# Incineration

of Hazardous, Toxic, and Mixed Wastes

# Incineration

## of Hazardous, Toxic, and Mixed Wastes

**A Textbook and Handbook**

**by James H. Gill and John M. Quiel**

Mr. Gill, a registered Professional Engineer, Consultant, and President of James H. Gill and Associates, Inc., Consulting Process Engineers, in St. Louis, Missouri, is an authority on incinerators and on regulatory constraints for incineration systems.

Mr. Quiel, Combustion and Process Control Engineer, is based in the North American Manufacturing Company's Santa Barbara, California engineering office. He has had broad experience in combustion and pollution control as related to incineration processes.

Published by

NORTH AMERICAN Mfg. Co.
4455 East 71st Street
Cleveland, OH 44105 USA
216-271-6000

The North American Manufacturing Company engineers and makes burners, blowers, controls, and complete combustion systems.

INCINERATION OF HAZARDOUS, TOXIC, AND MIXED WASTES
by Gill and Quiel



PRINTED IN THE UNITED STATES OF AMERICA

First Edition, first printing, 1993 4000 copies

**Library of Congress Cataloging in Publication Data**

Gill, James H., 1941-
Incineration of hazardous, toxic, and mixed wastes / by James H. Gill and John M. Quiel. -- 1st ed.
p. cm.
Includes bibliographical references and index.
ISBN 0-960 1596-4-9 :
1. Hazardous wastes -- Incineration. 2. Kilns, Rotary. I. Quiel, John M., 1951- II. Title.
TD1062.G55 1993
628.4′457 --dc20 93-10653

ISBN 0-9601596-4-9

# PREFACE

The authors and the publisher of this book realize that hazardous wastes can be incinerated in a wide variety of incinerator designs and styles, such as rotary drum (kiln) incinerators, fixed or moving grates, fluidized beds, duct burners, and regenerative thermal oxidizers. However, the most common state-of-the-art system for solid or liquid wastes is the *rotary kiln incinerator*, which is therefore the principal thrust of this textbook and handbook. Much of the technical data herein is applicable to other incineration systems.

Mr. Gill's and Mr. Quiel's inputs have been edited and integrated by Richard J. Reed, North American's Technical Information Director, in consultation with Charles Schroer and Ted Jablkowski, District Managers of North American Mfg. Company's Chicago and New England Districts, respectively. These five experienced engineers have worked together in making this book helpful and useful to both new and experienced persons working with incineration of many kinds of wastes, in many kinds of incineration systems.

## ACKNOWLEDGEMENTS

Frontispiece -- Roy F. Weston, Inc.

Curves for soils incineration with various volumetric loadings -- Dwight Brown, President, International Waste Energy Systems (IWES), Division of ICF/Kaiser Engineers, Pittsburgh, PA

Drawings -- International Waste Energy Systems (IWES), Division of ICF/Kaiser Engineers, Pittsburgh, PA

*(Credits for all other information not originated by the authors or editors are listed where that information appears.)*

# CONTENTS

It is the elimination and utilization of waste, waste effort, waste time and material, the minimization of destruction and damage, wear and tear that produce the great results in the industrial world. There is no magic in these accomplishments. The leaders in action or thought are not magicians, but steady persistent workers.

THEODORE N. VAIL, 1845-1920

This is a textbook and handbook of engineering practice and recommendations -- all subject to local, state, and federal codes, and insurance requirements... plus good common sense.

No patent liability is assumed with respect to the use of information from this book. Every precaution has been taken in preparing it, but the publisher does not assume a repsonsibility for errors or omissions; nor is any liability assumed for damages resulting from use of this information.

WARNING:

Situations dangerous to personnel and property can arise from incorrect operation of any kind of equipment. North American Mfg. Co. urges strict compliance with all safety standards and with insurance underwriters' recommendations, plus care in operation.

L B SMITH

# 1. INTRODUCTION

*Definition*

Incineration is a combustion process that uses rapid oxidation, excess air and high temperatures to produce conditions whereby hazardous and/or toxic waste constituents are thermally broken down and destroyed. Resulting gaseous acidic compounds are neutralized and removed in the *off gas treatment* (OGT) system. If the remaining ash is inert or meets proper specifications, it can be sent to a landfill. If from "mixed" (radioactive) waste, and/or it does not meet requirements for land filling, the ash must be drummed and stored.

**OGT**

If only atomizable liquid type wastes are to be processed, then the incinerator facility is essentially a combustion chamber fitted with the proper number and type of burners, combustion and process controls, followed by the off gas treatment system.

If in addition to atomizable liquid wastes, solid wastes and sludges are to be processed, a *primary combustion chamber* (PCC), usually a rotary drum kiln, or a rotary hearth, is utilized in the incinerator facility followed by a *secondary combustion chamber* (SCC) and OGT system.

**PCC**

**SCC**

Incineration has been demonstrated to be a *best available control technology* (BACT) for the safe and efficient disposal of hazardous and toxic wastes.

**BACT**

The most prevalent incinerator system presently is the *Rotary Kiln Incinerator* (RKI). These RKI systems offer great flexibility because of their ability to process a wide variety of hazardous, toxic, or mixed wastes on a continuous feed stream basis with minimum maintenance. The OGT system may be a wet or dry type. (See Figure 1.)

**RKI**

The wet type begins by adiabatically saturating and cooling the incinerator off gas from the SCC, followed by coarse particulate removal (>1 micron size), followed by acid gas neutralization and removal, followed by fine particulate removal (<1 micron size).

The dry type begins by sensibly tempering the incinerator off gas from the SCC in a tempering chamber followed by one or more spray dry absorbers to effect acid gas neutralization followed by particulate removal. (See Figure 2.)

Both caustic solutions at 20 to 50% by weight of sodium hydroxide, NaOH, and lime slurries at 10 to 20% by weight of calcium hydroxide, $Ca(OH)_2$, are used as scrubbing and neutralization media.

Because incinerators operate at temperatures of 1500 F to 2300 F, a refractory lining is required to protect the combustion chamber's steel casing from thermal damage.

**Figure 1. A transportable incineration system** consisting of (1) 7′6″ diameter by 25′ long rotary kiln, (2) 8′ diameter by 33′ high afterburner section, (3) conditioning (cooling) tower, (4) fabric filters for particulate emissions control, and (5) packed bed scrubber for acid gas emissions control. This dual waste incinerator system is RCRA-permitted for treating liquid and/or solid hazardous wastes, and is TSCA-permitted for 13,900 pounds per hour of PCB-contaminated waste.

The rotary kiln operates between 1200 F and 2200 F, and the afterburner is designed to provide a minimum gas residence time of two seconds at greater than 2100 F. This photo and the night-view on the frontispiece are printed here by courtesy of owner/operator, Roy F. Weston, Inc.

**Figure 2. Rotary-drum kiln TSCA incineration system.**
*(Please see Acknowledgements on page vi.)*

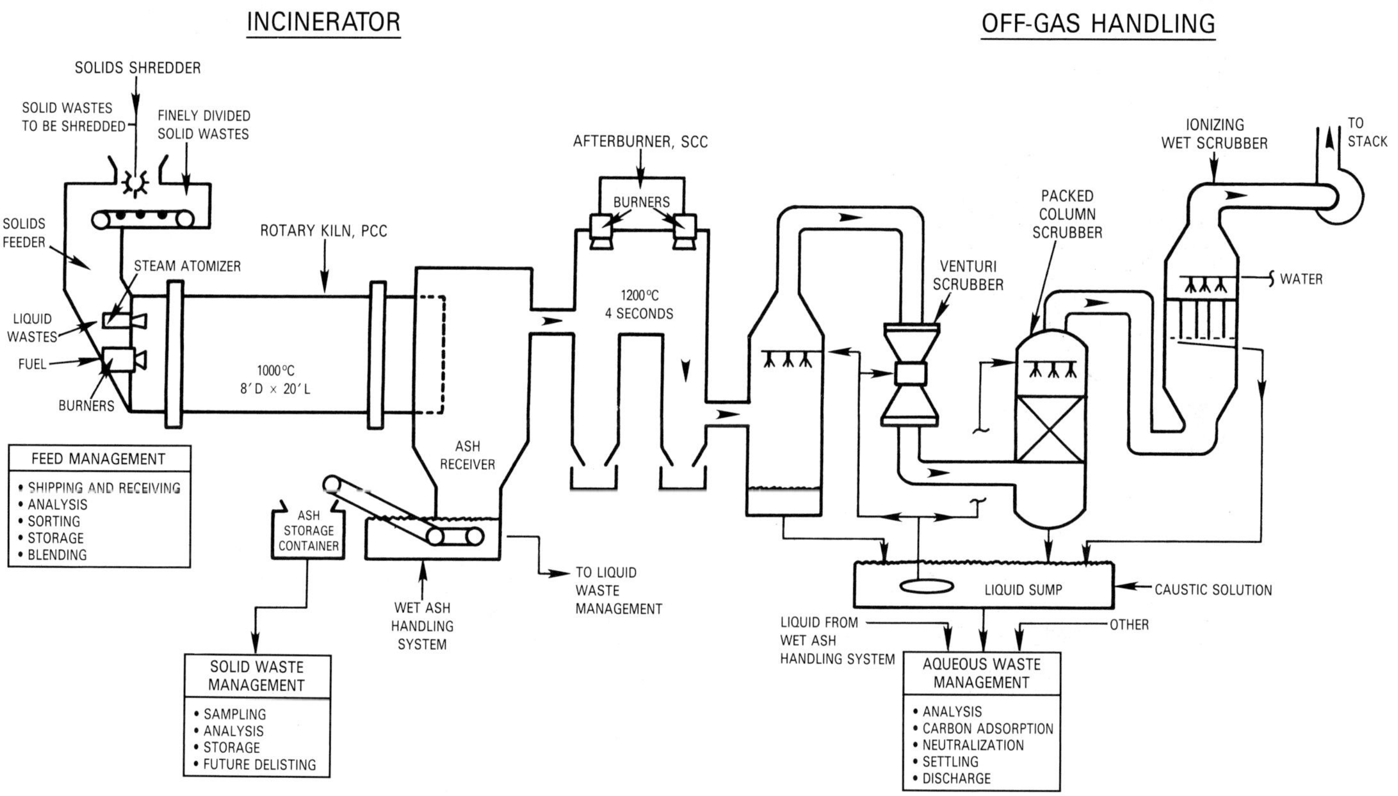

The refractory materials and any exposed internal metallic components must be selected to minimize corrosion attack from halogenated compounds and metal salts in the waste streams. They must also withstand the effects of temperature and slags. These problems are discussed at the end of this text.

Waste feed processing in the incinerator system cannot commence until the incinerator combustion chambers have attained prescribed regulatory permit temperature levels; so auxiliary fuels must be fired in the burner system to bring the incinerator system up to the required temperatures.

Hazardous and toxic waste incinerator systems are regulated and permitted for waste processing by the following:

**RCRA** Resource Conservation and Recovery Act (RCRA)
Hazardous Waste (toxic, corrosive, combustible) - 40 CFR 264.

**TSCA** Toxic Substance Control Act (TSCA -- controlled by EPA)
Toxic Waste (PCBs) - 40 CFR 761.

**NESHAP** National Emissions Standards for Hazardous Air Pollutants (NESHAP) -
Mixed waste (low level radioactivity).

**NPDES** National Pollutants Discharge Elimination System (NPDES) -
Liquid Effluents.

Air Test - Stack Emissions - for Particulates and Metals.

**Hazardous waste** is defined in the United States of America (USA, US) as
**EPA** any substance so listed by the US Environmental Protection Agency (EPA). (See Appendix B2.)

**Toxic waste** is defined in the US as any substance confirmed or suspected
**PCB** as being a carcinogen, e.g. polychlorinated byphenyls (PCBs), dioxins, and others listed by the EPA. Current lists of these wastes are in Appendix B2.

Some facilities are granted permits for incinerating both RCRA (Hazardous Wastes) and TSCA (Toxic Wastes). These types of incinerator facilities are referred to as a Dual Waste Incinerators.

*Low level radioactive wastes*

**Mixed waste.** Some incinerator facilities process RCRA and/or TSCA wastes and in addition, those that are classified as *low level radioactive wastes*. Such incinerator facilities are referred to as *Mixed Waste Incinerators*. "Low level radioactivity" is limited to < or = 1.0% by weight uranium and U 235 assay. Radionuclide sampling and analysis testing done at one mixed waste incinerator facility showed 97.0 % by weight of the uranium reported in the ash. High level wastes are sometimes blended with other materials to reduce their undesirable concentrations, allowing them to be incinerated.

Thermal incineration does not decrease nor eliminate the low level radioactivity present in a mixed waste. Even with this type of waste, however, volume reduction is achieved. That is the main benefit of thermal incineration. Also, liquid wastes are burned and principal organic hazard constituents are destroyed. These are discharged through the stack as essentially inert flue gases.

Low level radioactive solids and sludges, with their *principal organic hazard constituents* (POHC) volatized and destroyed, end up as an ash that is inert except for low level radioactivity. A 40% to 60% volume reduction is realized. **POHC**

In addition to complying with 40 CFR 761 (TSCA) and 40 CFR 264 (RCRA) waste processing regulations, incinerator facilities must comply with federal and state limits on particulate emissions, opacity, metals content (in ash, stack gas, and effluents), and the Occupational Safety and Health Administration (OSHA). **CFR** = *code of federal regulations* **OSHA**

Incinerator facilities must also follow the requirements and guidelines of the National Fire Protection Association (NFPA) and/or other insuring agencies for combustion systems. **NFPA**

While some operating conditions and permit levels are prescribed in 40 CFR 761 and 264, other operating conditions and permit levels are incorporated into the specific facility permit based on analysis and results in the trial burn report. Incinerator facilities must also comply with the requirements of 40 CFR 761.70, 264.345, and 264.347 which mandate operating and monitoring standards; i.e., stack emissions monitoring, waste feed monitoring, waste feed cut off parameters, hard copy strip charts, data logging/records keeping, and report files retention.

The EPA does allow RCRA type waste oils to be fired in certain non-incinerator processes. These processes should be inherently high temperature, 2000+ F, and achieve a hot gas residence time of 2.0+ seconds in normal operation. The basic criteria are:

a) Waste Oil - RCRA

b) Must demonstrate >99.99 *destruction and removal efficiency* (DRE) **DRE**

The cement industry is a major user of RCRA waste oils.

*Incineration Abbreviations and Acronyms* are listed in Appendix A1 of this book.

## 2. REGULATORY AND PERMIT CONSTRAINTS

**Minimum regulatory requirements** as of 1991 are listed below.

*Hazardous wastes*

*RCRA - Hazardous Wastes,*
per Code of Federal Regulations (CFR) 40, Part 264 (Figure 3):

Destruction & removal efficiency (DRE) = or >99.99% for each principal organic hazard constituent (POHC).

**HCL**

The allowable stack emission rate for HCL is the larger of 4 pounds/hr or 1.0% by weight of the total HCL exiting the SCC, as measured by stack test sampling. If this 1.0% total HCL is greater than 4.0 pounds per hour, the minimum acid gas removal efficiency (AGRE) shall be = or >99.0%.

**AGRE**

Particulate < or = 0.08 gr/dscf, corrected to 7% $O_2$ by volume, dry gas basis (dgb). [Expected to tighten to 0.015 or 0.010 gr/dscf within the next few years.]

Temperature in the SCC = 2192 F, ± 180 degrees (F) or 1200 C, ± 100 degrees (C).

**CGVI**

Combustion gas velocity indication (CGVI).
RCRA requires suitable, reliable instrumentation be provided, located in the system downstream of the SCC to provide indication of the system flow rate. A limiting value (not to exceed) shall be demonstrated during the RCRA trial burn and entered into the RCRA permit. The permit value insures acceptable gas residence time in the SCC. Exceeding the CGVI limit during incinerator operation shall automatically initiate a waste feed cutoff.

*Toxic wastes*

*TSCA - Toxic Wastes,*
per Code of Federal Regulations (CFR) 40, Part 761 (Figure 3):

DRE = or >99.9999% for each POHC

AGRE = or >99.9% (Stack emission = or < the greater of 4 pounds/hr HCL or 1.0% by weight of the HCL exiting the SCC, determined by stack testing and sampling).

Particulate < or = 0.08 grains per dry standard cubic foot (gr/dscf), corrected to 7% $O_2$ by volume, dry gas basis (dgb). [Expected to tighten to 0.015 or 0.010 gr/dscf within the next few years.]

**Figure 3. Code of Federal Regulations (CFR) 40, Part 264, relates to standards for owners and operators of hazardous waste treatment, storage, and disposal facilities (RCRA).** CFR 40 Part 761, deals with PCBs manufacturing, processing, distribution in commerce, and use prohibitions (TSCA).

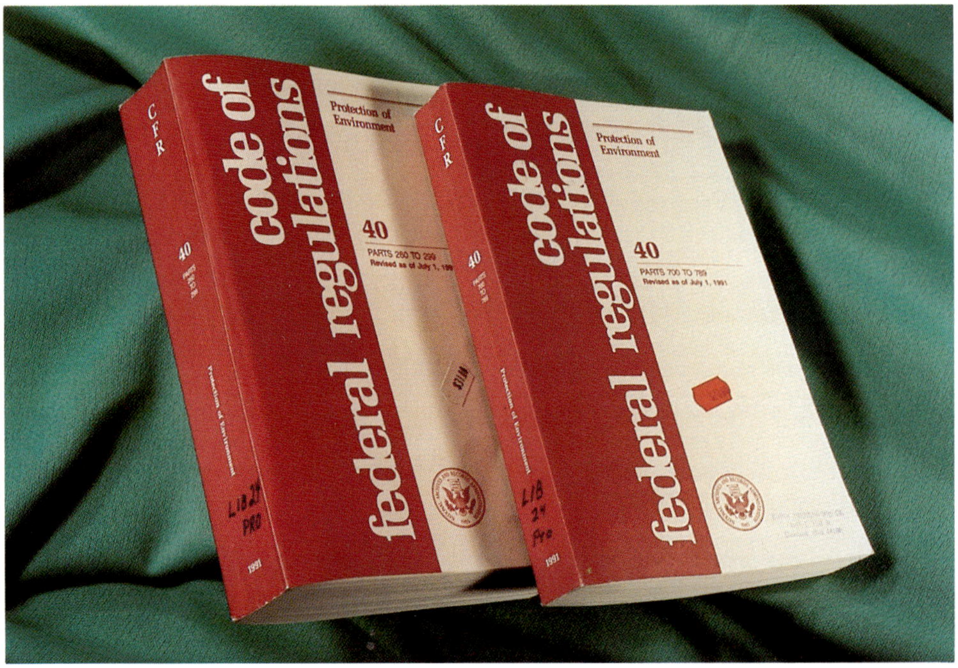

Temperature in the SCC = 2192 F, ± 122 degrees (F), or 1200 C, ± 50 degrees (C). <2070 F shall automatically initiate a waste feed cutoff.

Residence Time > or = 2.0 seconds at > or = 2192 F in the SCC.

Stack $O_2$ > or = 3.0% $O_2$, by volume (wgb). <3.0 % $O_2$ by volume (wgb) shall automatically initiate a waste feed cutoff.

*NESHAPS - Mixed Wastes:*

These standards apply to mixed waste incinerators and require continuous stack sampling and routine analysis, most often on a weekly basis, for site specific radionuclides and isotopes.

*Radioactivity*

Based on the stack gas and liquid effluent analyses, a dose calculation for the site is made and compared to acceptable standards. The allowable radioactive content in the waste feed is then established by blending it with other wastes to stay below the radioactive limit.

*NPDES - Liquid Effluents:*

**ppb**

Effluent content < or = 1.0 parts per billion (ppb) of polychloronated biphenyls (PCB)

*Air test for particulates, metals:*

Each state's environmental agency may require that a stack air test be made to demonstrate compliance with that state's allowable particulate emission level -- grains particulate concentration per dscf (gr Pc/dscf) -- and opacity.

*Wet and dry OGT*

Opacity - Stack discharge

- Dry off-gas treatment (OGT) System: 20% Ringleman
- Wet OGT System: Totally white plume by visual observation, performed by state authority.

As part of this air test, EPA requires that specified metals content in the stack shall be sampled and analyzed to demonstrate compliance with allowable emission rates. Usually, the metals with emissions rate constraints are:

Hazardous = lead (Pb)
Toxic = beryllium (Be), cadmium (Cd)

The final metal components to be sampled and analyzed are site-specific and a mutual agreement between the EPA and the Owner/Operator who sets the specific metal content and then EPA-listed allowable emission rates are applied.

**Compliance with permit values**. Although RCRA and TSCA mandate compliance within specified process and analytical requirements for waste processing, many other process conditions and process parameters (as listed in the EPA approved RCRA or TSCA trial burn plan) must be sampled and monitored during the actual trial burns.

*Monitoring required*

Upon analysis and review of the trial burn report results by the EPA and Owner/Operator, a permit constraint value ( = , <, or >) is established for each of the process conditions and process parameters. The values then become part of the final RCRA and/or TSCA permit for acceptable waste processing by the incinerator facility.

Non-compliance with these permit values, shall automatically initiate a waste feed cutoff; some instantaneously, some on a time delay basis (10 to 30 seconds), others on a rolling average basis (5 to 15 minutes).

The breadth and scope of these permitted waste feed cutoff process conditions and process parameters is shown by the following typical (but only partial) list:

OGT Venturi delta P*
OGT Effluent pH and conductivity
OGT Recycle flow rates
OGT Recycle pH*
OGT Ionizer amperage and voltage
OGT Slurry flow rates
Waste feed flow rates
Stack gas, CO*
Stack gas velocity (flow)
Kiln feed end pressures
Kiln discharge end pressures
Atomizing pressures

**CO** = *carbon monoxide*

In addition, running interlocks, as prescribed by NFPA, FM, or other insuring agency, are active and can initiate a waste feed cutoff.

Other final permit compliance limits are very site-specific and arrived at via discussions/negotiations between the Owner/Operator and EPA/State authorities.

* See Appendices A1, A7, or C1.

## 3. THE PERMITTING PROCESS

Following environmental assessments, impact statements and public hearings as required by the National Environmental Policy Act (NEPA), the RCRA permit process starts with the submittal by the incinerator facility Owner/Operator of a Part B Application. Most often an engineering firm (Designer/Contractor) is engaged to prepare the Part B Application based on an owner-supplied scope of work and/or overall facility requirements design basis.

**NEPA**

The Part B Application is an extensive document requiring about a 30% level of engineering content. 45 to 90 days should be scheduled for this level of effort.

The Part B Application must convey to the governing region of the EPA the following information:

a) Type of proposed incinerator facility: RCRA, TSCA, Mixed Waste.

b) Identify the wastes that the proposed incinerator will process; through-put of liquids, solids, and sludges; and the design thermal rating of the facility in millions of Btu/hour.

c) Thorough understanding of the incinerator's operation and control.

d) Demonstration that the incinerator design is safe to test and that the design will meet the performance requirements.

e) Demonstration that the incinerator test will accurately measure its performance for worst case operation over the expected operating range, and generate the process information required to establish permit conditions.

f) Demonstration that the operating protocol and process monitoring systems and equipment will guarantee operation of the incinerator on a continuous basis, within the required regulatory limits, and within the site-specific permit conditions.

To provide this level of information and documentation requires the following engineering activities:

*Documentation*

a) Process Description -- process flow sheets, heat and material balances.

b) Engineering Description -- piping and instrument diagrams (P & IDs), equipment list. Ideally, the P & IDs should be at the 90% completion level to maximize clear communications to the governing region.

c) Process Monitoring Description -- instrument list and index depicting ranges for flows, temperatures, pressures, type of device and output type (digital, analog), instrument device range, accuracy, calibration frequency.

d) Process Parameters List
e) Gaseous Emissions Monitoring Description
f) Automatic Waste Feed Cut-off Description
g) Waste Characterization Sampling and Analysis Description
h) Quality Assurance/Quality Control Description

Once compiled, the Part B Application should be submitted to the state environmental agency, if the state has primacy; otherwise, to the US EPA.

**NOD**

The governing authority will often respond to the Part B Application with a *notice of deficiency* (NOD), such as insufficient description or insufficient mechanical details. It may take 90 to 120 days before the NOD is issued to the owner/operator.

After receiving the NOD, the Owner/Operator has typically 30 to 60 days to address and correct the NOD and re-submit to the governing agency. The result at this point can be a second NOD. The permit process duration can be 5 months to a year, depending on how well it is prepared.

Obviously, the Part B must be prepared thoroughly and completely with as much specific detail information as possible. On some NOD items, this has required specifying items such as the burner type, model number, operating range, fuel pressure, atomizer type, and atomizing pressure.

## 4. PERFORMANCE TESTS

Performance tests are a series of tests mutually agreed to between the owner/operator and designer/contractor. The purpose of these tests is to demonstrate the process and mechanical integrity of the incinerator facility and compliance with RCRA and/or TSCA regulations.

A performance test plan, prepared by the designer/contractor, defines the process parameters to be sampled and monitored, acceptance criteria, sampling and analysis methods, and protocol and operating conditions.

*Sampling/ analysis*

An independent, EPA-approved firm should be engaged to perform the sampling and analysis and to prepare and issue the performance test report. The performance test report will be the criteria that the owner/operator will use to accept or reject the incinerator facility. Typically, a minimum of 2 performance tests are run. The tests should approach the upcoming trial burn conditions.

*Hierarchy rank*

Because the facility will not have been permitted during the performance tests, a designer/contractor liability question may arise if the performance tests are done using RCRA or TSCA wastes. Therefore, "Surrogate Wastes" are prepared from commercial chemicals. The "Surrogate" POHCs should be chosen from a list of commercial constituents that will equal or exceed the incinerability index of the most difficult expected RCRA and/or TSCA actual waste stream/compound. For example, the incinerability ranking of hydrogen cyanide is 1, which means that it is most difficult to incinerate; whereas the hierarchy ranking of endosulfan is 329, least difficult of the list to incinerate with 99% DRE.

Performance tests are long and expensive. If RCRA or TSCA type sampling protocol is utilized, up to 6 hours of continuous sampling for one test run may be required. Depending on the number of tests run, costs can be $100,000 to $250,000. The performance test plan must be a well thought out, realistic, detailed effort.

*Miniburn*

Sometimes a "miniburn" is done prior to the performance tests. This test utilizes straight commercial fuels (natural gas, No. 2 Fuel Oil) and is designed to operate the incinerator facility at or near the design point. The purpose of the test is to gain operator experience and familiarity along with systems and components reliability.

*Table 1* is a partial list of hierarchy ranking for surrogate POHCs. *Table 2* depicts a sample criteria for a performance test. *Table 3* depicts results of a sample performance test under RCRA conditions.

**Table 1. Thermal stability-based incinerability rankings for selected waste profile constituents and suggested trial burn constituents (surrogates).** *(For a complete list, see Appendix F.)*

| Hierarchy Ranking† | T99, C* | POHConstituent |
|---|---|---|
| CLASS 1– | | |
| 1 | >1500 | Hydrogen cyanide (hydrocyanic acid) |
| 3 | 1150 | Benzene |
| CLASS 2– | | |
| 38 | 895 | Toluene (methylbenzene) |
| CLASS 3– | | |
| 80-81 | 785 | Pyridine |
| 86-90 | 775 | Trichlorofluoromethane (Freon 11) |
| 118-119 | 730 | Methyl ethyl ketone (2-butane) |
| CLASS 4– | | |
| 145-146 | 655 | Nitrobenzene |
| 175-183 | 625 | Trichloromethane (chloroform) |
| 184-187 | 620 | Dichlorodiphenyltrichloroethane (DDT) |
| CLASS 5– | | |
| 229 | 545 | 1,1,1-Trichloroethane (methyl chloroform) |
| 240 | 511 | 1,2-Benzoisothiazolin-3-one,1,1-dioxide (saccharin) |
| CLASS 6– | | |
| 276 | 380 | Di-n-Octyl phthalate |
| 282-285 | 300 | (S)-3-[1-Methyl-2-pyrrolidinyl]pyridine (nicotine) |
| CLASS 7– | | |
| 329 | 100 | Endosulfan |

* T99, C means the temperature in degrees Celsius required to incinerate to 99% DRE (destruction and removal efficiency); so methyl ethyl ketone might be substituted as a surrogate during a trial burn for a waste material with a "T99" of 730 C.

† Explanation of rankings: The incinerability ranking of hydrogen cyanide is 1, which means that it is the most difficult (of this list) to incinerate and achieve 99% DRE; the hierarchy ranking of endosulfan is 329, the least difficult of this list to incinerate with 99% DRE.

**Table 2. Sample Performance Test Requirements.**

DRE = or >99.993%

AGRE = or >99.5% HCl
= or >99.2% Total Acidic Gases
($HCl + HF + H_2SO_4$)

Particulates = or < 0.06 gr/dscf corrected to 7% $O_2$ by volume (dgb)

**Table 3. Summary of Results for Critical Parameters, Performance Tests.**

| RCRA CONDITIONS | Test Number 1 | 2 | 3 |
|---|---|---|---|
| Waste feed rates, lb/hr | | | |
| Liquid | 562(S) | 889(P) | — |
| Solid | — | — | 687(P) |
| Thermal input, $10^6$ Btu/hr | 13.0 | 18.5 | 13.2 |
| Average chamber temp, F | | | |
| Primary | 730 | 1490 | 1480 |
| Secondary | 1920 | 1870 | 1840 |
| Stack gas data | | | |
| Flow rate, dscfm | 8950 | 8360 | 7940 |
| Oxygen, % (dgb) | 13.3 | 12.7 | 13.0 |
| Carbon monoxide, ppm | <10 | <10 | <10 |
| HCl removal efficiency, % (>99.5) | 99.97 | 99.88 | 99.97 |
| HCl emission rate (as chloride) lb/hr (<3.32) | 0.019 | 0.136 | 0.009 |
| Average particulate concentration, gr/dscf (<0.06) at 7% $O_2$ (dgb) | 0.059 | 0.045 | 0.040 |
| Target POHCs | TCFM | TCFM | $CCl_4$ |
| DRE value, (% >99.993) | 99.9999 | >99.9999 | >99.9999 |

Numbers in parentheses indicate the allowed minimums.
Letters in parentheses indicate location and/or type of feed:
P = primary chamber; S = secondary chamber; A = aqueous.

## 5. TRIAL BURNS

Trial burns are performed to demonstrate to the EPA and state environmental agency that the incinerator facility complies with RCRA and/or TSCA regulations. As with performance tests, trial burn tests are long and expensive.

Each trial burn, RCRA or TSCA, requires 3 separate runs with 2 to 6 hours of continuous sampling required per run. A fourth, "Archive" run is normally performed to ensure extra samples in case of shipping, handling or laboratory losses.

*Time limits*

720 hours of pre-trial operation is allowed for a RCRA trial burn at the RCRA trial burn operating conditions on the RCRA trial burn waste materials. An additional 720 hours is available upon request.

*Surrogate waste*

The RCRA trial burn waste may be either a "Surrogate" waste, an actual facility site RCRA waste, or (the best) an actual facility site RCRA waste spiked with "Surrogates".

The latter option affords the incinerator facility Owner/Operator the greatest flexibility in the final RCRA permit. The essential criteria are as defined in the RCRA trial burn plan. Its composition becomes the final permit limits as to incinerability hierarchy ranking, POHC content, organic chlorine concentration, ash content, heating value, and feed rate.

The RCRA trial burn must be thorough, complete and detailed. The cost of conducting a RCRA trial burn is substantial due to handling, spiking, blending of large volumes of the waste materials, engaging an independent certified sampling and analysis contractor, and analytical laboratory and manpower requirements.

*Shakedown*

The incinerator facility must have thorough shakedown runs and operators must be thoroughly trained and knowledgeable about the instrumentation and control systems before a trial burn is attempted.

*Interruptions*

While small-duration sampling interruptions (minutes) due to process upsets or mechanical difficulties are allowed, any interruption approaching an hour's duration leads to an invalid test, scrubbing the run, and a total test re-start. Reliability of the combustion system (burners, blowers, fuel train components, burner stability, flame configuration, atomizer) operation is absolutely essential.

With the exception of higher operating temperatures and more stringent DRE requirements, the TSCA trial burn is similar to the RCRA trial burn in sampling and analysis requirements, waste feed preparation, continuous sampling durations, and costs. For a TSCA trial burn, the Owner/Operator may, for the first time, burn PCBs. Dioxins and furans, if included in the waste characterizaton plan, may also be burned.

Cost

Trial burns may cost between $100,000 and $200,000. *Table 4* depicts sample TSCA trial burn results along with the TSCA regulatory requirements.

**Table 4. Summary of Results for Critical Parameters, TSCA Trial Burn Tests.**

| | Test Number | | |
|---|---|---|---|
| | 1 | 2 | 3 |
| Date | 5-12-88 | 5-13-88 | 5-25-88 |
| Waste feed rates, lb/hr | | | |
| Liquid | 622(S) | 606(P) | — |
| Solid | — | — | 693 |
| Thermal input, $10^6$ Btu/hr | 13.9 | 19.8 | 15.2 |
| Average chamber temp, F | | | |
| Primary | 1130 | 1820 | 1820 |
| Secondary | 2210 | 2210 | 2200 |
| Stack gas data | | | |
| Flow rate, dscfm | 7610 | 7740 | 7520 |
| Oxygen, % (dgb) | 12.1 | 9.9 | 10.8 |
| Carbon monoxide, ppm | <10 | <10 | <10 |
| HCl removal efficiency, % (>99.5) | 99.90 | 99.90 | 99.93 |
| HCl emission rate (as chloride) lb/hr (<3.32) | 0.166 | 0.176 | 0.012 |
| Average particulate concentration, gr/dscf (<0.06) at 7% $O_2$ (dgb) | 0.019 | 0.030 | 0.003 |
| DRE value, (% >99.9999) | | | |
| Total chlorobenzenes detected | 99.99996 | 99.99996 | 99.99993 |
| Combustion efficiency, % (>99.95) | >99.99 | >99.99 | >99.99 |
| Residence time, sec (>4) | NA | 4.2 | 4.4 |

Numbers in parentheses are allowable minimums.
Letters in parentheses indicate location of feed:
P = primary chamber; S = secondary chamber.

## 6. WASTE CHARACTERIZATION

Waste characterizaton is a process involving the following activities:

a) Sampling and analysis determination of:

Chemical composition, including metals and metallic salts
Heating value
Viscosity
Ash and water content

b) Compatibility analysis on as-received wastes. Wastes are usually blended to meet a heating value requirement as a primary or secondary waste. Compatibiliy analysis insures that when different wastes are blended, mixed, and agitated, no unknown adverse mixing reactions are initiated, resulting in the formation of highly exothermic, highly corrosive, or volatile compounds.

c) Record Keeping:

Shipping manifests
Receiving manifests
Storing manifests
Blending manifests
Blending data sheets
Compatibility data sheets
Burn sheets
Ash disposal manifests
Aqueous liquid disposal manifests

**Table 5. Example of a waste stream.** 14 650 tons/year liquid waste stream: 7.99% by weight halogenated solvents; 32.4% Marpol oil; 29.4% mineral oil; 9.22% non-halogenated solvents; 4.37% paint waste; 6.83% tank cleaning sludges; plus catalyst waste, dyestuff waste, fuel oil, interceptor and treatment sludges, miscellaneous chemical waste, mixed organic compounds, oil/water mixtures, PCBs, pharmaceuticals, phenol and derivatives, polymerization precursors and production wastes, painting waste, tar and pitch.

| | **Units** | **Exothermic Liquids** | **Sludge** |
|---|---|---|---|
| Heat content: | kJ/kg | 27 900 to 46 500 | 2330 to 23 300 |
| | kcal/kg | 6 670 to 11 110 | 557 to 5 570 |
| | Btu/lb | 11 990 to 19 990 | 1000 to 10 000 |
| Analysis: | | | |
| C, carbon | % wt | 45-70 | 5-50 |
| H, hydrogen | % wt | 5-15 | 3-10 |
| Cl, chlorine | % wt | 0-20 | 0-5 |
| O, oxygen | % wt | 0-10 | 5-15 |
| N, nitrogen | % wt | 0-3 | 0-3 |
| $H_2O$, moisture | % wt | 0-6 | 10-50 |
| Ash | % wt | 0-3 | 20-50 |
| Viscosity: | SSU | <100 | >100 |
| Maximum Particle Size: | mm | < or = 3 | < or = 6 |

## 7. TYPICAL WASTE STREAMS

**Table 6. Typical waste stream analyses.** Tabulated below are data for four typical primary wastes and one typical secondary waste. See also Table 5.

The higher heating values (HHV) of **HHV**
*primary wastes* range from 4 000 to 10 000 Btu/pound;
*secondary wastes* range from 10 000 to 20 000 Btu/pound.

*PCC Liquid Waste*

| Component/Property | % by weight | Range |
|---|---|---|
| Carbon | 27.60 | 22.86 - 48.61 |
| Hydrogen | 3.92 | 2.29 - 6.42 |
| Nitrogen | 3.00 | 1.87 - 6.95 |
| Oxygen | 34.98 | 4.00 - 11.48 |
| Chlorine | 19.00 | 1.17 - 23.50 |
| Fluorine | 1.00 | 0.00 - 2.00 |
| Sulfur | 2.00 | 0.31 - 3.00 |
| Water | 3.50 | 5.02 - 42.63 |
| Ash | 5.00 | 0.50 - 17.39 |
| HHV = 6000 Btu/pound | | 4,500 - 10,000 |
| VISCOSITY = 2.8 cp | | 1.0 - 4.0 |
| SPECIFIC GRAVITY = 1.03 | | 0.92 - 1.2 |

*PCC Aqueous Waste*

| Component/Property | % by weight | Range |
|---|---|---|
| Carbon | 5.73 | 0.00 - 5.58 |
| Hydrogen | 0.76 | 0.00 - 0.66 |
| Nitrogen | 1.58 | 0.00 - 14.10 |
| Oxygen | 0.99 | 0.00 - 6.90 |
| Chlorine | 0.04 | 0.04 - 6.60 |
| Fluorine | 0.50 | 0.00 - 1.00 |
| Sulfur | 0.10 | 0.00 - 0.08 |
| Water | 86.30 | 47.20 - 92.72 |
| Ash | 4.00 | 1.00 - 24.12 |
| HHV = 1200 Btu/pound | | 0 - 1200 |
| VISCOSITY = 1.0 cp | | 1.0 - 2.0 |
| SPECIFIC GRAVITY = 1.0 | | 0.95 - 1.0 |

**Table 6. Typical waste stream analyses** (concluded)

*PCC Pumpable Sludge Waste*

| Component/Property | % by weight | Range |
|---|---|---|
| Carbon | 24.35 | 20.52 - 41.40 |
| Hydrogen | 3.22 | 1.61 - 5.87 |
| Nitrogen | 0.00 | 0.00 - 3.47 |
| Oxygen | 0.00 | 0.00 - 4.73 |
| Chlorine | 0.00 | 1.17 - 29.48 |
| Fluorine | 0.00 | 0.00 - 1.00 |
| Sulfur | 0.00 | 0.07 - 1.00 |
| Water | 62.43 | 4.00 - 55.80 |
| Ash | 10.00 | 12.63 - 25.00 |
| HHV = 5400 Btu/pound | | 3,200 - 9,500 |
| VISCOSITY = 500 cp | | 200 - 1,000 |
| SPECIFIC GRAVITY = 1.2 | | 1.0 - 1.71 |

*PCC Solid Waste*

| Component/Property | % by weight | Range |
|---|---|---|
| Carbon | 13.80 | 3.39 - 20.70 |
| Hydrogen | 1.95 | 0.43 - 2.97 |
| Nitrogen | 3.25 | 0.00 - 7.80 |
| Oxygen | 25.00 | 0.00 - 7.62 |
| Chlorine | 15.00 | 0.76 - 16.00 |
| Fluorine | 0.00 | 0.00 - 1.00 |
| Sulfur | 1.00 | 0.00 - 1.00 |
| Water | 15.00 | 10.00 - 22.00 |
| Ash | 25.00 | 48.33 - 58.00 |
| HHV = 3000 Btu/pound | | 700 - 4,500 |

*SCC Liquid Waste*

| Component/Property | % by weight | Range |
|---|---|---|
| Carbon | 57.30 | 48.61 - 75.90 |
| Hydrogen | 7.56 | 6.42 - 10.88 |
| Nitrogen | 22.52 | 0.58 - 28.61 |
| Oxygen | 6.00 | 7.48 - 10.26 |
| Chlorine | 1.47 | 1.10 - 4.68 |
| Fluorine | 0.01 | 0.00 - 1.00 |
| Sulfur | 0.01 | 0.00 - 1.00 |
| Water | 4.21 | 0.00 - 5.02 |
| Ash | 0.92 | 0.54 - 1.00 |
| HHV = 12,000 Btu/pound | | 10,000 - 20,000 |
| VISCOSITY = 15.2 cp | | 1.0 - 20.0 |
| SPECIFIC GRAVITY = 0.95 | | 0.92 - 1.10 |

## 8. INCINERATOR SYSTEM DESCRIPTION

The primary focus of this book will be on Rotary Kiln Incinerator (RKI) systems because they have the widest application and feasibility. The RKI system will be described in detail in order to insure a good understanding of the major system components and the principal operating parameters. The description under secondary combustion chambers (SCC) should suffice to familiarize the reader with liquid-waste-only incinerator facilities. The four major equipment components of an RKI System as shown in Figure 4 are:

(a) the primary combustion chamber,

(b) the mixing chamber,

(c) the secondary combustion chamber, and

(d) the off-gas treatment system.

## 9. PRIMARY COMBUSTION CHAMBER (PCC)

Most often the PCC is a rotary kiln operated in the non-slagging mode, over a temperature range from 1500 F to 1850 F. Typical kiln sizes have shells ranging from 7′ to 12′ ID and from 25′ to 65′ in length. Most kilns are co-current fired, although some incinerator facilities, specifically with horizontal SCCs, are counter-current fired. On co-current-fired kilns, the auxiliary fuel burner(s) and liquid waste burner(s) are mounted through the kiln face plate at the feed end. [Co-current fired, or parallel flow, means that the products of combustion (poc) and the charged load move in the same direction. Counter-current or counterflow means that the poc and charged load move in opposite directions.] See Figures 5 and 6.

*Co-current*

*Counter-current*

Some incinerator facilities combine the auxiliary fuel firing and liquid waste firing into a single burner body equipped with multiple fuel atomizers (guns), or a combination gas nozzle/fuel gun, or others switch over from auxiliary fuel burners to waste fuel burners after the permitted temperature has been attained. See Figures 22 and 24, in Section 15.

*Liquid wastes*

Aqueous wastes and pumpable sludges are atomized and fired through specially designed nozzles and lances that are also mounted through the kiln face plate.

**Figure 4. Major incinerator system equipment components:** (a) Primary combustion takes place in a rotary kiln incinerator, preceded by a shredder and/or other feed management equipment.

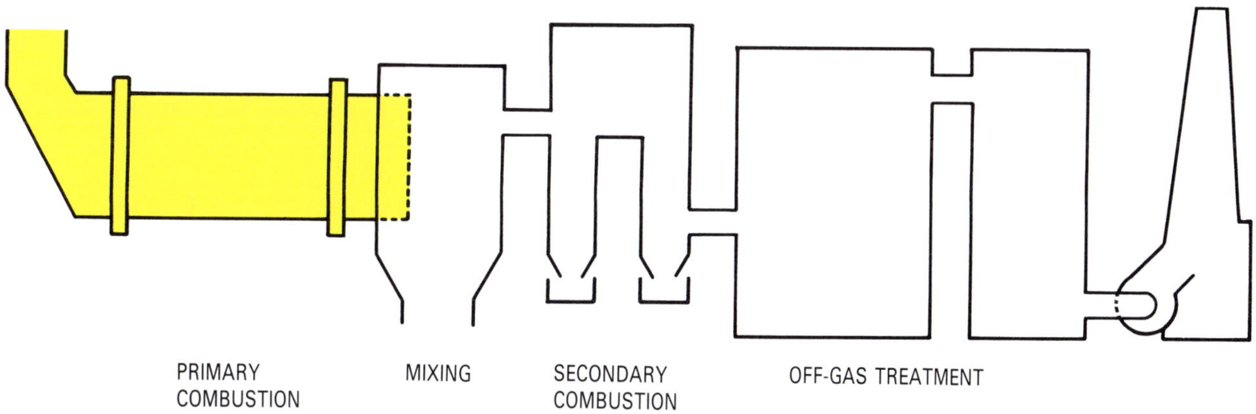

*Solid wastes*

Solid wastes, contaminated soils, shredded solids, non-pumpable sludges, and shredded drum containers, are fed into the kiln at the feed end via a hydraulically driven reciprocating ram feeder.

The solids and sludges may be energetic (exothermic organics and combustible constituents), or non-energetic (inert, or endothermic) with no appreciable heating value, but contaminated.

**XSAir** = *excess air*

Auxiliary fuel burners, and liquid waste burner(s) are usually fired at 0-20% excess air. Total XSAir in the PCC may be as high as 100% when processing energetic solids and sludges or when temperature maintenance at the low end of the kiln operating temperature range (1500 F) is desired. The additional % XSA may be introduced into the kiln feed end via a tertiary air duct and damper.

*Tertiary air*

An extension on the tertiary air duct, above the tertiary air damper location, serves as an explosion vent to relieve the kiln gases if a detonation should occur within the kiln. This pressure surge vent should be equipped with a gravity flapper type damper to minimize uncontrolled air leakage into the kiln during normal operation. The pressure in the PCC should be maintained negative ( – 0.3″ to – 0.8″wc) to prevent out-gassing of fugitive emissions, partially destroyed organics, and dust particulate.

**Figure 5. Typical temperature profiles in a counter-current-fired rotary drum incinerator.**

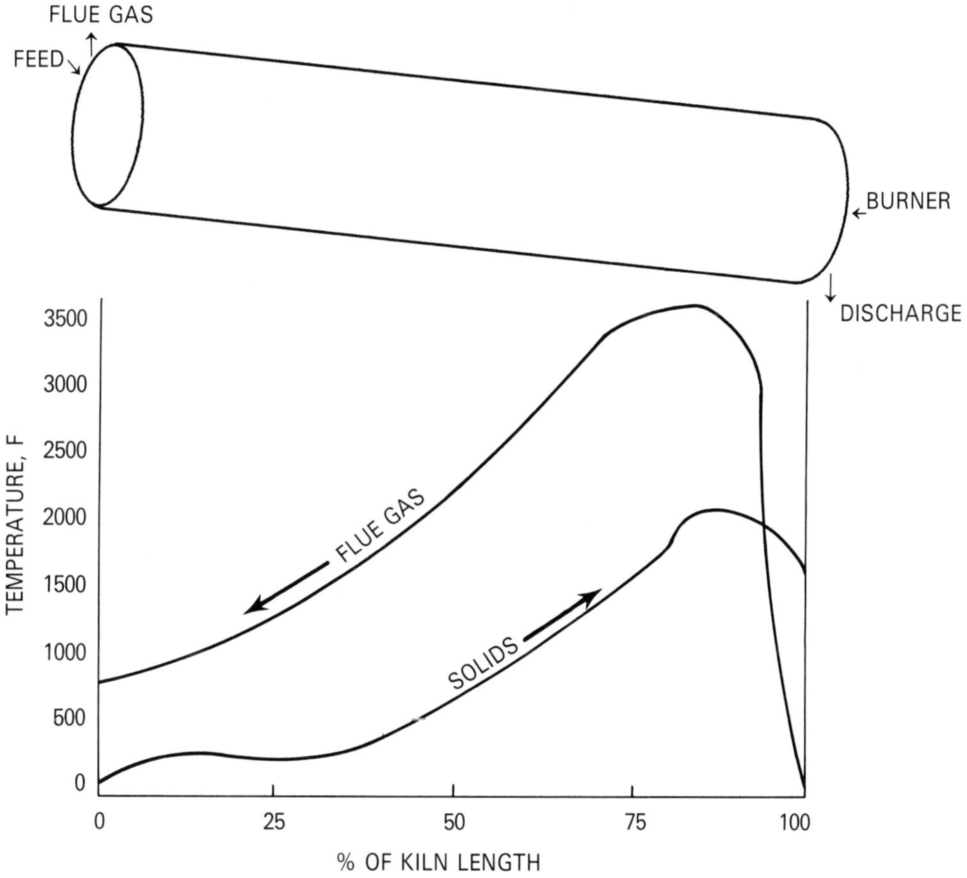

**Figure 6. Typical temperature profiles in a co-current-fired rotary drum incinerator.**

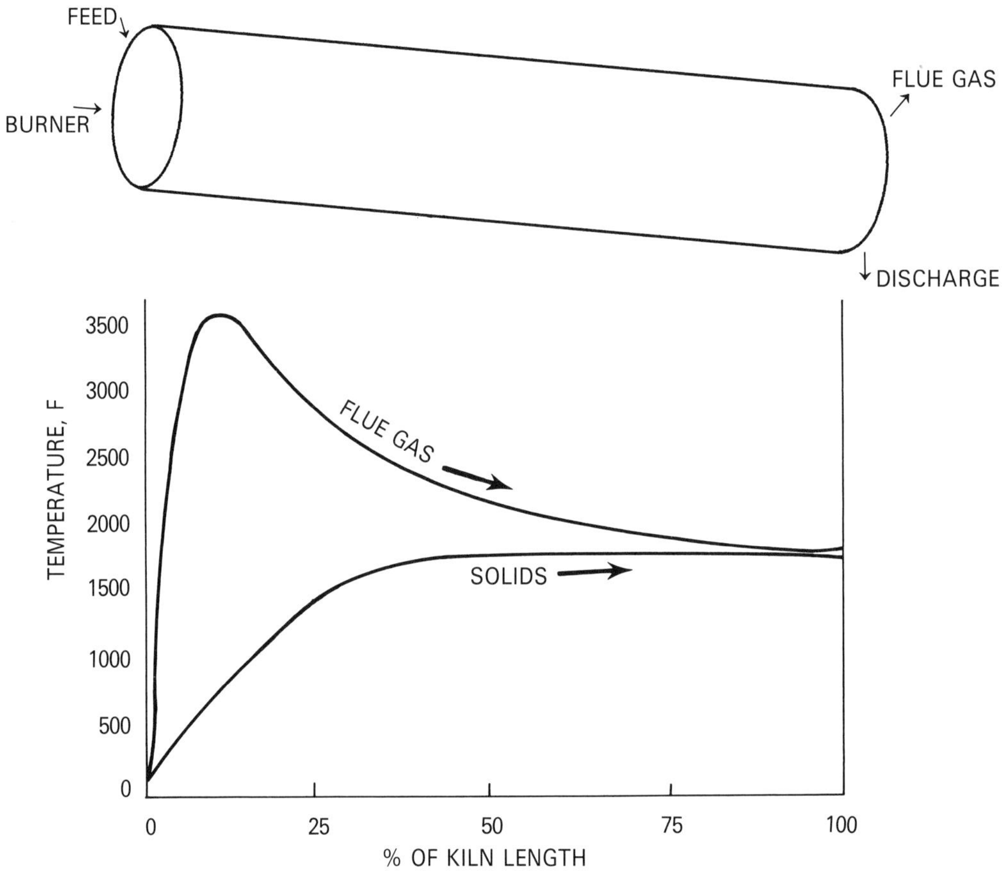

*Refractory lining*

Refractory hot face lining thicknesses range from 9″ to 12″ with or without a layer of backup insulation. If an insulating backup lining is used, it is typically an insulating board or insulating brick material. The hot face lining material is most often a refractory brick shape of 60% or 70% alumina, $Al_2O_3$.[*] Castable usage is to be avoided if possible, because chlorine and/or fluorine constituents in the waste feeds and flue gases may attack the lime binder in the castable refractory.

**PCC Design Parameters.**

Kiln length/diameter Ratio (L/D) = 3:1 to 5:1

Kiln slope = 1 degree for non-slagging, 3 degrees for slagging

Kiln speed = 0.5 to 3.0 rpm

Solids retention time = 30 to 90 minutes

Kiln solids loading = 6% to 12.5%; 10% typical average

*Retention*

The solids retention time for a trial kiln design is calculated from the following relationship:

$$T = (1.77)\,(\text{theta})^{0.5}\,(L)\,(\text{factor}) \,/\, (S)\,(D)\,(N) \qquad \text{[EQUATION A]}$$

where:

T = Solids (ash) retention time in the kiln, minutes
Theta = Solid material angle of repose, degrees (Typical range is 35 to 40 degrees.)
L = Kiln length, feet
S = Kiln slope, degrees
D = Kiln refractory lining ID, feet
N = Kiln rotational speed, rpm
Factor = Adjustment for lifters and suspension chains. (Usually 1.0 for incinerator kilns.)

*Loading*

The solids loading can be calculated as follows:

$$\%LDG = (100)\,(cfh) \,/\, (60)\,(fpm)\,(A) \qquad \text{[EQUATION B]}$$

where:

%LDG = Solids % loading, the percent of kiln cross sectional area (based on refractory ID) theoretically filled with material.
cfh = Volumetric flow of solid material through kiln, cubic feet/hour.
fpm = Solid material travel rate through kiln = shell length divided by solids retention time, L/T.
A = Kiln cross sectional area, $ft^2$, based on refractory ID.

**PCC Process Parameters.** Volumetric heat release = 7000 to 14 000 Btu/ft$^3$. (Typical average is 11 500 Btu/ft$^3$.) Kiln flue gas exit velocity = 15 ft/sec, maximum. (Typical average is 10 ft/sec.)

**Kiln Design Procedure Sequence.** (See Figures 7, 8, and 9.)
Start with T = 90 minutes and N = 0.5 rpm.
With a trial D, calculate L from Equation A.
Check if resultant L/D is within 3:1 to 5:1 range.
Calculate kiln volume = (kiln design heat release, Btu/hr) divided by (average volumetric heat release of 11 500 Btu/ft$^3$hr).
Adjust D to match kiln volume at L.
Choose compromise D and L values to get L/D = 3:1 to 5:1.
Using Equation B, check to see that % LDG is between 6% to 12.5% with selected D and L.
Estimate flue gas acfh at design heat release + auxiliary fuel input.
At kiln cross sectional A, based on D, check if kiln exit flue gas velocity is <15 ft/sec.
Adjust D and L as required to meet PCC design and process parameter limits.

EXAMPLE illustrating use of Figure 7.

*Given:* KILN -- 1347 ft$^3$ volume, 1 360 000 Btu/hr heat loss, 1° slope, 3 rpm max, 37.23° solids repose angle, 10% volumetric loading, no effect of lifters, dam or baffle adequate for volumetric loading.

*$c_p$ = specific heat at constant pressure*

LOAD -- $c_p$ of solid = 0.25 Btu/lb°F, solids at 60 F, 100 lb/ft$^3$ solids bulk density, zero HHV in solids, 20% moisture in solids, 30 minutes required retention time at 1000 F.

FUEL, FIRING, TEMPERATURES -- 60 F reference temperature,
#2 fuel oil with 18 993 Btu/lb HHV,
fuel and air at 60 F,
$h_V$ = 7000 Btu/hr ft$^3$,
humidity = 0.013 lb $H_2O$/lb dry air
$t_g - t_s$ = 600°(F)
maximum heat release = 25 million Btu/hr.

*Solution:* Select Figure 7. Enter at 20% moisture, and proceed vertically until either of the 1000 F curves is reached, or until the 30-minute horizontal line is reached, whichever is reached first (lowest).

In this case, find that the 1000 F curve for *heat release* is reached first; so at that point, read 21 600 lb/hr on the left scale. At the same point, read that operating at 21 600 lb/hr will afford nearly 37 minutes of retention time.

*Comments, Alternate Situations.* If the required retention time had been 42 minutes rather than 30, the 42-minute horizontal line would have been reached before either of the 1000 F lines. In that instance, the feedrate would have been 19 500 lb/hr as controlled by the needed residence time.

If the solid had a bulk density of 120 lb/ft³ in this latter case, the maximum feedrate would have been 19 500 (120/100) = 23 400 lb/hr. If the maximum volumetric loading were limited to 6%, then the feedrate would be reduced to 23 400 (0.06/0.10) = 14 040 lb/hr.

**Figure 7. Data for one design of rotary drum incinerator processing contaminated soils, 10% volumetric loading.** See example illustrating its use. *(Acknowledgements on page vi.)*

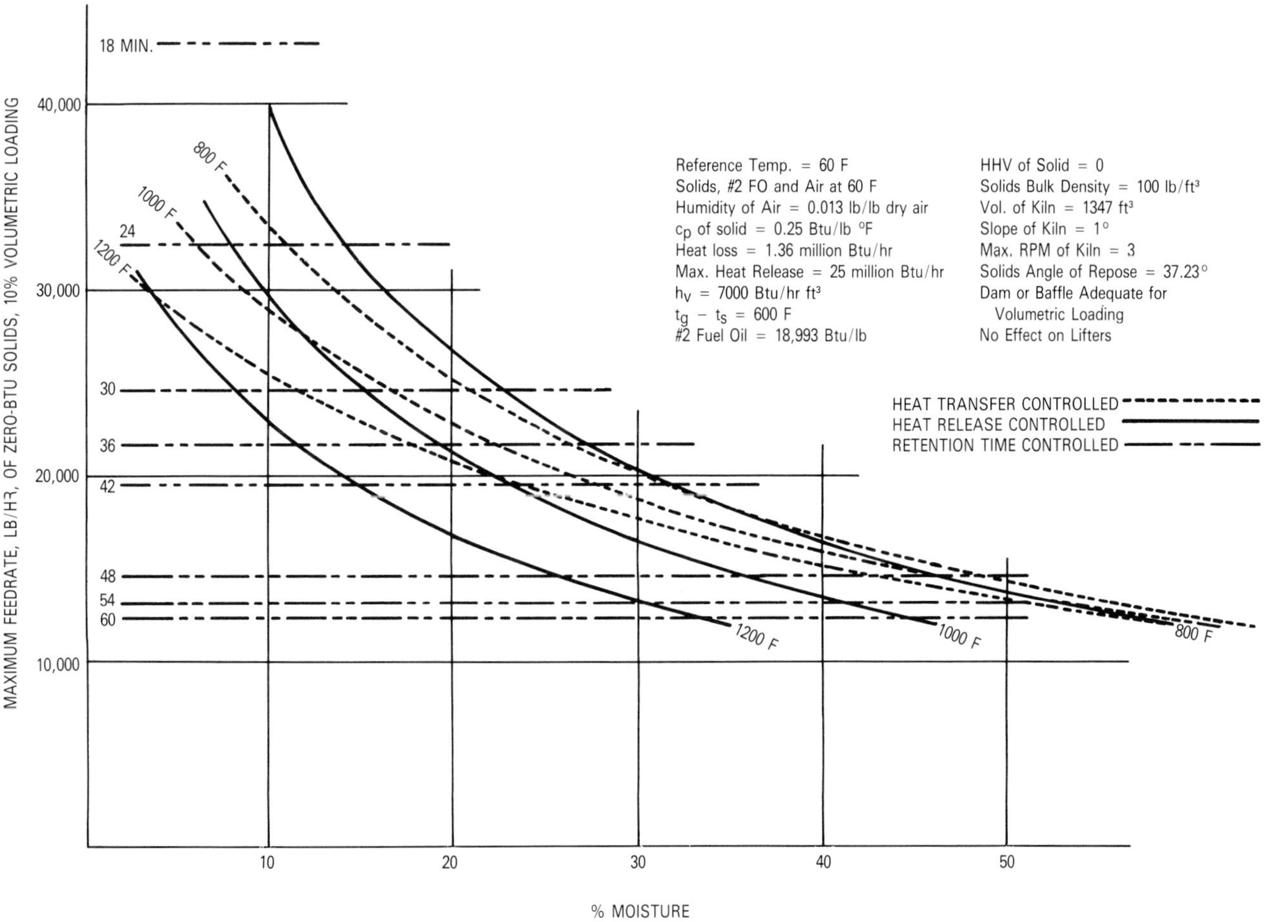

**Figure 8. Data for one design of rotary drum incinerator processing contaminated soils, 7.5% volumetric loading.** *(Acknowledgements on page vi.)*

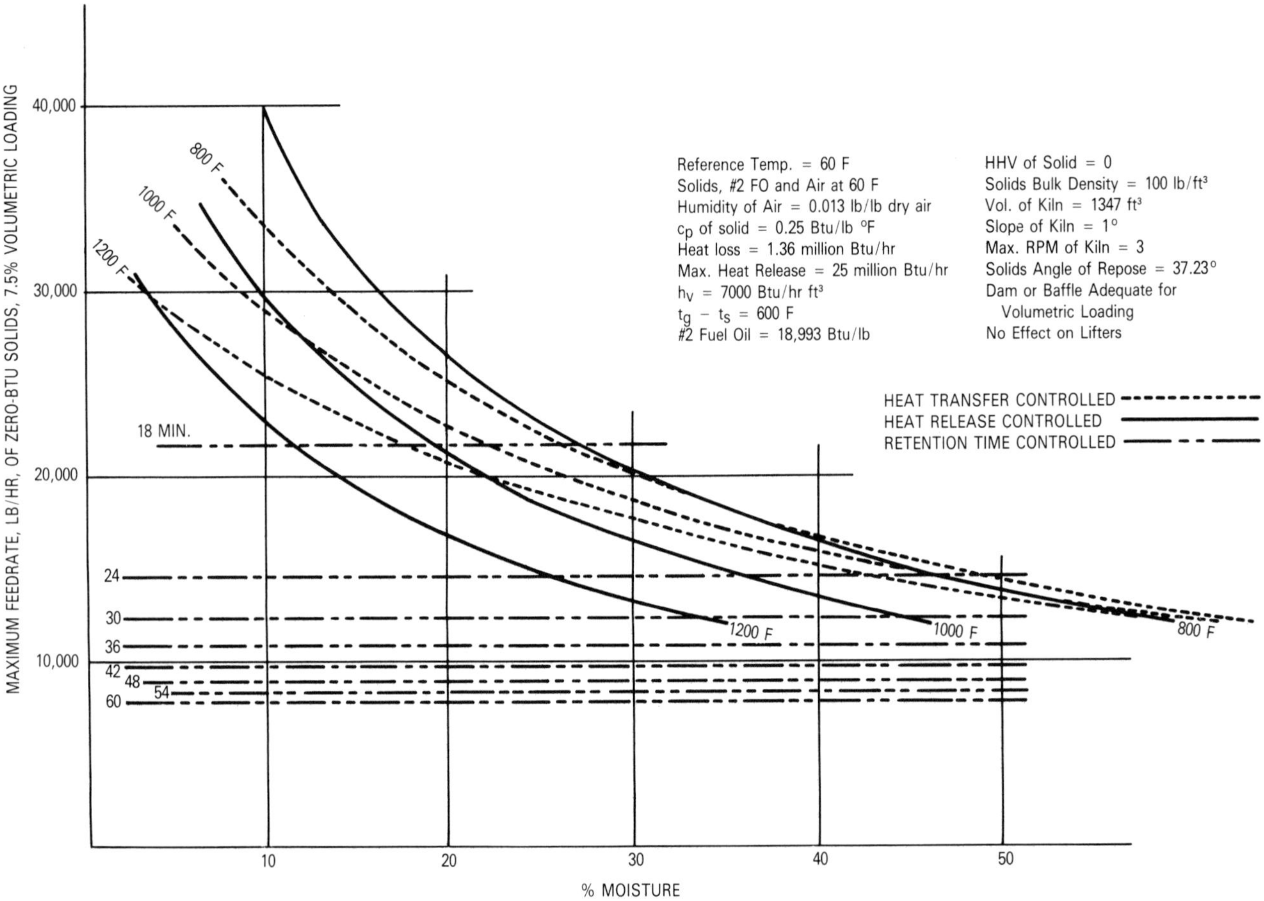

**Figure 9. Data for one design of rotary drum incinerator processing contaminated soils, 5% volumetric loading.** *(Acknowledgements on page vi.)*

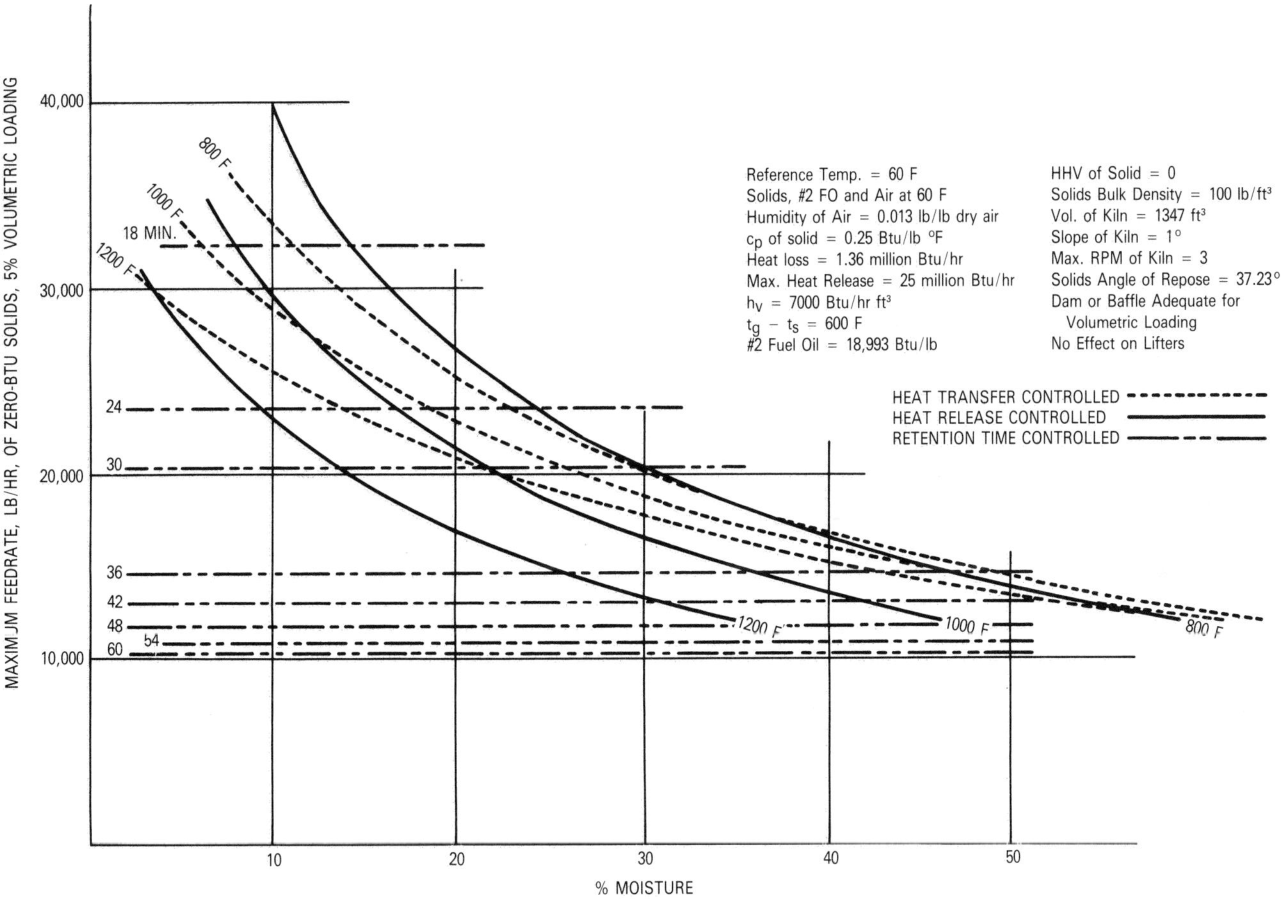

## 10. SLAGGING KILNS

The present practice of operating PCCs in the slagging mode is on the decrease worldwide. Originally championed in European facilities and later in a few facilities in North America, this mode of operation has fallen into disfavor based on results of actual sampling and testing of the quenched ash.

The slagging mode is achieved by operating the kiln above 2200 F and with high energetic solids so that the ash melts, producing a molten slag. The molten slag flows through the kiln to a reservoir pool near the kiln discharge end. The overflow from this reservoir pool flows out the kiln discharge end into the mixing chamber water pool where it is quenched, the intent being that the quick quenching of the molten slag by the water pool would achieve an impervious glassy encapsulation of the metals constituents and any residual organics, thus rendering the ash inert and safe for land filling.

Actual sampling and testing of the ash, however, revealed that the quenching of the molten slag mass was fracturing the glassy encapsulation layer thus exposing the metal constituents and/or residual organics on the ash surface. The level of this exposure has been demonstrated to exceed the allowable Toxicity Characterization Leaching Procedure (TCLP) limits; so the ash must still be considered a hazardous or toxic waste requiring further process treatment. **TCLP**

Furthermore, to ensure proper slag fluidity and to avoid localized "freezing" of the slag mass with attendant "damming" and back spill over the kiln feed dam, the feed end of the kiln has to be operated at 2200 F. At that temperature, the useful life of stainless steels and high nickel alloy components (i.e. ram feeder chute, waste nozzles, and lances) is reduced to months. Because of this temperature requirement and the combined erosion, abrasion and chemical attack of the molten slag, useful refractory life can be increased from 1 to 5 years by switching to a non-slagging mode of operation.

Some work is currently under way with flux agent additions to preserve the encapsulation through the quenching action and to achieve a proper slag fluidity at lower kiln firing temperatures. Unless a fluxing agent is found that is readily available and usable in its natural state at virtually no cost, the economics of flux additions do not appear to be viable.

## 11. MIXING CHAMBER

The mixing chamber is a refractory-lined transition section between the PCC (Rotary Kiln) and the Secondary Combustion Chamber (SCC). It accomplishes the following functions:

a) The mixing chamber causes large size particulate dropout from the kiln exit flue gases by decreasing the flue gas velocity. It then turns the flue gas path through a 90 degree change in direction before entering the SCC.

b) The mixing chamber acts as a sump to contain the water pool in which solids (ash) from the kiln are cooled for subsequent removal into drums or ash tote boxes.

   The ash may be disposed of as landfill if verified as inert by sampling and analysis (total burn out of organics and POHCs has been effected in the kiln), and if it meets the Toxicity Characteristic Leaching Procedure (TCLP) requirement, currently at less than 50 ppb of PCBs and metals (Be, Pb, Hg, Cd). If the above criteria are not met, the ash must be drummed and stored. At present, any ash resulting from processing low level radioactive wastes must be drummed and stored. Investigations are currently underway concerning the feasibility of delisting this material.

c) Some mixing chambers are designed with a "pinch point" throat section to accelerate and promote mixing of the flue gas prior to entering the SCC proper. Dry ash systems have the benefit of generating less aqueous, evaporative load on the SCC; but dust control and in-leakage control become major problems. *Throat*

**Figure 10. Major incinerator system equipment components:** (b) Mixing occurs in a mixing chamber or connecting sections adjacent to the ash receiver.

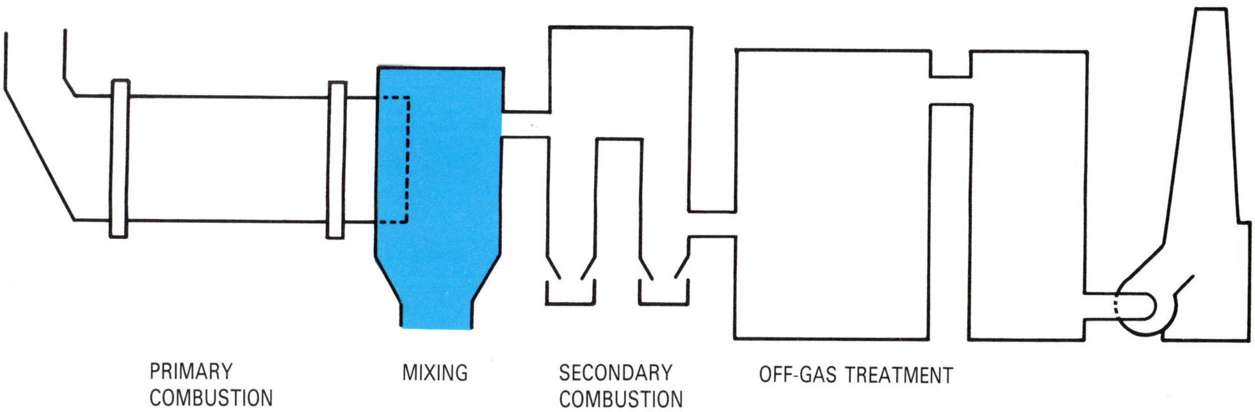

## 12. SECONDARY COMBUSTION CHAMBER (SCC)

The SCC is most often a vertical cylindrical refractory-lined vessel equipped with burners to fire auxiliary fuels and high Btu liquid wastes (HHV >10,000 Btu/pound). Some SCCs are equipped with a dedicated auxiliary fuel burner plus dedicated liquid waste burner(s). See Figures 12 and 13.

Other SCCs are equipped with a "Master Pilot Burner", one or more dual fuel burners, and dedicated liquid waste burners.

The "Master Pilot Burner" may have a heat release of 1 to 3 million Btu/hr, firing commercial hydrocarbon fuels such as natural gas, distillate fuel oil (No. 2), or residual fuel oil (No.6). The "Master Pilot Burner" should be equipped with flame supervision and an interrupted pilot.

The dual fuel burner(s) fire commercial hydrocarbon fuels to attain the permit temperature in the SCC, then switch over to waste fuel firing along with any dedicated waste burners after the waste permit interlocks are satisfied.

**Figure 11. Major incinerator system equipment components:** (c) Secondary combustion proceeds in an afterburner chamber.

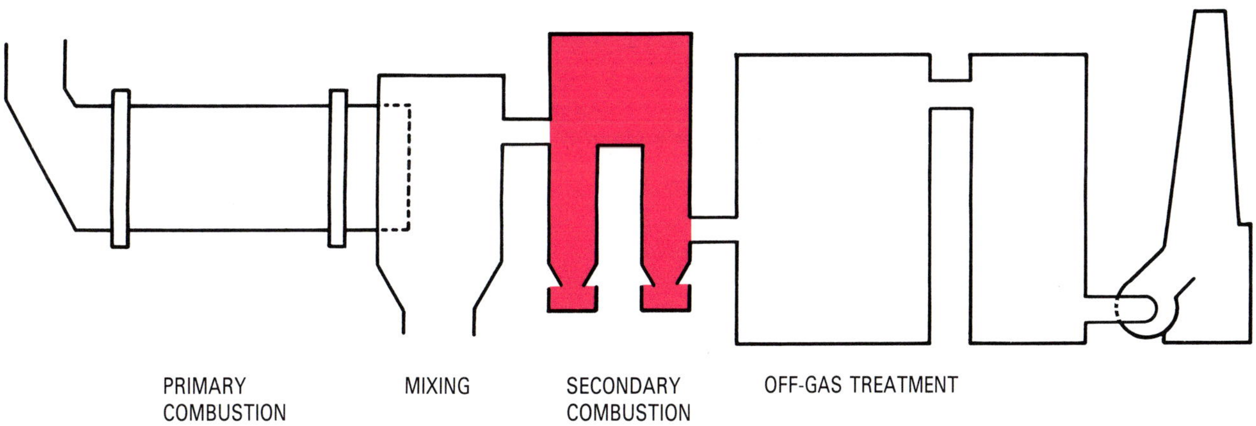

In addition to destroying secondary wastes, the SCC also ensures complete destruction (after-burning) of any residual POHCs contained in the PCC flue gas.

Satisfactory mixing and turbulence levels of the flue gas volume are set by the diameter of the SCC, inlet geometry, and burner location. The (hot) residence time is set by the length (usually height) of the flue gas flow path through the SCC.

In order to satisfy the residence time requirement and to avoid an excessively tall SCC, some SCCs are configured with an upcomer section, crossover section, and downcomer section (an inverted J or U design).

The actual residence time is figured, at the hot flue gas velocity, over a flow path from the center line of the uppermost burner to the location of the SCC control thermocouple.

**IFB**

Refractory linings for the SCC are 6″ to 9″ of 60% to 70% $Al_2O_3$ brick shapes as the hot face lining, with 3″ of insulating fire brick (IFB) as the backup lining.

**SCC Design Parameters.** Length-to-diameter ratio = 6:1 to 8:1. Flue gas hot velocity = 20 ft/sec maximum; 7 ft/sec minimum. Residence time = 2 to 4 seconds (hot).

**SCC Process Parameters.** Operating temperature: 1800 F to 2250 F

**TRV**

The top of the SCC should be equipped with a thermal relief vent (TRV) to vent the hot flue gases to atmosphere in the event of excess temperature or loss of water in the off-gas treatment (OGT) system. An auxiliary fuel burner or master pilot burner in the SCC must stay on for 30 minutes minimum during a TRV event and maintain the SCC at 1800 F.

**Figure 12. Three views of primary and secondary combustion chambers of a large incineration system in Texas, capable of incinerating liquid wastes, drummed wastes, and bulk solids.** A robotic arm sorts the bulk solids for the crusher or grinder. *(Courtesy of Chemical Waste Management.)*

**Figure 13. Liquid waste incineration.** Most modern systems inject liquid waste through the burner. Older-style arrangements have 4 peripheral atomizers in the extended tile section. Liquid fuels are usually easier to handle and burn than are liquid wastes because the fuels are more likely to have consistent characteristics.

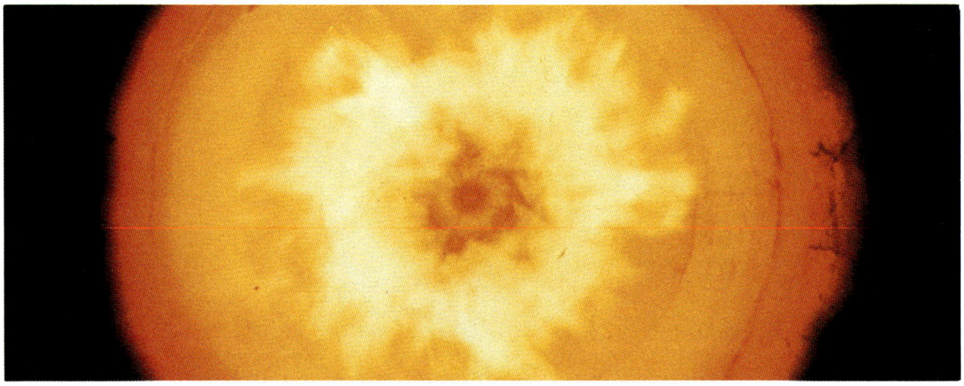

**Figure 14. A large integrated incineration system may have heat recovery, providing combustion air preheat to the SCC burner, in place of a cooler prior to the exit gas cleanup equipment.** *(Courtesy of ABB Raymond, Lisle, IL.)*

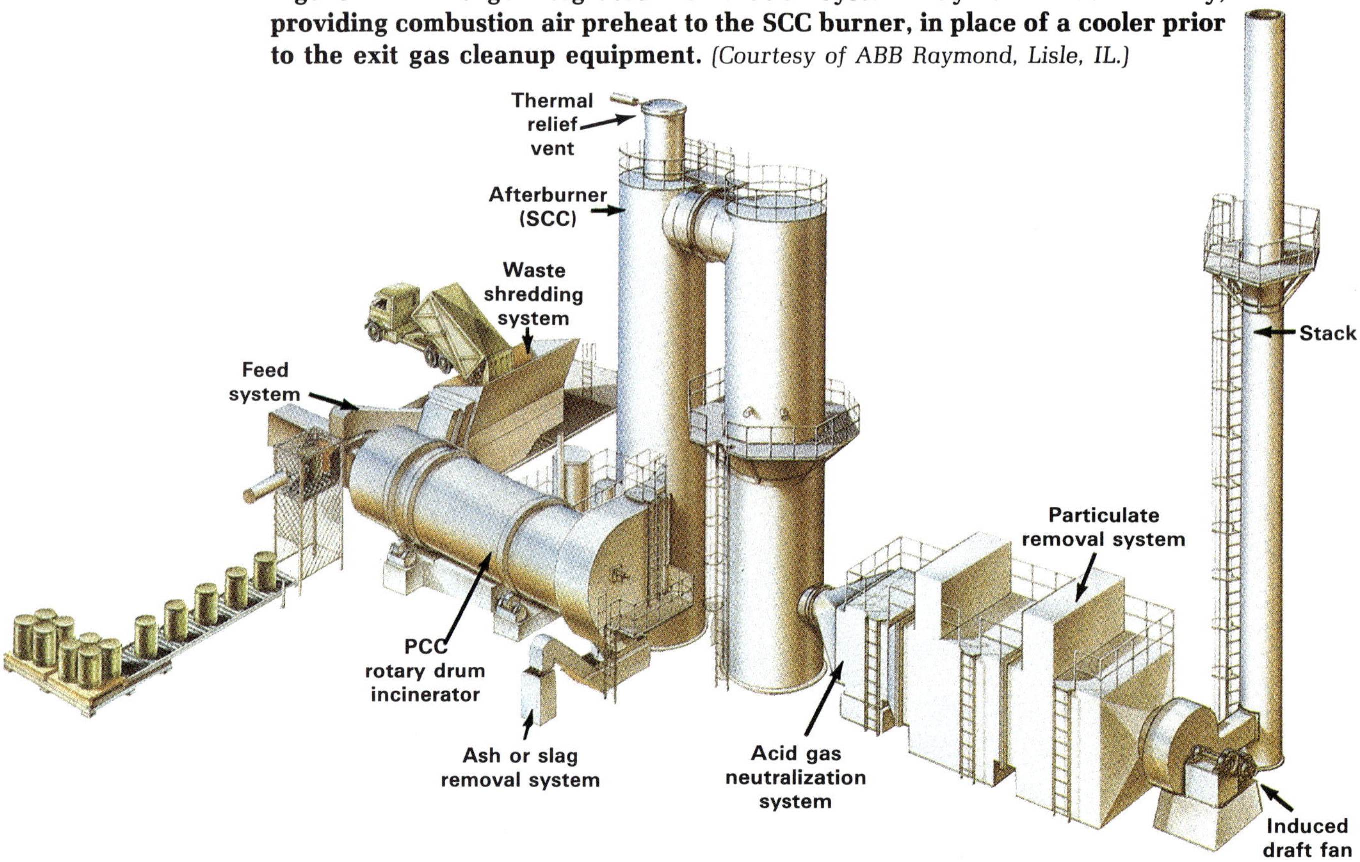

**Figure 15. Temperature required for 99.99% destruction efficiency, with various residence times.** *From J.L. Torres, Thermal Decomposition of Chlorinated Benzenes, Univ. of Dayton, 1982.*

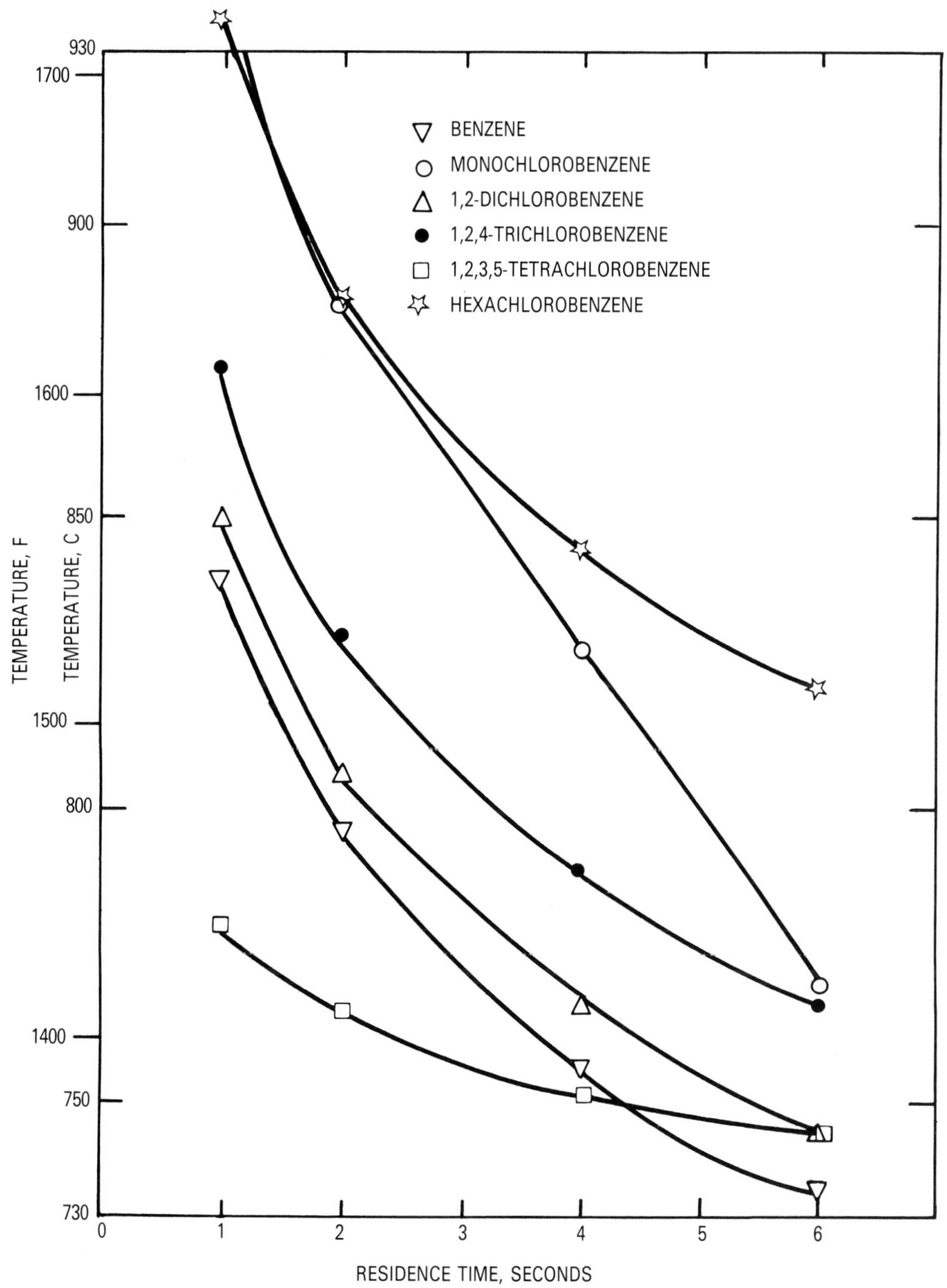

The lower 1800 F temperature is allowed, even for a TSCA waste, because the TRV event will have initiated a waste feed cutoff with subsequent decrease in the flue gas volume and hence an increase in the residence time. Under TRV event conditions, the residence time typically increases to 6-8 seconds; so at 1800 F, adequate DREs are maintained. System natural draft calculations need to be made by the designer to ensure adequate net stack effect.

## 13. OFF-GAS TREATMENT SYSTEM (OGT)

The purpose of the off-gas treatment system is to remove acidic gas constituents and particulates from the SCC off-gas so as to comply with EPA, state, and local limits. There are two (2) OGT systems in use: wet and dry. **SCC OGT**

**A Wet Off-Gas Treatment System** (Figure 17) consists of the following components:

a. *Adiabatic saturation quench chamber.* The quench chamber is lined with thermal shock resistant refractory in the upper third of its height and acid-proof refractory in the lower two thirds. The quench chamber cools and saturates the SCC off-gases from 1800-2250 F down to 160-190 F. Both fresh water (top nozzle) and recycled water (lower nozzles) are sprayed into the chamber. The chamber is sized for 0.75 to 1.00 seconds gas residence time and a water recirculation rate of 7 to 9 gallons per 1000 cubic foot of gas. See Table 7. *Cooling*

In most installations, caustic solution is introduced into the quench chamber via metered injection into the recycle spray loop. Significant amounts of the entering HCl are removed in the quench chamber; so the pH of the effluent stream may range from 3 to 5.5. *pH*

**Figure 16 Major incinerator system equipment components:** (d) Off-gas treatment may be accomplished in a number of units (which will include scrubbers with an aqueous waste management system).

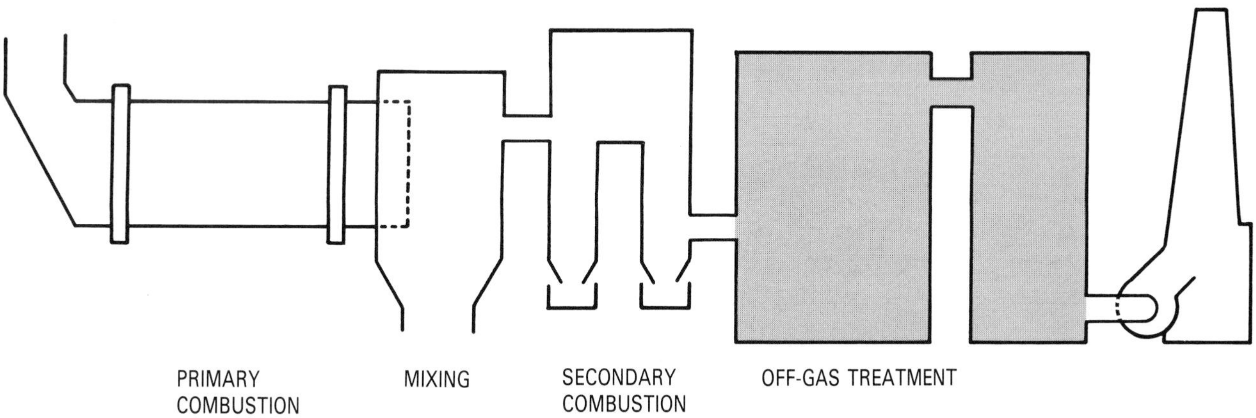

b. *Venturi Scrubber* -- low energy type. The delta P (pressure drop) across the throat ranges from 8″wc to 10″wc. The scrubber removes 90% of the >1 micron size particulate. Recirculation rates are 8 to 10 gallons/1000 acf gas.

c. *Cross Flow Scrubber* -- packed bed type with demister section. This unit removes the final acid gas constituents via adsorption onto the packing; then washes the resultant salts and carbonates to effluent removal. Caustic solution is introduced in the cross flow scrubber to maintain the recirculation stream at 6 to 7 pH. Recirculation rates are 7 to 9 gallons/1000 acf of gas.

*Particulates*

d. *Wet Electrostatic Precipitator or Ionizing Wet Scrubber* -- accomplishes final fine particulate (<1 micron size) removal. Recirculation rates range from 25 to 30 gallons/1000 acf of gas. Blowdown pH is maintained at 6.5 to 7.5.

Table 7. Effects of inlet process variables of a wet gas cleaning system on the outlet saturation temperature and humidity.

Inlet process variables:

Temperature ($T_i$), F
Pressure ($P_i$), ″wc
Humidity ratio ($H_i$), pounds $H_2O$/pound dry gas

Outlet process variables:

Saturation temperature ($T_{sat}$), F
Humidity ratio ($H_o$), pounds $H_2O$/pound dry gas

| **Case** | **Constants** | **Variable** | **Resultant Changes** |
|---|---|---|---|
| Case 1 | $P_i$, $T_i$ | $H_i$ increases | $T_{sat}$ increases, $H_o$ increases |
| Case 2 | $H_i$, $T_i$ | $P_i$ increases | $T_{sat}$ increases, $H_o$ increases |
| Case 3 | $H_i$, $P_i$ | $T_i$ increases | $T_{sat}$ increases, $H_o$ increases |
| Case 4 | $P_i$, $T_i$, $H_i$ | Gas composition deviates, % $CO_2$ increases | $T_{sat}$ increases, $H_o$ increases |

e. *Induced Draft (ID) Fan* -- establishes the negative pressure, and conveys the flue gases through the incineration and off-gas treatment systems. The ID fan is most often a centrifugal type with a radial tip, or forward curved blade configuration, and equipped with inlet volume control. The pressure rise (delta P) across the fan is typically 15 to 22″wc. Typical fan speeds are 1100 to 1400 rpm.

*Fan*

f. *Stack.* The main exhaust stack delivers the acceptably cleaned waste gases to the atmosphere at a safe and convenient elevation. It should be sized for an average velocity of 18 to 20 fps, with 100 ft of height. It also contains the sensor probes for the Stack Gas Monitoring (SGM) system, currently required to check CO, $CO_2$, and $O_2$. Proposed EPA requirements may add HCl, THC (Total Hydrocarbons), and NOx to the SGM requirements.

*Stack velocity*

**SGM**

**THC**

The stack must also be fitted with at least two EPA sample port connections at a common elevation and 90 degrees opposed, to serve as the sampling points of the stack gases during performance tests and the trial burns.

*Sampling*

In the wet type OGT system, the equipment units are fabricated of *fiber reinforced plastic* (FRP) with a fire retardant resin. The FRP used has a maximum use limit temperature of 250 F.

**FRP**

Components in the system which must be metallic (Venturi throat section, ionizing wires, charged collector plates) are constructed of high alloy nickels, such as Hastalloy C or Alloy 20.

*Materials*

**A Dry Off-Gas Treatment System** (Figure 18a) consists of the following:

a. *Tempering Chamber* -- a refractory-lined vessel that uses spray water to sensibly cool the SCC off-gases from 1800-2250 F down to 650-750 F. The tempering chamber is most often the downflow type.

b. *Spray Dry Absorption Tower(s)* remove acid gas constituents from the flue gas. The 650-750 F flue gas from the tempering chamber enters the top of the towers where a lime slurry (10% to 20% by weight) is sprayed into the gas stream. The acid gas constituents are removed at the tower bottom as calcium salts and carbonates in a dry powder form. The flue gases leave the tower(s) at 300-400 F.

*Acid gas*

c. *Fabric Filter.* The fabric filter reduces particulate emission from the flue gas to the required compliance levels. Currently, that required level is 0.08 grains per dscf corrected to 7% $O_2$ by volume for federal EPA compliance, but can be as low as 0.02 grains per dscf in California and some other states.

*Particulate*

**Figure 17. Block flow diagram of a TSCA incinerator process with wet off-gas treatment.** *(See Acknowledgements, page vi.)*

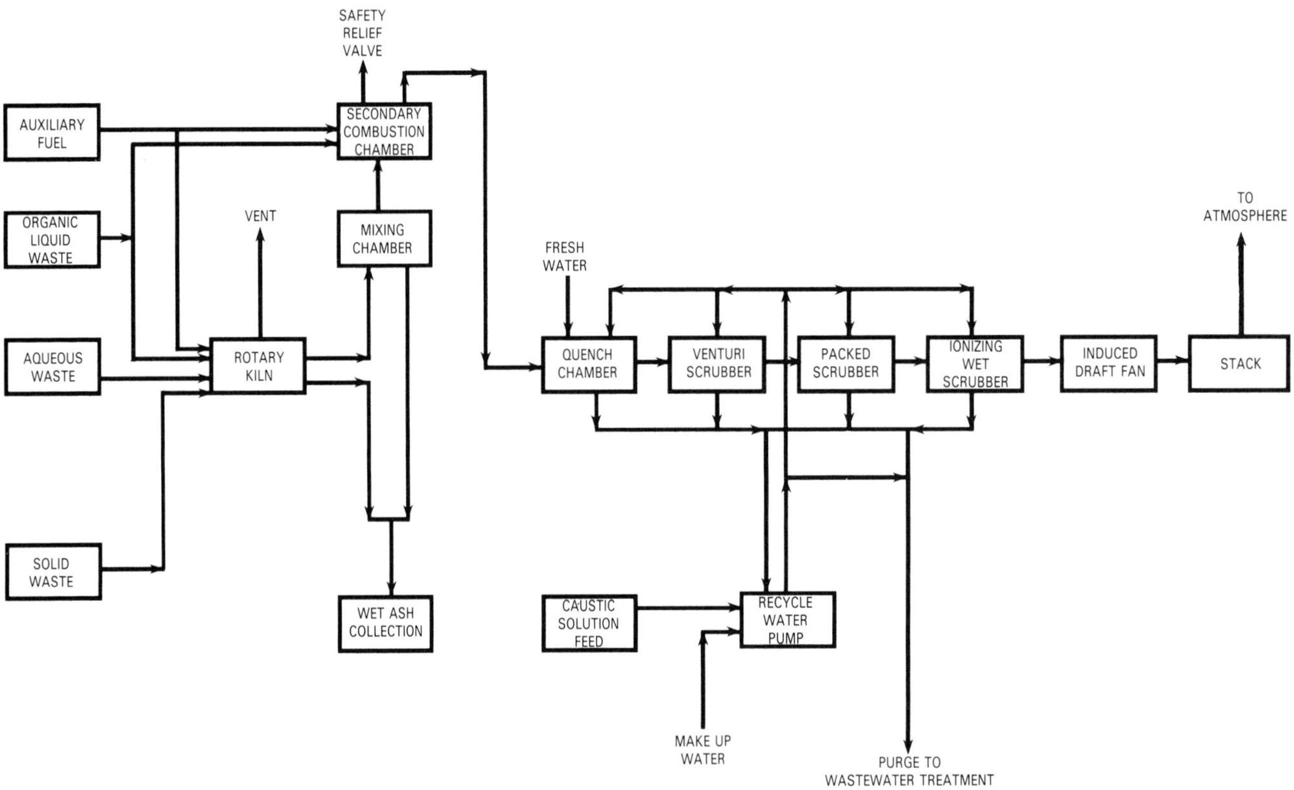

As with any fabric filter type collector, a relatively narrow temperature window is required for proper operation to avoid thermal damage to the bag material from too high a flue gas temperature and to avoid condensation, with attendant bag blinding, from too low a flue gas temperature. The usual temperature range for proper operation is >250 F but <450 F. Good combustion control is required to avoid a fire in the fabric filter from unburned hydrocarbon particulates.

*Temperature window*

d. *Dry Off-gas Induced Draft (ID) Fan.* The induced draft fan is a centrifugal type of carbon steel construction with radial tip blade configuration. It should develop a pressure rise (delta P) of 10-20″wc (hot) at 1750 rpm, and it should be equipped with inlet volume control.

e. *Dry Off-gas Stack.* Design parameters and height are the same as those for the wet OGT stack, including the SGM and EPA sample port requirements. One difference is that the dry OGT stack is carbon steel material.

**Figure 18a. Block flow diagram of an incinerator process with dry off-gas treatment.** *(See Acknowledgements, page vi. See Figures 18b and 18c for legend and data.)*

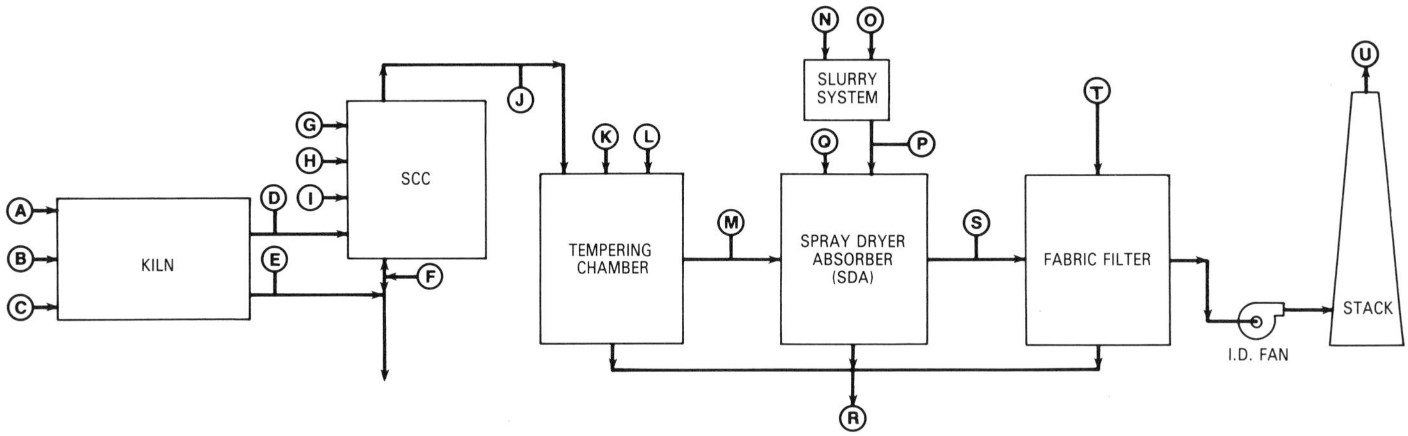

**Figure 18b. Legend and data for Figure 18a.** *(continued on Figure 18c)*

| | A Kiln Waste Input | B Aux. Fuel Input | C Combustion Air Input | | D Kiln Outlet | E Hot Ash Output | F Water Input | G SCC Outlet Input | H Aux. Fuel Input | I Combustion Air Input | |
|---|---|---|---|---|---|---|---|---|---|---|---|
| | | | From Fans | In Leakage | | | | | | From Fans | In Leakage |
| Carbon | — | 1,150 | — | — | — | — | — | — | 752 | — | — |
| Hydrogen | — | 167 | — | — | — | — | — | — | 109 | — | — |
| Nitrogen | — | — | 23,076 | 13,672 | 36,749 | — | — | — | — | 15,736 | — |
| Oxygen | — | — | 6,972 | 4,130 | 6,674 | — | — | — | — | 4,754 | — |
| Chlorine | — | — | — | — | — | — | — | — | — | — | — |
| Sulfur | — | — | — | — | — | — | — | — | — | — | — |
| Water | — | — | 391 | 231 | 2,117 | — | — | — | — | 266 | — |
| Ash | — | — | — | — | — | — | — | — | — | — | — |
| $CO_2$ | — | — | — | — | 4,213 | — | — | — | — | — | — |
| HCl | — | — | — | — | — | — | — | — | — | — | — |
| $SO_2$ | — | — | — | — | — | — | — | — | — | — | — |
| $Ca(OH)_2$ | — | — | — | — | — | — | — | — | — | — | — |
| $CaCl/CaSO_2$ | — | — | — | — | — | — | — | — | — | — | — |
| Total lb/hr | 0 | 1,317 | 30,438 | 18,034 | 49,753 | 0 | 0 | 0 | 861 | 20,756 | 0 |
| Millions Btu/hr | 0.00 | 25.01 | 0.64 | 0.38 | 23.45 | 0.00 | 0.00 | 0.00 | 16.36 | 0.44 | 0.00 |
| Btu/lb | 0 | 18,993 | — | — | — | — | — | 0 | 18,993 | — | — |
| gpm/acfm | — | — | 6,630 | 3,928 | 43,506 | — | 0 | 0 | — | 4,521 | — |
| Temp, F | 90 | 90 | 90 | 90 | 1628 | 400 | 90 | 90 | 90 | 90 | 90 |

**Figure 18c. Legend and data for Figure 18a.** *(concluded)*

| | J SCC Outlet | K Water Input | L Atom. Air Input | M SDA Inlet | N Water Input | O Lime Input | P Lime Slurry | Q Atom. Air Input | R Solids Discharge Output | S SDA Outlet | T Pulse Air & Leakage Input | U Flue Gas Output |
|---|---|---|---|---|---|---|---|---|---|---|---|---|
| Carbon | — | — | — | — | — | — | — | — | — | — | — | — |
| Hydrogen | — | — | — | — | — | — | — | — | — | — | — | — |
| Nitrogen | 52,484 | — | 2,662 | 55,147 | — | — | — | 1,421 | — | 56,568 | 4,612 | 61,179 |
| Oxygen | 8,540 | — | 804 | 9,345 | — | — | — | 429 | — | 9,774 | 1,388 | 11,162 |
| Chlorine | — | — | — | — | — | — | — | — | — | — | — | — |
| Sulfur | — | — | — | — | — | — | — | — | — | — | — | — |
| Water | 3,361 | 13,867 | 45 | 17,273 | 6,798 | — | 6,798 | 24 | 36 | 24,059 | — | 24,032 |
| Ash | — | — | — | — | — | 34 | 34 | — | 32 | — | — | 2 |
| $CO_2$ | 6,968 | — | — | 6,968 | — | — | — | — | — | 6,968 | — | 6,968 |
| HCl | — | — | — | — | — | — | — | — | — | — | — | — |
| $SO_2$ | — | — | — | — | — | — | — | — | — | — | — | — |
| $Ca(OH)_2$ | — | — | — | — | — | 568 | 568 | — | 568 | 426 | — | — |
| $CaCl/CaSO_3$ | — | — | — | — | — | — | — | — | — | — | — | — |
| Total lb/hr | 71,353 | 13,867 | 3,512 | 88,732 | 6,798 | 602 | 7,400 | 1,874 | 636 | 97,794 | 6,000 | 103,344 |
| Millions Btu/hr | 37.25 | 0.42 | 0.08 | 36.83 | 0.20 | 0.00 | 0.21 | 0.04 | 0.04 | 36.72 | 0.05 | 36.62 |
| Btu/lb | — | — | — | — | — | — | — | — | — | — | — | — |
| gpm/acfm | 66,765 | 28 | 98 | 50,980 | 14 | — | 18 | 52 | — | 42,821 | 1,307 | 41,823 |
| Temp, F | 1774 | 90 | 100 | 753 | 90 | 90 | 90 | 100 | 400 | 402 | 90 | 377 |

## 14. THERMAL ENHANCEMENT METHODS

**Combustion Air Preheat (CAPH).** Preheating combustion air has not found wide application in the hazardous, toxic or mixed waste incinerator field. Even though the flue gases exit the SCC at 1800 F to 2250 F, which provides more than enough thermal head for combustion air preheating, there are four considerations, all detrimental to some extent to combustion air preheating.

a) *Temperature.* At 1800 F to 2250 F, high temperature heat resistant alloys are required for useful service life. While commercial alloys are available to withstand these temperatures, they are expensive and at these temperatures, tertiary creep affects will govern the final finite service life of the alloys.

*Materials*

b) *Corrosion.* While high temperature heat resistant alloys will solve the temperature consideration and do so at an economical service life, the alloys may not survive the corrosion attack of HCl, HF and $SO_2$. At 1800 F to 2250 F service temperatures, the corrosion attack could accelerate stress corrosion cracking, resulting in short term failure of the alloys. A solution to this problem is to use ceramic materials for the preheater internals. While ceramic materials are commercially available to withstand flue gas temperatures to 2700 F and resist corrosion attack from gaseous HCl, HF and $SO_2$, such a CAPH unit will be expensive and somewhat fragile.

*Acid attack*

**CAPH**

c) *Economic utilization.* CAPH is only viable when applied to the auxiliary fuel burner system. While the firing and stability characteristics of some waste fuels might be enhanced on preheated air, this benefit is offset by the reduction in burner capacity when changing from cold combustion air firing to CAPH, which in turn reduces the waste fuel firing rate. The use of a dedicated waste burner designed and sized for a specific heat release and waste fuel firing rate solely to operate with combustion air preheat should be evaluated. These considerations relate to CAPH on the auxiliary fuel system only.

*Justification*

With the use of CAPH on an auxiliary fuel burner system, there are some fine details that need to be considered. Because a CAPH unit will operate with the total system flue gas volume at 1800-2250 F, the CAPH unit may generate hotter preheated combustion air than the conventional 700-1000 F. To effectively utilize this, the burner must be capable of handling 1400 F to 1600 F preheated air, i.e. with internals of high temperature heat resistant alloys and ceramic materials.

*Burner*

At air preheat temperatures of 1400 F to 1600 F, consideration must be given to the effects of increased flame temperature on the heat available, necessitating automatic reduction of auxiliary fuel consumption; but that is a gain. PCC and SCC temperature controller set points can control auxiliary fuel input.

*Fouling*

d) *Slagging/plugging.* By far the most serious problem with CAPH in incinerator facilities to date has been particulate carry-over from processing solids and sludges in the PCC. At 1800 F to 2250 F, the particulates may exhibit a sticky slag type characteristic, which has completely plugged up the CAPH unit, blocking flue gas flow in as short a time as 3 to 4 hours. Some success has been achieved in deslagging the CAPH unit by cutting off the solid and/or sludge feeds and raising the SCC exit flue gas temperature, but the required frequency was prohibitive.

*Air heater location*

For an incinerator facility operating with a wet OGT system, the only viable location for the CAPH unit is between the SCC and quench chamber. At this location, however, the first three above considerations need to be thoroughly evaluated, and if the PCC is processing solids and/or sludges the fourth consideration needs to be thoroughly evaluated also.

For an incinerator facility operating with a dry OGT system with the CAPH unit located between the SCC and tempering chamber, the above considerations also need to be thoroughly evaluated.

The possibility of locating a CAPH unit downstream of the tempering chamber in a dry OGT system, with 650 F to 750 F inlet flue gas temperature, appears to be worth evaluating. The particulate behavior characteristics at 650 F to 750 F are not presently well understood, and the low amount of thermal head would require substantial CAPH unit surface area.

Additional investigations leading to the availability of sound design data needs to be done in the area of particulate behavior characteristics versus flue gas temperature along with the amount and size of particulate carryover.

**Waste Heat Recovery.** For many of the same reasons as with CAPH (temperature, corrosion, and slagging), waste heat recovery has not been widely applied to incinerator facilities. There are a few waste heat boilers presently in operation on small incinerator units at private industrial sites.

**WTE** = *waste to energy*

There has been a major effort to build steam generators fueled with municipal waste. Because of the variable rate of waste material available, these units are best coordinated with other steam-generating or electricity-generating units that can take up the slack, or cut back in event of excess availability. For private or industrial incinerator/boiler combinations, a demand for the substantial amounts of steam capable of being generated from the incinerator facility flue gases must be readily available on site.

In the final analysis however, more knowledge and experience is needed to accurately predict the particulate behavior characteristics at the flue gas temperatures that would be encountered in the various boiler sections.

*Cogen*

Co-generation, either from waste heat steam generation or direct-fired gas turbine/generator systems followed by exhaust gas cleanup and treatment for compliance emission levels is in its infancy principally because of:

(a) Unknown particulate behavior characteristics at temperatures encountered.
(b) Capital and maintenance costs associated with the machinery and equipment involved.
(c) Ability to handle only relatively clean waste streams.
(d) Time and temperature requirements for required DRE.
(e) The necessity for committment from long term paying customers for the electricity generated.
(f) Impact of corrosive nature of the products of combustion on metallic components in the turbine and steam generator.

**Oxygen Enrichment.** $O_2$ Enrichment is a promising method for reducing auxiliary fuel usage and cost, and for reducing the required OGT system flue gas volume (OGT system cost).

*Less poc volume*

In other high temperature thermal processing systems, principally the metallurgical processing industries, application of oxygen ($O_2$) enriched burners have achieved 20% to 40% reduction in flue gas volumes by using air streams with oxygen enriched to 25-50% oxygen instead of the 20.9% oxygen in normal air.

*Payback*

Depending on the local cost of oxygen, there may or may not be a positive payback; net dollars saved by auxiliary fuel reduction over the cost of oxygen. Limited studies done to date on the use of oxygen enrichment through auxiliary fuel burners at one incinerator facility resulted in the break-even points listed in Table 8.

**Table 8. Requirements for minimum reduction in fuel input using oxygen costing \$0.20/100 ft$^3$.**

| Natural gas cost | % Oxygen | Fuel saving required to break even |
|---|---|---|
| \$3.29/million Btu | 50% | 10.7 million Btu/hr |
| | 25% | 7.2 |
| \$5.00/million Btu | 50% | 7.2 |
| | 25% | 6.8 |

The net cost analysis and break even analysis are very sensitive to the actual cost of oxygen over the life of the incinerator facility, net reduction in heat fired, and the resultant saving in auxiliary fuel at cost. The attendant reduction in the flue gas volume alone may constitute a favorable economic evaluation as follows:

For a planned new incinerator facility, the resultant decrease in the OGT system required size and hence lowered capital cost with oxygen-enriched burners compared to the size and capital cost of a conventional OGT system might be justification for construction of the incinerator, especially considering potential reductions in operating costs such as lowered quench water usage and smaller effluent volumes and treatment costs.

*$O_2$ plusses*

For an existing incinerator facility, the application of oxygen-enriched burners could achieve the following:

a. Allow an increase in waste fuel firing rate and/or waste feed processing rates to achieve the original design flue gas volume to the OGT system without a decrease in residence times or increase in unit equipment gas velocities. This condition and evaluation would need to be within any existing RCRA and/or TSCA compliance permit constraints.

b. A significant increase in performance efficiencies, e. g. combustion efficiency, acid gas removal efficiency, and particulate collection efficiency because of increased residence time and reduced packing wear and deterioration.

c. The potential for an existing particulate collection system, if of the fractional collection characteristic type, to achieve 0.01 or 0.02 gr/dscf with the reduced flue gas volume, could have significant cost saving consequences in an existing OGT system.

*$O_2$ negatives*

As with CAPH, it is necessary to make a careful evaluation of the effects of increased flame temperature on materials requirements, NOx emissions, sticky slags, heat available on the system absorbed duty, and PCC and SCC exit flue gas temperatures.

## 15. INCINERATOR BURNERS

Nozzle mix type burners are used in incinerator applications. Some general, universal, required performance characteristics for incinerator burners are described in the following sections.

a) *Auxiliary Fuel Burners* must be capable of stable operation over a wide range of combustion air inputs, from 200% excess air (lean) down to 70% of theoretical air (rich). Figures 19 and 20 are graphs showing % available heat (best possible efficiency) for three liquid wastes with varying amounts of excess air, and with a wide range of exit gas temperatures.

*Available heat*

When firing auxiliary fuels, dedicated auxiliary fuel burners (dual-fuel or combination fuel burners -- Figure 22) normally operate from 0% to 20% XSA, during the incinerator heat up sequence to the waste feed permit temperatures in the PCC and SCC. However, there are certain periods of operation (e.g. initial waste feed firing), when the firing rate of these burners must be turned down to maintain the PCC and SCC set point temperatures. Under these conditions and until thermal equilibrium is re-established, the burners may have to be operated with constant fast mix air flow, so the % XSA through the burners may reach 200%.

**XSA**

*Fuel rich operation*

In the PCC, it may sometimes be desirable to run the auxiliary fuel burner *sub-stoichiometric* (i.e. rich) to compensate for excess ambient air in-leakage through the rotary kiln feed end and discharge end seals. Excess in-leakage may result from (a) operating the rotary kiln at a temperature lower than that to which the seals were previously tensioned, or (b) normal wear on the seals.

The burner should be capable of a *turndown ratio* of 10:1 when burning gas, and of 5:1 on liquid fuels.

b) *Waste Fuel Burners* must be able to accomodate widely varying waste fuel firing rates; so the burner's *turndown ratio* must be capable of accomodating a range from high Btu liquid at low fuel flow rate to low Btu liquid at higher fuel flow rates.

*Liquid wastes*

The burner design should have broad-range *stability*, incorporating internal flame retention/stabilizing features. Some waste liquids, primarily PCC waste liquids, have relatively low HHV, and they contain inerts and water. With waste liquids of such variable composition, the auto-ignition characteristic can be adversely affected over the operating range of fuel and air flow rates.

**Figure 19. Available heat curves for primary liquid wastes, not corrected for dissociation.**

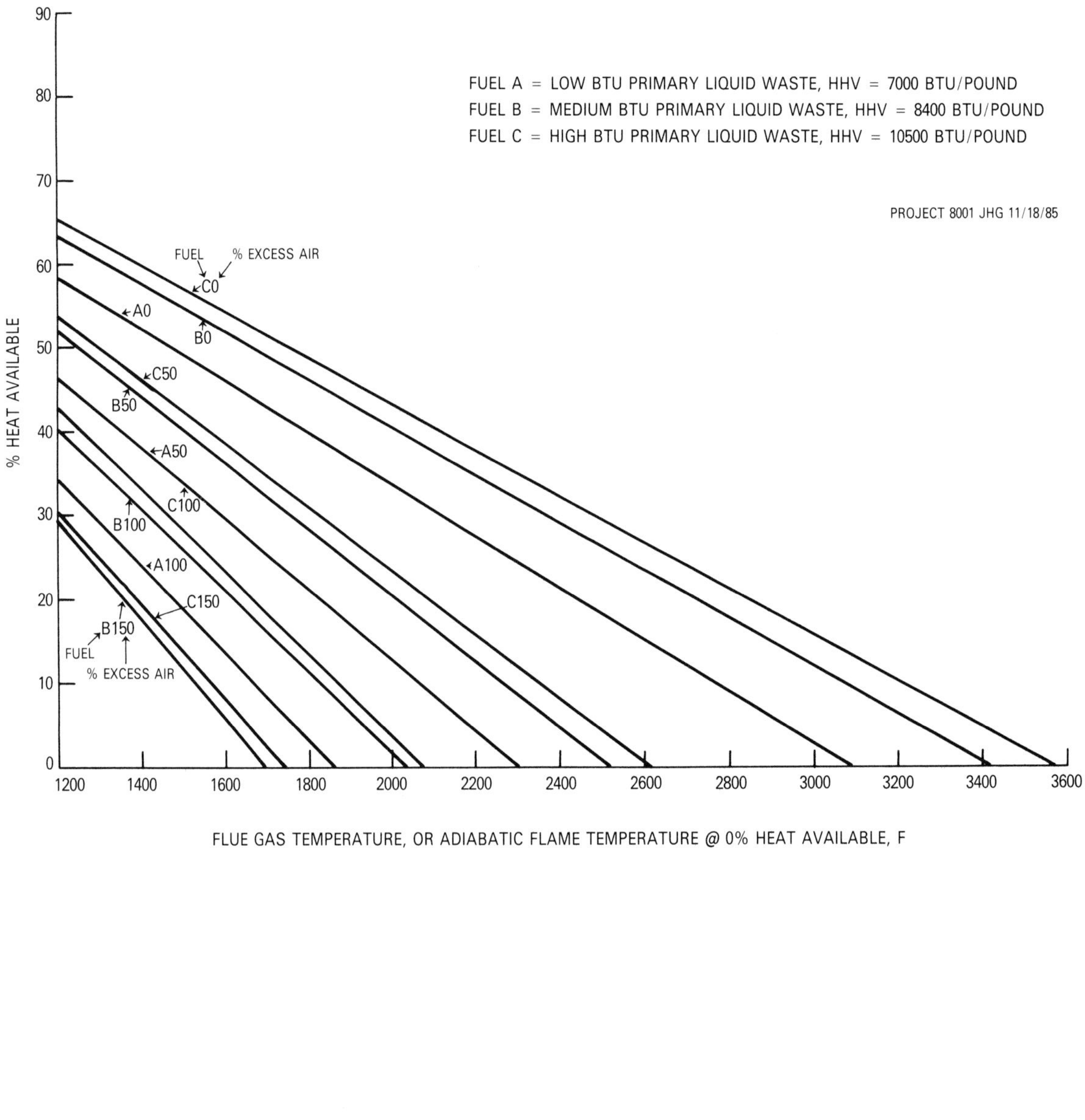

**Figure 20. Available heat curves for secondary liquid wastes, not corrected for dissociation.**

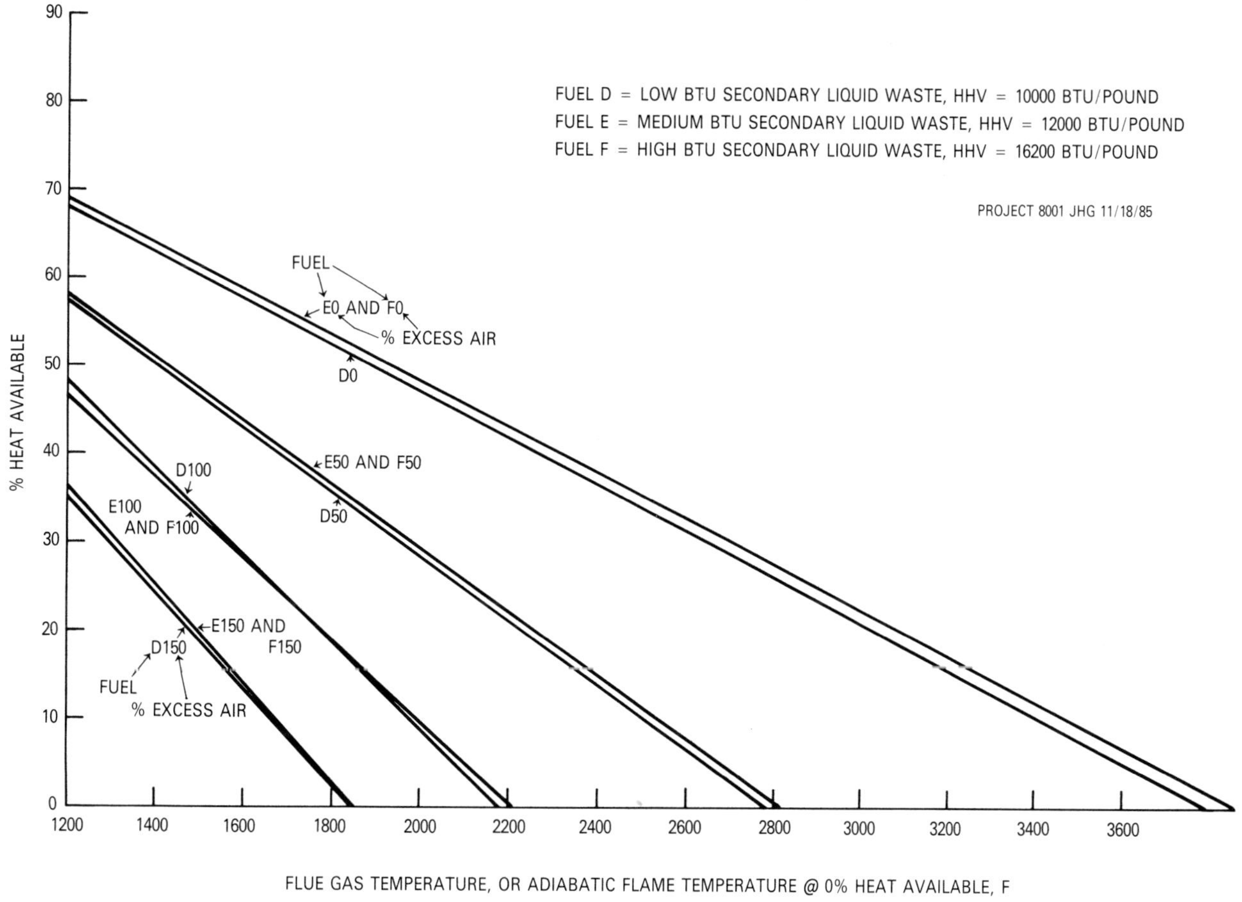

Pilots

Sturdy, stable, well-designed intermittent type pilots on the waste burners have minimized flame-outs when firing waste fuels. (An intermittent pilot operates from startup to shutdown of the waste fuel burner that it is designed to light.) For safety, both the pilot and waste fuel burner *must* have separate flame monitors (neither of which can detect the other flame) to minimize the flame-out tendency when firing waste fuels.

c) *Atomizers.* The burner should be capable of accommodating an internal type atomizer for clean waste fuels or an external type atomizer for dirty waste fuels with significant ash content. Two different type atomizers for dirty waste fuels are shown in Figure 23.

Flame throw

**Burners for Primary Combustion Chambers.** Figures 13, 21, 22, 23. For a PCC Rotary Kiln with significant solids throughput, ability to vary the flame throw over the length (or portion of the length of solids) is a beneficial characteristic to ensure complete burnout of any organic constituents.

In a rotary kiln, well defined, tight, forward flames are desired to avoid flame impingement on the refractory walls and lifters. Multiple fuel gun burners, dual-fuel (combination) burners, and Fuel-Directed® burners have been used successfully on incinerator PCCs.

**Burners for Secondary Combustion Chambers.** In SCCs, many small heat release burners are often used. The attendant fuel train piping/flame supervision complexity is not desirable. To keep the number of burners to a minimum, the heat release of the burners is often increased, resulting in the flame throw from standard nozzle mix burners exceeding the SCC diameter with flame impingement/raw fuel burning and/or accumulation on the refractory walls. A fast mixing type burner (Figures 24 and 25) eliminates these problems. Both high pressure and low pressure atomizers are utilized.

Flame impingement

**Basics for Most Burners.** New fuels, including many waste and by-product materials, make it essential that incineration engineers and managers understand some basics of burner design.

A burner usually conforms to the principles that govern the behavior of free jets. Their fundamental requirement is to provide enough energy at the burner orifices to project the combustion air and fuel to the desired flame tip position. The injection energy must equal the kinetic energy at the flame tip. The total injection energy is the sum of the injection energies of the fuel, air, and entrained jet stream, each calculated separately from:

Injection jets

$$\text{Injection energy, ft-pounds/sec} = \frac{(\text{flow rate, pounds/sec})\ (\text{velocity, ft/sec})^2}{(2)\ (32.2\ \text{ft/sec}^2)} \qquad \text{[EQUATION C]}$$

**Figure 21. Left to right: fuel train, pilot train, atomizing air, main air, and tertiary air supplies to a North American forward flame type burner on a rotary drum kiln in Kansas.**

**Figure 22. Tri-fuel incinerator burner for destroying hazardous wastes in a rotary kiln -- burns gas and/or fuel oil, and a waste stream.** Top left view shows the burner's hot face with its liquid waste atomizer at the 3 o'clock position and the oil atomizer and gas injector on burner centerline.

**Figure 23a. Fuel-Directed® Burner with two adjusting screws for changing proportions of radial and forward gaseous fuel, provides an adjustable thermal profile.** Lower views show high pressure atomizers (external mixing and internal mixing) that may be used with this and several other types of burners.

**Figure 23b. Typical external nozzle mix type atomizer (top); typical internal full emulsion type atomizer (bottom).**

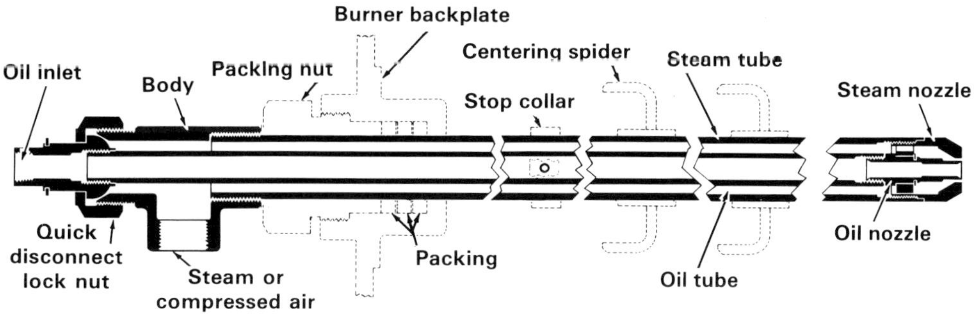


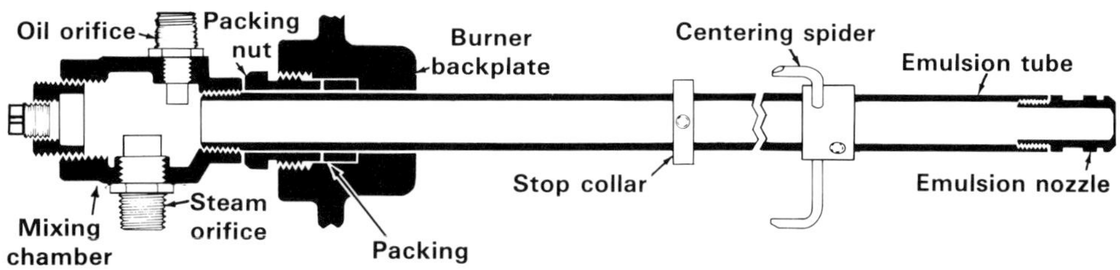

**Figure 24. A fast-mixing dual-fuel burner.** Progressing along the burner body from right to left, main air is supplied at 3 o'clock (near side), auxiliary gas or waste gas at 12, atomizing medium (steam or air) and liquid waste or fuel oil at the back end.

**Figure 25. A fast-mixing dual-fuel burner with a short flame that helps avoid impingement on chamber walls.** Top photo shows fast mix burner firing natural gas at 16 million Btu/hr (flame dimensions 5′ dia. × 6′ long). Lower photo shows same burner firing #2 fuel oil at same 16 million Btu/hr rate and same flame dimensions.

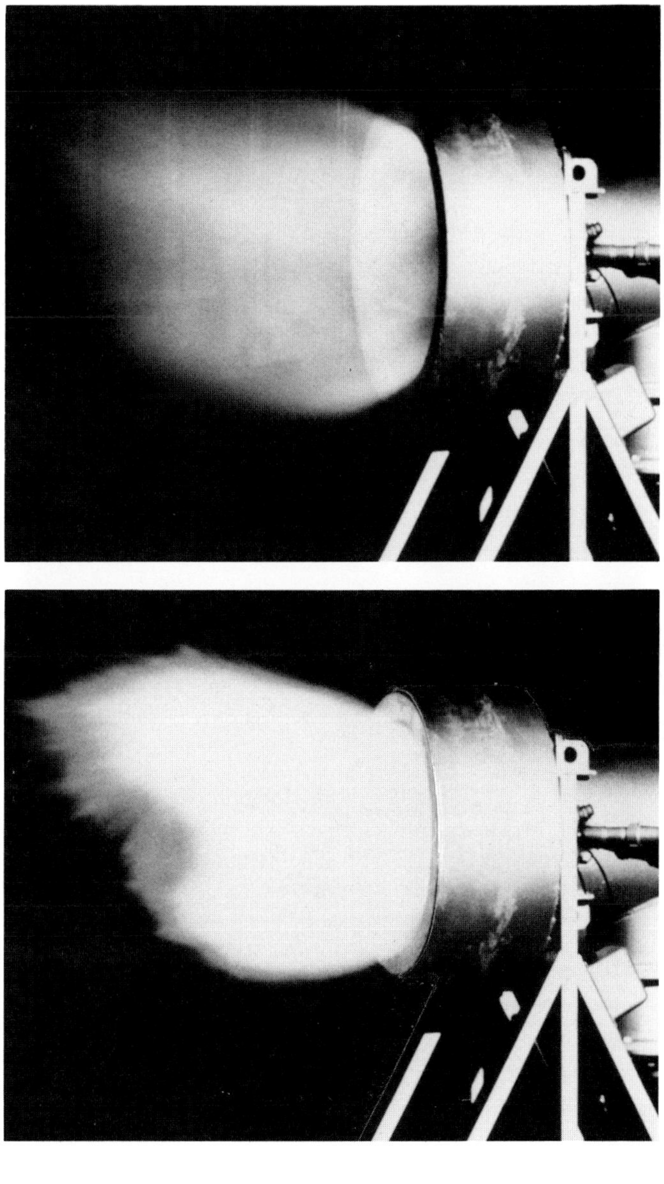

The jet stream entrainment may be estimated by EQUATION D:

Entrainment $ft^3/sec = (0.32\ L\ Q_0/D') - Q_0$

where L = flame length, ft

$D'$ = 0.5 (flame diameter, ft)

$Q_0$ = air + gas injected through the burner, $ft^3/sec$

Entrainment pounds/sec =

($ft^3/sec$) (density from equation G below, at estimated mean temperature) [EQUATION D]

Air velocity is calculated on the basis of effective net free area in the burner air orifice. Because of the area of an oil gun collar, air flow velocity for an oil burner is sometimes estimated at 60% of the tile open area. When calculating the fuel energy, it is important to realize that the maximum velocity of a gas or a vapor may be the critical or sonic velocity. Oil atomized with steam or compresssed air is injected as a gas or vapor.

For standard nozzle-mix type burners for gas or well-vaporized oil, the flame throw and flame diameter may be estimated from the following EQUATION E:

Flame throw, in feet = $2.67\ R^{0.625}$ for type A flames*;

Flame diameter, in feet = $1.375\ R^{0.2275}$ for type A flames;

Flame throw, in feet = $2.67\ R^{0.307}$ for type C flames;

where R = heat release rate in millions of Btu/hr. [EQUATION E]

The mean flame velocity at any point along the flame length can be approximated by EQUATION F:

$$\text{Mean fps} = \frac{\text{\#fuel/sec} + \text{\#air/sec} + \text{\#entrained/sec}}{\text{(cross sectional area, sq ft) (flame density, \#/ft}^3\text{)}}$$

[EQUATION F]

The mean density of the gases in the flame depends on the estimated flame temperature at the point of the cross-sectional area used in Equation F. That density can be approximated by EQUATION G:

Density, $\#/ft^3 = (0.076)(520) / (460 + \text{flame temp, in F})$ [EQUATION G]

* Slow-burning particulates may result in considerably longer flames. Type A flames have zero swirl, and an L/D ratio from 5:1 to 8:1. Type C flames have considerable swirl, and an L/D ratio of about 1:1. See North American's Handbook Supplement 230, or Part 6 of Volume II of the North American COMBUSTION HANDBOOK.

To better predict the firing characteristics of waste fuels and to achieve the best burner for the application, the North American Manufacturing Company's "Stoic" computer program may help determine the % heat available with various flue gas exit temperatures and excess air rates. It also prints out the flue gas analysis (which may help in evaluating the corrosion potential of the flue gas), and stp volume flow rates. The limitations of this program are that its input accepts only 74 gaseous compounds or 5 elements and water for solid or liquid wastes; which may not be inclusive enough for some waste streams. North American is attempting to expand the program capability to include a greater number of liquid compounds.

*Computer program*

## 16. COMBUSTION CONTROL

Combustion control systems on incinerator facilities usually incorporate the requirements and devices specified or suggested by NFPA or the insuring organization.

*Permissives*

All required pre-ignition permissives -- purge cycle times, trial for ignition periods, and running interlocks -- are provided. Where applicable, approved devices from an approved manufacturer are provided.

*Flame monitoring*

Flame supervision (flame scanner and programmer) is provided on all burners, but sometimes not on nozzles or lances. Interruptible type pilots are provided on the primary auxiliary fuel ignition source burner in the SCC and PCC.

The combustion control system is often furnished on a pre-assembled, pre-piped, and pre-wired rack. A local control panel, which houses the flame programmer and all required hard-wired relays and timers, is usually mounted onto the rack.

*Interlocks*

The minimum pre-ignition permissive parameters and devices applicable with auxiliary, liquid, gaseous, and waste fuels are:

- Low fuel pressure
- High fuel pressure
- Low atomizing medium pressure
- High atomizing medium pressure
- Low atomizing/fuel differential pressure
- Low combustion air pressure
- Low fire position for primary auxiliary ignition source fuel flow control valves for SCC and PCC auxiliary burners
- Fuel shut-off solenoid valves, proof of closure
- Induced draft fan damper, prove open
- Induced draft fan, prove running.

*Programmed startup*

All of the devices must be proved to be in the proper positions (open or closed, contacts made or not made), and proved that parameters are at acceptable levels before the purge cycle may be initiated.

Once the purge cycle has been successfully completed, the trial for ignition may be initiated for the pilot on the SCC auxiliary fuel burner.

The purge cycle time is set to accomplish 4 to 5 volume changes in the PCC and SCC on cold air, but for most incinerator facilities, a minimum of 5 minutes is set for the purge cycle.

The *trial for ignition time* should be set in the burner programmer in accordance with the insuring authority's specifications: usually

direct spark . . . . . . . . . . . . . . . . . . . 5 seconds
gas pilot . . . . . . . . . . . . . . . . . . . . . . 10 seconds
main flame, gas or light oil . . . . . . 10 seconds
main flame, heavy oil . . . . . . . . . . . 15 seconds.

*Fuel shutoff*

There should be redundant automatic fuel shutoff valves in a "double block" series arrangement. Some insuring organizations still require a vent solenoid valve and vent line through the roof for gaseous auxiliary fuels at over 400,000 Btu/hr heat release per burner.

With the exception of the following three pre-ignition permissives, the balance of the pre-ignition permissives become running interlocks upon proof of main flame ignition:

a) low fire position for primary auxiliary fuel ignition source fuel flow control valves supplying all the PCC and SCC auxiliary burners,

b) proof of closure of automatic fuel shutoff valves,

c) induced draft fan damper proved open.

In addition, the following process parameters should become running interlocks:

a) high temperature limit in the PCC,

b) high temperature limit in the SCC,

c) loss of flame via burner flame monitor.

Loss of one of these running interlocks on a particular waste burner, nozzle, or lance, should initiate a shutdown of that specific burner, nozzle, or lance.

*Shutdowns*

If a loss of a running interlock occurs on an auxiliary fuel burner in the PCC, then a shutdown of that specific burner and all PCC waste burners, nozzles, and lances and waste feed should occur immediately.

If a loss of a running interlock occurs on the SCC auxiliary fuel burner or SCC master pilot burner, then a total combustion system shutdown should occur (all waste burners, nozzles, lances, and PCC auxiliary fuel burner or PCC burner on auxiliary fuel and waste solids feeds to the PCC).

The pre-ignition permissives and running interlocks should be as specified by Owner/Operator's individual insuring agency and local codes.

NFPA requirements incorporate the insuring agency requirements in total and require in addition the following pre-ignition permissives:

*Prior to prepurge and postpurge* --

a) Combustion air damper(s) proved 100% open

b) Induced draft fan damper proved 100% open

c) All fuel (waste and auxiliary) flow control valves at their low fire position

d) ID fan and combustion air blower(s) proved running via motor contact(s).

*Prior to pilot trial for ignition* --

a) Combustion air dampers and valves at their low fire position

b) All fuel (waste and auxiliary) flow control valves at their low fire position

*Prior to main flame trial for ignition* -- same as pilot trial for ignition, plus

a) All fuel shutoff valves proved open.

Only the last three additional NFPA pre-ignition permissives on page 70 become running interlocks upon proof of main flame ignition.

*Startup program sequence*

A typical ignition sequence for an incinerator facility is described below.

*First Burner Start* -- SCC auxiliary burner or SCC master pilot burner or designated SCC dual-fuel burner *on auxiliary fuel.*

a) SCC auxiliary burner or SCC master pilot burner or designated SCC dual-fuel burner on auxiliary fuel.

b) Pre-ignition permissives satisfied.

c) Purge cycle completed.

d) Trial for gas pilot ignition completed, i.e. flame scanner proves pilot flame within prescribed 10 seconds.

e) Main auxiliary fuel automatic block valves open admitting auxiliary fuel to the burner with the fuel flow control valve at the low fire position.

f) Pilot flame establishes main flame ignition.

g) After the prescribed 10 or 15 second trial for main flame igntion, pilot flame is programmed off. If the flame monitor does not then detect presence of main flame, all automatic fuel valves trip shut, operators should locate the trouble, purge, and start the burner lighting procedure again from the beginning.

h) If main flame is proved, the SCC temperature control loop is then placed in the "auto" mode, and it begins increasing the SCC temperature to the RCRA or TSCA operating temperature according to refractory heat-up schedule.

*Second Burner Start* -- PCC auxiliary burner or PCC designated dual-fuel burner *on auxiliary fuel.*

a) Pre-ignition permissives satisfied.

b) Purge cycle completed.

c) Trial for pilot ignition completed, i.e. flame scanner proves pilot flame within the specified 10 seconds.

d) Main auxiliary fuel automatic block valves open admitting auxiliary fuel to the burner with the fuel flow control valve at low fire position. See Figure 26.

e) Pilot flame establishes main flame ignition.

f) After a prescribed 10 or 15 second trial for main flame ignition, the pilot flame is programmed off. If the flame monitor does not then detect presence of main flame, all automatic fuel valves trip shut, operators should locate the trouble, purge, and start the burner lighting procedure again from the beginning.

g) If main flame is proved, the PCC temperature control loop is then placed in "auto" mode and it begins increasing the PCC temperature to the RCRA or TSCA operating temperature according to the refractory heat-up schedule.

After the SCC is at the RCRA or TSCA operating temperature, the ignition procedure for the *SCC waste burner(s)* may commence.

After the PCC is at the RCRA or TSCA operating temperature, the ignition procedure for the *PCC waste burner(s)* may be started. Waste nozzles, sludge lances, and solids feed may also be started.

*Waste burners*

The ignition sequence for the waste burners is similar to that for the auxiliary fuel burners. One notable exception, however, is that the pilots on the waste burners may sometimes be the intermittent type (rather than interrupted), i.e. the pilot stays on as long as the burner is on...*if* separate flame scanners for pilot and main flame are located so neither scanner can detect the presence of the other's flame.

All system running interlocks are active. Failure of the SCC auxiliary burner or master pilot burner due to a loss of a running interlock will initiate a total combustion system shut down. A system re-purge cycle is required to re-start.

Figures 27, 28, 29 and 36 illustrate typical combustion control *piping and instrument diagrams* (P&ID).

**Figure 26. Automatic shutoff (block, trip) valves for auxiliary fuel are located in the lowest pipe (inlet, from right) of this pre-packaged fuel train designed and built by North American Manufacturing Co.**

**Figure 27. Left (input) side of a 3-part P&ID drawing, continued on Figure 28.** Legend is on Figure 29. *(Courtesy of ABB Raymond Combustion Engineering.)*

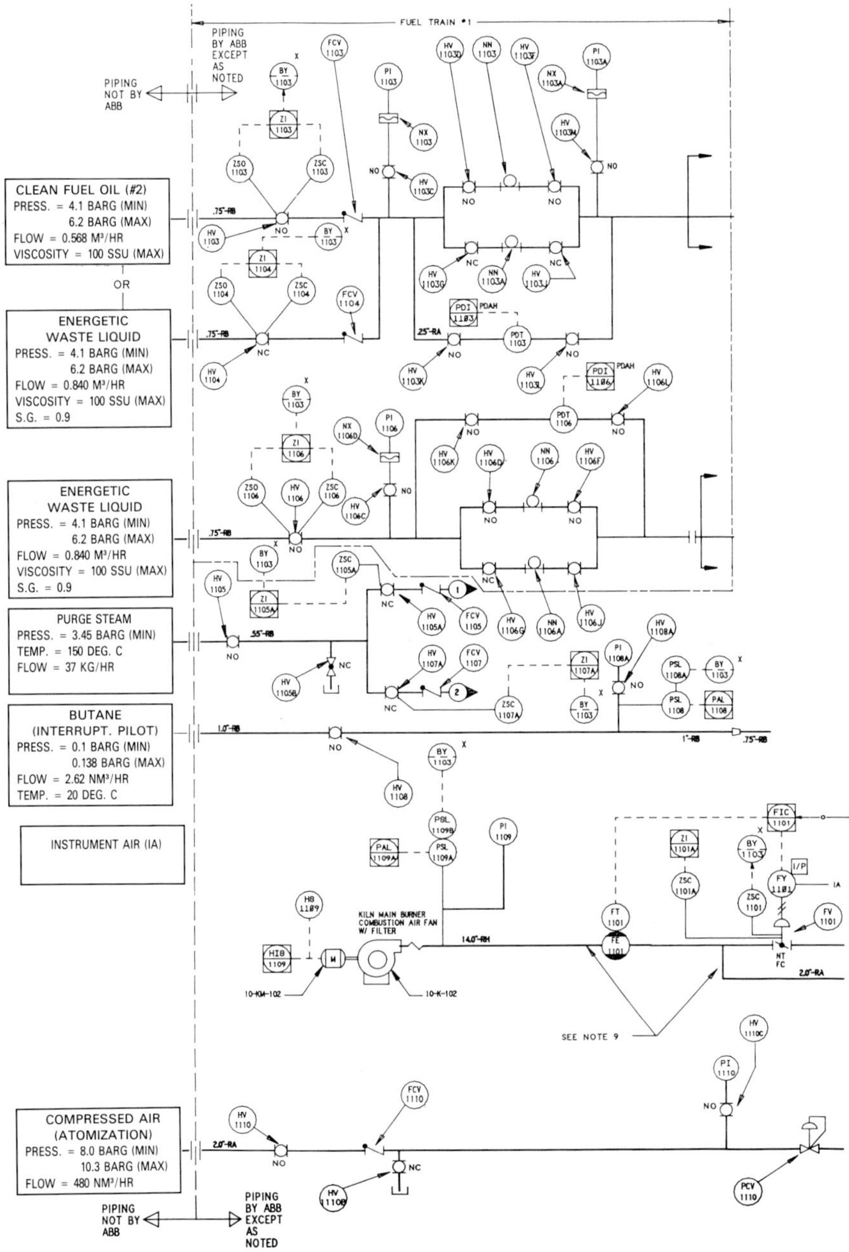

**Figure 28. Right (burner) side of a 3-part P & ID drawing, concluded from Figure 27.** Legend is on Figure 29. *(Courtesy of ABB Raymond Combustion Engineering.)*

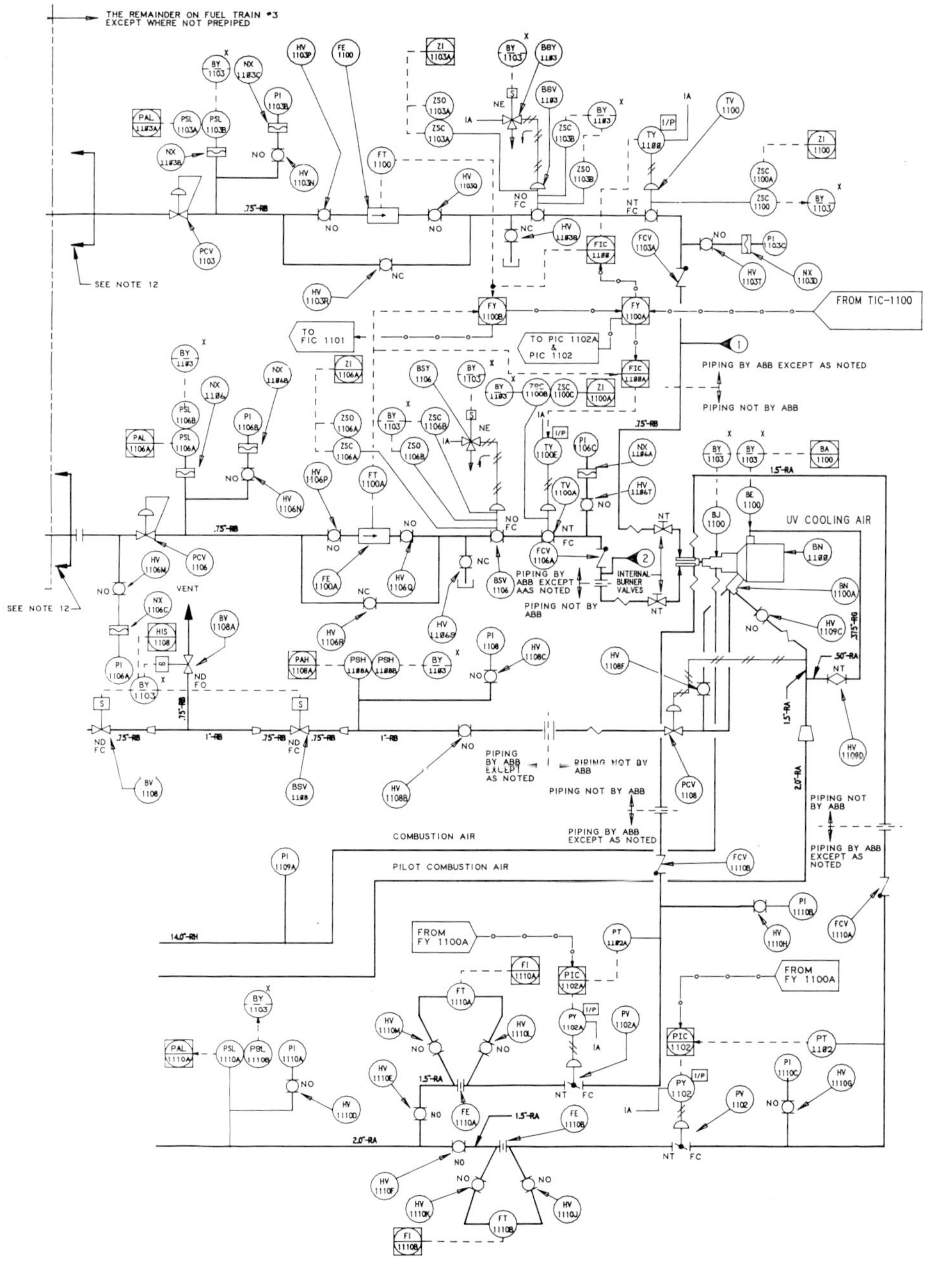

**Figure 29. Legend for P&ID drawing on Figures 27 and 28.** *(Courtesy of ABB Raymond Combustion Engineering.)*

BE Burner Flame Scanner
BJ Burner/Ignition Transformer
BN Burner
BSV Burner Safety Valve
BSY Burner Safety Valve Relay
BV Burner Valve
BY Burner Flame Safety System
FE Flow Sensor (Element)
(FE) Thermal Dispersion Flow Meter (Mass)
FI Flow Indicator
FIC Flow Indicating Controller
FT Flow Transmitter
FV Flow Control Valve
FY Flow Converter (Compute)
I/P Electro-Pneumatic Converter
MV Manually Operated Valve (Hand Valve)
NN Strainer Filter
NX Chemical Seal
NXX Lubricator
PCV Pressure Control Valve
PDT Differential Pressure Transmitter
PI Pressure Indicator
PIC Pressure Indicating Controller
PSL Low Pressure Switch
PT Pressure Transmitter
TIC Temperature Indicating Controller
TV Temperature Control Valve

Notes:

1. See sequence of operation, logic diagrams, and control wiring diagrams for a description of interlocking and control.
2. All equipment shown on this drawing by ABB except as noted.
3. Heat lines and tracing not by ABB.
4. Piping insulation not by ABB.
5. See above for legend.
6. Fluids may have to be heated to reach the proper viscosities. The heating system is not by ABB. Do not heat liquid to a temperature which could cause an unsafe condition or to a temperature which causes damage to the equipment.
7. Vents, drains, and steam traps not by ABB.
8. All pressure sensing lines are constructed with 0.25″ carbon steel schedule 40 pipe.
9. Pipe and elbows, furnished by ABB, are to be field cut, routed, and installed by others. Extra pipe and elbows will be furnished by ABB. These lines are not prepiped. This note applies where indicated.
10. (BY/1163) X Indicates the kiln main burner flame safety system, which is located in the flame safety panel.
11. For more detailed descriptions of control loops 1100, 1101, 1102, and 1112, see flow instrument diagram (drawing D-FS-2072Q).
12. This section of piping, not prepiped and not by ABB.

## 17. PROCESS CONTROL

A complete process control description of an incinerator facility cannot be attempted herein due to the number and complexity of the required control loops. Also, the facility process control is dependent on the type of OGT system. Some common features of an incinerator facility process control system, however, are described hereafter.

The two most important process control parameters are temperature and pressure.

Separate, independent *temperature control* loops are provided on the PCC and SCC. The temperature control loop consists of the following instrument devices:

a) Temperature Indicating Controller (TIC)
   Control mode is "auto".
b) Auxiliary fuel, Flow Indicating Controller (FFIC)
   Control mode is "cascade remote".
c) Waste fuel, Flow Indicating Controller (FFIC)
   Control mode is "cascade remote".
d) Combustion air, Flow Indicating Controller (FFIC)
   Control mode is "auto".
e) % XSA, ratio device (FY).
f) Temperature, split range device (TY).

*Set points*

The desired set point temperature for the PCC or SCC is set in the TIC. The TIC output is directed to both the auxiliary and waste fuel flow controllers, through the temperature split range device, TY. This output signal acts as a remote set point to the two fuel flow controllers and adjusts the auxiliary fuel and waste fuel flow rates to maintain the set point temperature on a split range basis; i.e. it maximizes the amount of waste fuel fired and minimizes the amount of auxiliary fuel fired.

The output from the fuel flow controllers is directed to the % XSA ratio devices, one for auxiliary fuel combustion air and one for waste fuel combustion air. The desired % XSA set point values (adjustable) are operator-set in the FYs. The output signal from the FYs is directed to the two combustion air flow controllers for the correct amount of combustion air at the % XSA set point to be delivered with the auxiliary fuel and waste fuel firing rates, respectively.

*Fuel flows*

Auxiliary fuel flow rates are measured by conventional in-line metering orifice runs. Waste fuel flow rates are measured by in-line mass flow (Coriolis type) meters because of their varying densities.

**Figure 30. Prepiped gas train on a rack.** High pressure supply gas enters at lower left. Controlled low pressure gas goes to the burner at upper left. Two transducers at top center translate pressure drops across gas and air orifices to 4-20 milliamp signals to an electronic comparator for automatic flow-balanced air/gas ratio control.

**Figure 31. Prepiped strainer rack for three liquid waste streams.** Pressure sensors warn operators when strainers are becoming plugged.

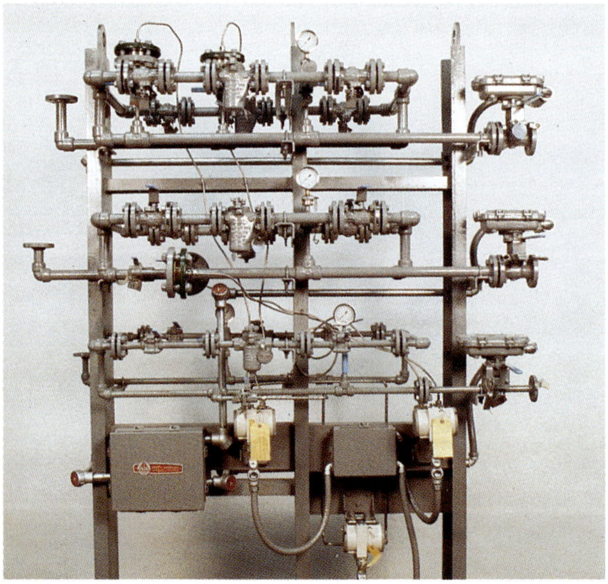

**Figure 32. Rotary drum PCC, SCC with relief vent, and stack of an incineration system in Tennessee.**

**Figure 33. Closeup of thermal relief vent atop the SCC.**

**Figure 34. Control/annunciator panel and fuel/waste trains.**

**Figure 35. Control room for entire incinerator system.** At the far right are opacity and oxygen monitors.

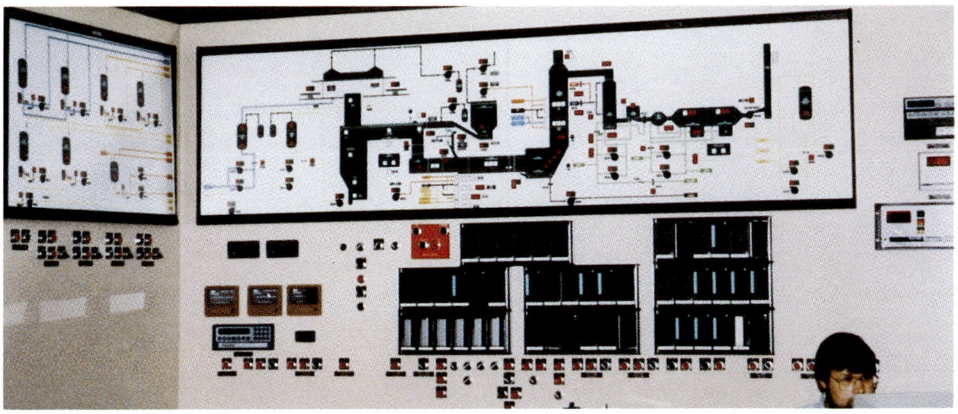

**Figure 36. Typical process control piping and instrument drawing for a primary combustion chamber.** *(See Acknowledgements, page vi.)*

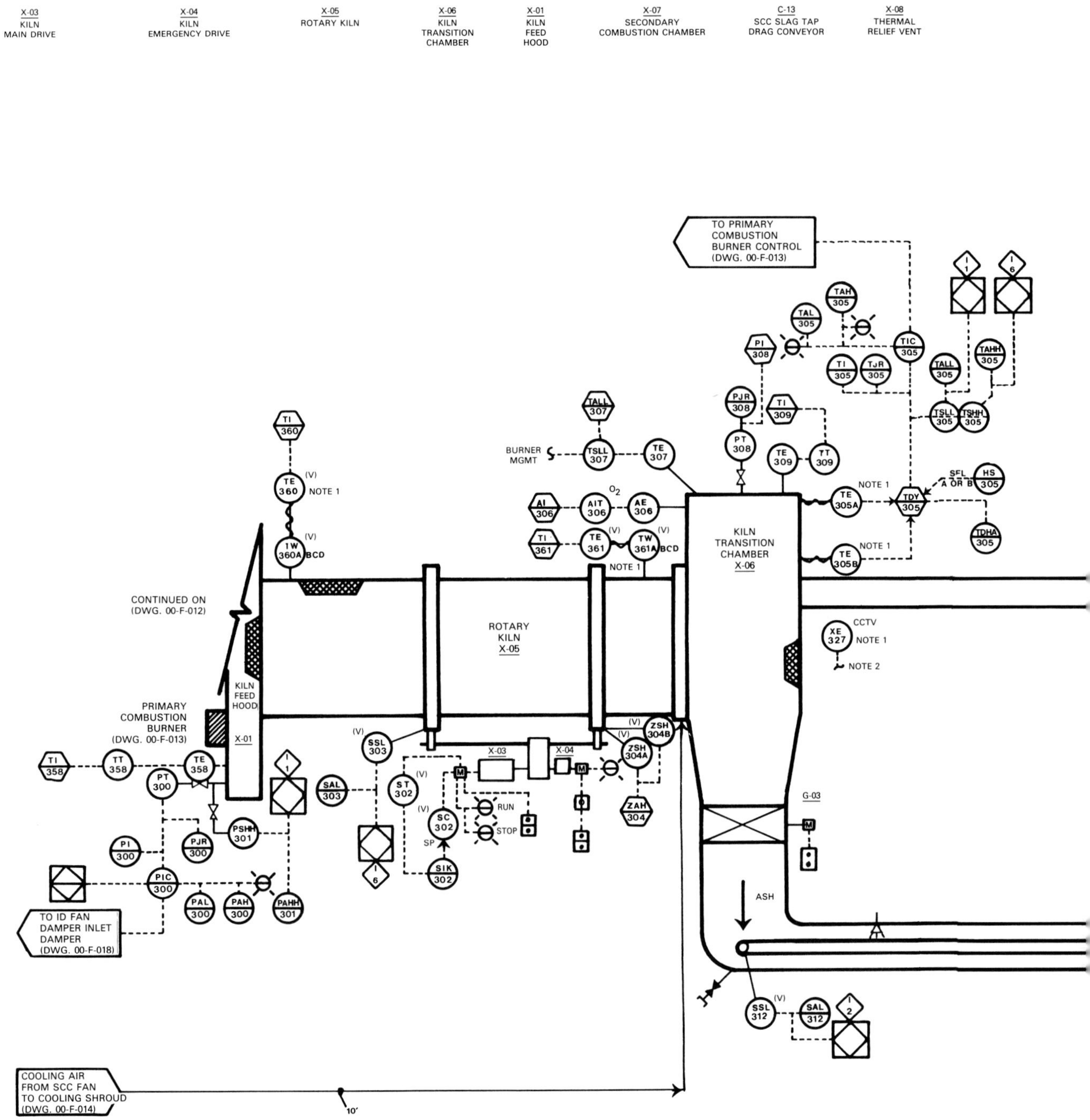

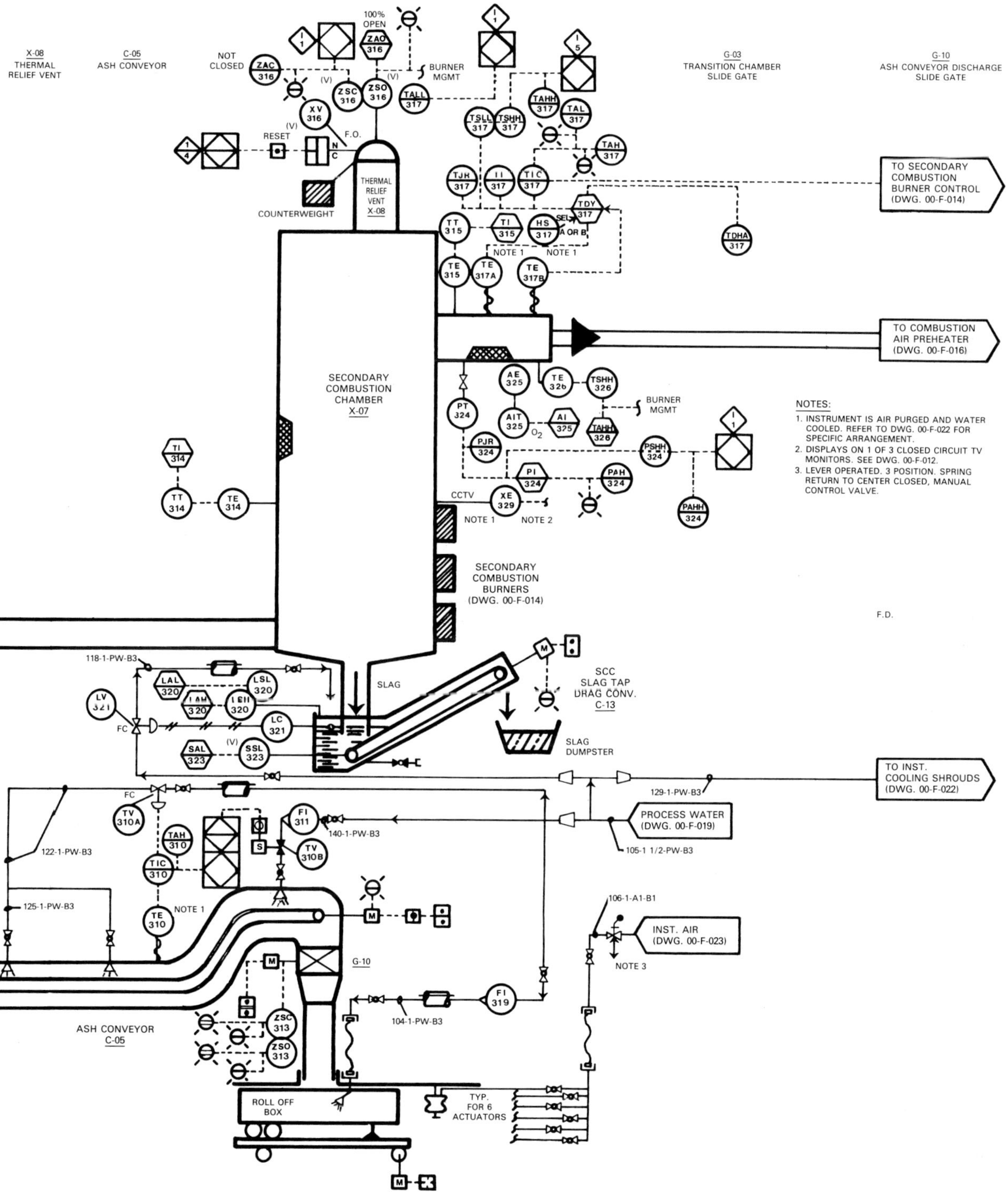
X-08
THERMAL RELIEF VENT
C-05
ASH CONVEYOR
NOT CLOSED
100% OPEN
BURNER MGMT
G-03
TRANSITION CHAMBER SLIDE GATE
G-10
ASH CONVEYOR DISCHARGE SLIDE GATE
RESET
F.O.
COUNTERWEIGHT
THERMAL RELIEF VENT X-08
TO SECONDARY COMBUSTION BURNER CONTROL (DWG. 00-F-014)
NOTE 1
SEL A OR B
TO COMBUSTION AIR PREHEATER (DWG. 00-F-016)
SECONDARY COMBUSTION CHAMBER X-07
BURNER MGMT
NOTES:
1. INSTRUMENT IS AIR PURGED AND WATER COOLED. REFER TO DWG. 00-F-022 FOR SPECIFIC ARRANGEMENT.
2. DISPLAYS ON 1 OF 3 CLOSED CIRCUIT TV MONITORS. SEE DWG. 00-F-012.
3. LEVER OPERATED. 3 POSITION. SPRING RETURN TO CENTER CLOSED, MANUAL CONTROL VALVE.
CCTV
NOTE 1
NOTE 2
SECONDARY COMBUSTION BURNERS (DWG. 00-F-014)
F.D.
118-1-PW-B3
SLAG
SCC SLAG TAP DRAG CONV. C-13
SLAG DUMPSTER
FC
TO INST. COOLING SHROUDS (DWG. 00-F-022)
129-1-PW-B3
140-1-PW-B3
PROCESS WATER (DWG. 00-F-019)
105-1 1/2-PW-B3
122-1-PW-B3
125-1-PW-B3
NOTE 1
106-1-A1-B1
INST. AIR (DWG. 00-F-023)
NOTE 3
G-10
104-1-PW-B3
ASH CONVEYOR C-05
ROLL OFF BOX
TYP. FOR 6 ACTUATORS

Liquid auxiliary fuel and waste fuel flow rates are reported in units of pounds per hour (lb/hr). Gaseous auxiliary fuel and combustion air flow rates are reported in units of scfh and scfm.

*Pumps*

Auxiliary liquid fuels and waste fuels are pumped via rotary gear and rotary lobe type positive displacement pumps in a recirculating loop under back pressure control at the combustion skids.

*Filter/strainers*

Because of the unpredictable nature of liquid wastes, it is advisable to have strainer/filters in each liquid waste line to prevent blockage of control valves, regulators, pumps, and atomizers. A preventive maintenance (PM) program is a "must" to check the strainers/filters for plugging. Figure 31 shows a strainer rack (prepiped by North American) with pressure sensors that sound an alarm and annunciator readout to warn operators when the pressure drop, due to plugging, across a strainer exceeds a practical limit.

*Pressure control* of the incinerator facility is accomplished by using the mixing chamber pressure ( – 0.5 to – 0.8″wc) as the set point in a pressure indicating control (PIC) loop. The output signal from the PIC adjusts the inlet damper on the induced draft (ID) fan to maintain a slightly negative pressure throughout the incinerator and OGT systems.

A weighted relief vent atop the SCC opens in response to abnormally high pressure in the SCC. See Figures 1, 32, 33, and Frontispiece.

*Process loops*

Other process control loops are:

a) pH control via caustic flow control or lime slurry flow control.

b) Level control in OGT equipment sumps and recycle tank.

c) Rotary kiln speed.

d) Ram feeder cycle/feed rate control.

Most of the process control loops incorporate a process parameter high alarm and low alarm and a process parameter high high shutdown interlock and low low waste feed cutoff interlock.

Figure 36 is a *piping and instrument diagram* (P & ID) of the process controls for a typical incinerator facility.

## 18. WASTE FEED PERMISSIVES AND CUTOFF INTERLOCKS

In addition to NFPA and insuring agency pre-ignition permissives and running interlock requirements, the incinerator facility must comply with waste feed permissives and have an Automatic Waste Feed Cut-Off System on line.

*Automatic waste feed cutoff*

The process parameters involved and the final permit parameter permissive and cut-off values are established by discussions and negotiations between the Owner/Operator and the EPA/State environmental agency, based on parameter result values as contained in the RCRA trial burn report and/or the TSCA trial burn report. The final permit conditions for waste feed permissives and the Automatic Waste Feed Cut-off System can, however, involve almost every single process parameter in the incinerator and OGT systems, such as:

a) pH levels, max/min

b) pressures, max/min

c) temperatures, max/min

d) waste flow rates, max/min

e) stack gas flow rate, max

f) OGT recirc and recycle flow rates, min

g) ionizer current and voltage, min

h) stack gas composition, max/min

Some of the common process parameter/process equipment permissives are:

*Process parameters*

a) OGT system in operation:
    recirculation and recycle pumps running
    recirculation and recycle flow rates to equipment acceptable
    pH and conductivity levels acceptable
    ionizer current and voltage acceptable

b) SCC auxiliary burner on:
    stack $O_2$ level $>3.0\%$ wgb
    stack CO level $<100$ ppm
    thermal relief vent (TRV) closed
    combustion efficiency $>99.9\%$

All the permissive minimum values must be satisfied to initiate waste feed. Loss of any one of the permissive minimum values or exceeding any one of the maximum values shall initiate the Automatic Waste Feed Cut-Off System.

## 19. MATERIALS OF CONSTRUCTION

This area in the incineration field may be considered the most undefined to date.

The halogenated constituents in the waste feeds (chlorine and flourine compounds) tend to steer the designer away from the carbon steels and austenitic stainless steels, and toward the more exotic high nickel/high chromium alloys (such as Hastalloys and Inconels).

*Refractories*

The metals compounds in some waste feeds, such as particulates of vanadium, sodium, and potassium in the flue gases at elevated temperature tend to steer the designer away from the conventional high alumina refractory shapes (60% to 70% alumina) with mullite bond to the more expensive basic type refractory shapes with their spinel bonds. Yet, to date in a number of incinerator facilities, carbon steels, austenitic stainless steels, and 60% $Al_2O_3$ refractory shapes have all given acceptable service life in incineration environments at recognized industry limits on chlorine, fluorine, and metallic salts.

There have been some notable exceptions, particularly with 60% $Al_2O_3$ refractory shapes, which failed over a six week period. Upon subsequent investigation it was learned that a waste feed with a preburn, somewhat borderline sodium content, broke down other compounds during combustion resulting in 30% by weight sodium in the flue gas, which "took out" the refractory shapes from the kiln face plate to the bottom of the quench chamber.

The following sections list materials of construction used successfully in specific areas of incinerators, along with the best available industry limits on concentrations.

**Materials for Wet Off-Gas Treatment Systems.** The shells and casings for this system have been constructed entirely of *fiberglass reinforced plastics* (FRP).

FRP materials have provided good corrosion resistance and service life (4 years plus) at saturation temperatures of 160 F to 190 F and with chlorine concentrations of 1% by weight in the flue gases. Above 240 F, these materials are subject to blistering and de-lamination, both of which result in a significant loss in corrosion resistance.

*Alloys*

High wear areas, such as venturi throat sections, venturi throat dampers, induced draft fan inlet dampers, and certain electrical components (ionizer wires and collector plates) need to be constructed from high nickel/high chromium alloys such as Hastalloy C-276 or Alloy 20.

It is necessary to carefully evaluate the structural integrity of FRP materials used as casings and roofs under negative pressures, particularly the *equivalent fan static pressures* (EFSPs) developed during the purge cycle on cold air.

**EFSP**

Highly stressed, rotating elements, such as induced draft fan wheels of FRP materials exposed to wet gaseous chlorine and fluorine at saturation temperatures of 160F to 190F need to be (a) inspected every six months for signs of de-lamination, or (b) made from corrosion resistant metals.

*Refractory lining*

**Materials for PCC and SCC Refractory Hot Face Linings.** Refractory hot face linings of 60% alumina ($Al_2O_3$) refractory shapes have given service lives of five years plus in incinerator atmospheres at 1500 F to 2300 F. See Figures 37 through 47.

Recognized refractory industry limits in the flue gas are:

1% Chlorine by weight
1% Fluorine by weight
1% Vanadium by weight
1% Sodium by weight

Recent investigations have led to a criteria limit of 5.0% by weight of (vanadium + sodium + potassium) in any slag formed.

Monolithic hot face linings using plastic refractory materials with 85% to 90% $Al_2O_3$ have given good resistance and service life for five plus years.

**Table 9. Typical incinerator refractory heat-up rates.**

| Rate | To | Rate | To |
|---|---|---|---|
| 50-75°F/hr | *to 1000 F* | 100°F/hr | *to 1850 F* PCC |
| 100°F/hr | *to 1000 F* | 200°F/hr | *to 2250 F* SCC |

*Burners*

**Materials for Burner Internals.** For *auxiliary fuel burners*, carbon steel fuel gun assemblies, atomizer tips, and air distributor disks have proven satisfactory if cooled with some minimum amount of combustion air or atomizing medium when the fuel gun is not in service.

For *waste fuel burners*, stainless steel fuel gun assemblies and high nickel/high chromium heat-resistant and corrosion-resistant alloys such as Hastalloy C-276 have provided good service.

Port tiles of standard 3000 F castable materials have given reliable service even in waste burners. In waste burner service, however, these port tiles do suffer inter-granular deterioration, but to date, have held their configuration and continued to perform the function of the port tile. An important consideration is careful attention during maintenance activities to avoid impact on these port tiles and subsequent disintegration, requiring replacement. See Figure 48.

**Figure 37. Suspended refractory roof in a chemical waste incinerator, looking up at the base of the stack leading up to the thermal relief vent.** *(Courtesy of Merkle Engineers, Inc.)*

**Figure 38. Anchored refractory sidewalls, roof and hearth in a traveling grate incinerator, with a rounded bullnose divider visible at the far end.** The waste feed direction is from the photographer's back as he stands on a downwardly-inclined hearth. *(Courtesy of Merkle Engineers, Inc.)*

**Figure 39. Detail of a partially-installed bullnose illustrating anchoring steel and patented refractory shapes (in this case high alumina).** When completed, all metal piping and structural shapes will be protected by refractory.
*(Courtesy of Merkle Engineers, Inc.)*

**Figure 40. Refractory installation in a rotary kiln, looking uphill toward the feed chute (left).** A North American burner is at the top right, and an aqueous waste injector is at the bottom right, in the stationary refractory end wall. Four refractory lifters built into the rotating cylindrical wall stir and elevate the solids to provide better contact with the hot kiln gases.
*(Courtesy of Gillespie and Powers, Inc.)*

**Figure 41. Refractory installation in a rotary kiln and mixing chamber.** Hot gases exit the rotary kiln from right to left, entering a disengaging and mixing chamber. Cast iron sections, for retaining the rotary kiln refractory, are visible at the top right. *(Photo courtesy of Gillespie and Powers, Inc.)*
[The reader's orientation of Figures 40-47 may be helped by consulting Figures 2 and 14.]

**Figure 42. Ash discharge from a rotary kiln.** (Just below the view of Figure 41.) Ash from the discharge of the rotary kiln falls into water in the ash pit at the bottom of this photo. The water level will be 6″ above the bottom of the diagonal water-resistant refractory ledge. Ash is removed from the pit by conveyor. The sloped floor at the bottom corner of this picture is the beginning of the angled refractory baffle (Figure 43) that directs gases up into the "pinch point". *(Photo courtesy of Gillespie and Powers, Inc.)*

**Figure 43. Pinch point entry under a bullnose.** After mixing in the ash discharge disengaging chamber above the ash pit, the hot gases are deflected upward around a bullnose and through a pinch point in the SCC. In this view, the camera is pointed 180 degrees from Figures 41 and 42. Ports at the bottom admit a thermocouple and a TV camera lens aimed through the rotary kiln.
*(Courtesy of Gillespie and Powers, Inc.)*

**Figure 44. SCC entrance.** Looking up through the pinch point, which increases velocity for better mixing of the kiln gases with poc from the afterburners. The sloping baffle is at the lower left; the bullnose at right. The refractory quarl of one of the afterburners is partially visible at center.
*(Courtesy of Gillespie and Powers, Inc.)*

**Figure 45. Refractory installation in a secondary combustion chamber.** One of two North American burners firing horizontally and tangentially across the lower end of the uptake of the vertical cylindrical SCC. In the small port at left is an aqueous waste injection nozzle. Many installations now use the oil nozzle of a Dual-Fuel burner to inject aqueous waste.
*(Courtesy of Gillespie and Powers, Inc.)*

**Figure 46. Crossover plenum at the top of the U-shaped SCC.** A thermocouple port checks temperature as the gases turn down through the second cylindrical leg of the secondary chamber, toward the quench. The letters X on larger refractory pieces mark anchor brick sets. *(Courtesy of Gillespie and Powers, Inc.)*

**Figure 47. Refractory installation at the top of the SCC.** Looking up through the first (up-flow) leg of the U-shaped SCC. Burners are not visible because they are at camera level. The port at the top of the crossover is thermal relief vent (TRV). *(Courtesy of Gillespie and Powers, Inc.)*

**Figure 48. Burner internals, refractory quarl, and the combustion chambers themselves are subjected to severe thermal and corrosion stresses.** Stored heat in the chamber walls may damage parts of a burner if air is not allowed to flow through the burner after fuel shutoff.

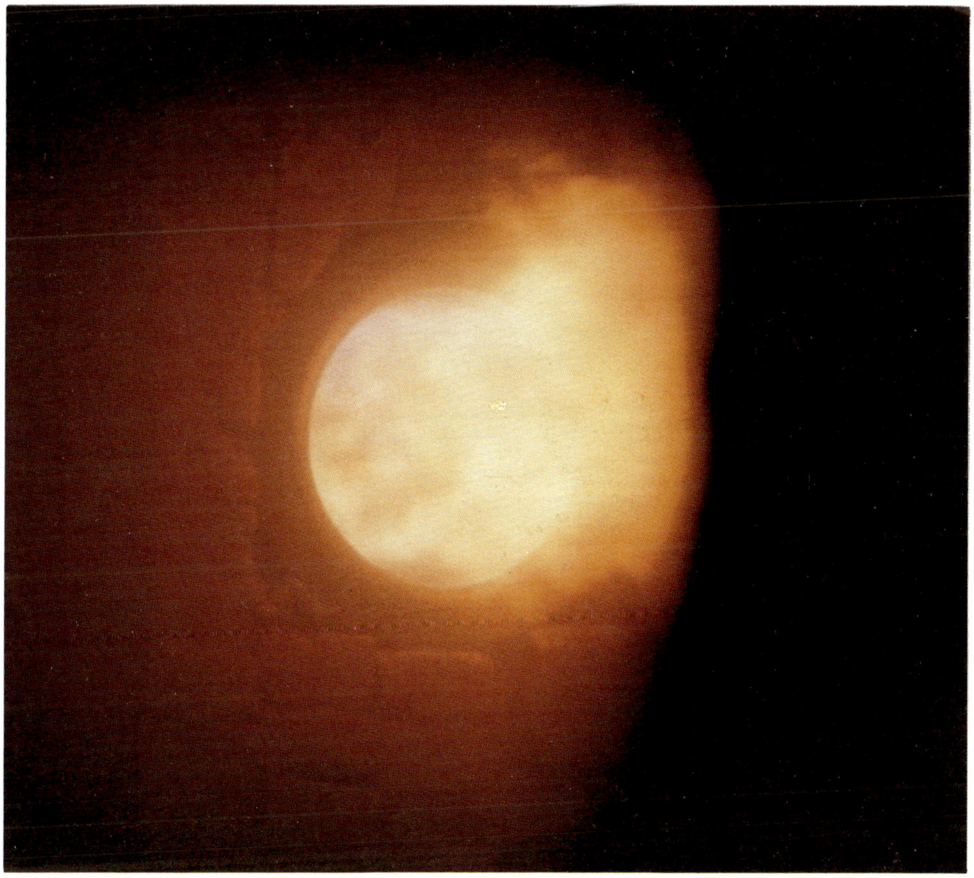

**Materials for Fuel Train Piping Components.** In general, fuel train and recirculation pipes of schedule 40 carbon steel pipe (schedule 80 at less than 1″ bore), and equivalent fittings and flanges have not shown signs of accelerated deterioration and have provided long-term service. However, a well-documented and implemented waste characterization plan is a must to avoid unforseen exothermic reactions and/or severe corrosive concentrations during waste blending operations.

*Valves*

Flow control valves with standard cast iron/steel bodies, with stainless steel trim have provided good service. Flow control valves with V ball type ports have given the best repeatability at typical incinerator waste fuel flow rates. Waste fuel flow control valves should be furnished with positioners.

*Switches*

Pressure switches in waste fuel service have provided good service when equipped with Teflon diaphragms as chemical seals.

Standard fuel shutoff solenoid valves of cast iron/cast steel construction have been used extensively with no serious problems to date. They should be leak-checked as part of routine maintenance. On some projects, where severe corrosion of waste fuel shutoff valves is expected, insuring authorities have allowed substitution of an automatic ball valve with stainless steel internal trim for fuel shutoff service. The approval however is specific to and only to the service waste fuel in a specific incinerator facility.

In waste fuel service, any internal or wetted components of brass, bronze, neoprene or buna-N materials should be avoided.

# Part 20. INCINERATOR ANALYSIS COMPUTATIONS

*This section is exerpted, with permission from Mr. Calvin R. Brunner's book, INCINERATION SYSTEMS, Selection and Design (Reference 2.)*

## THE MASS BALANCE

The flow into an incinerator must equal the flow of products leaving the incinerator. Input includes waste fuel, air (including humidity entrained within the air), and supplementary fuel. The flow exiting the incinerator includes moisture and dry gas of the exhaust as well as ash, both in the exhaust as fly ash and exiting as bottom ash.

Table 9-1, the mass flow table, provides an orderly method in establishing a mass balance surrounding a combustion system. Of initial interest is waste quality: its moisture content, ash content (non-combustible fraction), and heat content. Of prime importance is the generation of moisture and dry flue gas from the combustion process.

For the purpose of this example the following waste will be assumed, at the indicated firing rate and other combustion parameters:

8000 pounds per hour of paper waste.
5250 Btu per pound as fired.
22.5% moisture as fired.
5% ash as fired.
7.50 pounds of dry flue gas generated per 10 000 Btu released.
0.51 pounds of moisture generated per 10 000 Btu released.
Fired with 125% excess air.

Following Table 9-1:

*Wet Feed*, as received charging rate, is 8000 pounds per hour.

*Moisture*. Percent moisture, by weight, of the wet feed, is 22.5%. The moisture rate is 22.5% of the wet feed rate, 0.225 × 8000 lb/hr = 1800 lb/hr.

*Dry Feed* equals total wet feed less moisture, 8000 lb/hr − 1800 lb/hr = 6200 lb/hr.

*Ash* is percent of total dry feed that remains after combustion. From the data provided, 5% of the total wet feed, 0.05 × 8000 lb/hr = 400 lb ash/hr. The ash fraction of the dry feed is 400 lb/hr ÷ 6200 lb/hr = 0.0645.

*Volatile* is that portion of dry feed that is combusted. It is found by subtracting ash from dry feed: 6200 lb/hr − 400 lb/hr = 5800 lb/hr.

*Volatile Heating Value* is the Btu value of the waste per pound of volatile matter. The total heating value as charged is 8000 lb/hr × 5250 Btu/lb = 42,000,000 Btu/hr. The total waste heat value is 42.00 million Btu/hr. On a unit volatile basis, with 5800 lb volatile/hr charged, the heating value per pound volatile is 42 million Btu/hr ÷ 5800 lb/hr = 7241 Btu/lb.

**Table 9-1. Mass Flow.**

| | **Example** |
|---|---|
| Wet Feed, lb/hr | 8 000 |
| Moisture, % | 22.5 |
| lb/hr | 1 800 |
| Dry Feed, lb/hr | 6 200 |
| Ash, % | 6.45 |
| lb/hr | 400 |
| Volatile | 5 800 |
| Vol. Htg. Value, Btu/lb | 7 241 |
| Vol., million Btu/hr | 42.00 |
| Dry Gas, lb/10 kB | 7.5 |
| lb/hr | 31 500 |
| Comb. $H_2O$, lb/10 kB | 0.51 |
| lb/hr | 2 142 |
| DG + Comb. $H_2O$, lb/hr | 33 642 |
| 100% Air, lb/hr | 27 842 |
| Total Air Fract. | 2.25 |
| Total Air, lb/hr | 62 645 |
| Excess Air, lb/hr | 34 803 |
| Humid/Dry Gas (Air), lb/lb | 0.01 |
| Humidity, lb/hr | 626 |
| Total $H_2O$, lb/hr | 4 568 |
| Total Dry Gas, lb/hr | 66 303 |

*Dry Gas* produced from combustion of the waste is given for this example as 7.5 lb/10 000 Btu. The dry gas flow is this figure multiplied by the Btu released or (7.5 lb/10 000 Btu) × 42 million Btu/hr = 31 500 lb dry gas/hr.

*Combustion $H_2O$* is the moisture generated from burning the waste, given in this example as 0.51 lb/10 000 Btu. Therefore, the combustion moisture flow rate is (0.51 lb/10 000 Btu) × 42 million Btu/hr = 2142 lb moisture/hr.

*DG + COMB $H_2O$* is the sum of the dry gas and the moisture products of combustion, 31 500 lb/hr + 2142 lb/hr = 33 642 lb/hr. This figure is convenient in order to obtain the amount of air required for combustion.

*100% Air* is the dry gas and moisture weights that are produced by combustion of the volatile component, which equals the weight of the volatile component plus the weight of the air provided. Likewise, the air requirement is equal to the sum of the dry gas and moisture of combustion less the volatile component. This air requirement is the stoichiometric air requirement, that amount of air necessary for complete combustion of the volatile component of the waste. Using the above figures, the value for 100% air is as follows:

$$33\ 642 \text{ lb/hr (DG + } H_2O) - 5800 \text{ lb/hr (volatile)} = 27\ 842 \text{ lb/hr}$$

*Total Air Fraction* is the air required for effective combustion. This is basically a function of the physical state of the fuel (gas, liquid, or solid) and the nature of the burning equipment. In this example, 125% air is required, providing 1.25 + 1.00 = 2.25 total air fraction.

*Total Air* is the stoichiometric requirement multiplied by the total air fraction, 27 842 lb/hr × 2.25 = 62 645 lb/hr.

*Excess Air* provided to the system is the total air less the stoichiometric air requirement, 62 645 lb/hr − 27 842 lb/hr = 34 803 lb/hr.

*Humid/Dry Gas (Air).* The humidity of the air entering the incinerator will have a significant effect on the heat balance to be calculated further in this chapter. In this case, assume a humidity of 0.01 pound of moisture per pound of dry air.

*Humidity* is the flow of moisture into the system with the air supply. Knowing the air flow and the fractional humidity, the humidity can be calculated as 62 645 lb air/hr × 0.01 lb $H_2O$/lb air = 626 lb moisture/hr.

*Total $H_2O$* is that moisture exiting the system. It is equal to the sum of three moisture components: moisture in the feed plus moisture of combustion plus humidity, or 1800 lb/hr + 2142 lb/hr + 626 lb/hr = 4568 lb/hr total $H_2O$.

*Total Dry Gas* exiting the system is equal to the sum of the dry gas generated by the combustion of the volatiles, with stoichiometric air, plus the flow of excess air into the system. Thus, 31 500 lb dry gas/hr + 34 803 lb excess air/hr = 66,303 lb/hr total dry gas.

## HEAT BALANCE

Heat, like mass, is conserved within a system. The heat exiting a system is equal to the amount of heat entering that system. Table 9-2 presents a quantitative means of establishing a heat balance for an incinerator. The heat balance must be preceded by a mass flow balance, as was discussed previously. The result of a heat balance is a determination of the incinerator outlet temperature,

**Table 9-2. Heat Balance.**

| | Example w/Fuel Oil | Example w/Gas |
|---|---|---|
| Reference *t*, F | 60 | 60 |
| Cooling Air Wasted, lb/hr | 9 000 | |
| F | 450 | |
| Btu/lb | 94 | |
| million Btu/hr | 0.85 | |
| Ash, lb/hr | 400 | |
| Btu/lb | 130 | |
| million Btu/hr | 0.05 | |
| Radiation, % | 1.5 | |
| million Btu/hr | 0.63 | |
| Humidity, lb/hr | 626 | |
| Correction (at 970 Btu/lb), million Btu/hr | − 0.61 | |
| Losses, Total, million Btu/hr | 0.92 | |
| Input to the system, million Btu/hr | 42.00 | |
| Outlet heat content, million Btu/hr | 41.08 | |
| Dry Gas, lb/hr | 66 303 | |
| $H_2O$, lb/hr | 4 568 | |
| Temperature, F | 1 899 | |
| Desired Temperature, F | 2 000 | |
| Heat content of wet gas stream, million Btu/hr | 43.26 | |
| Net, million Btu/hr | 2.16 | |
| Fuel Oil, Air Fraction | 1.20 | 1.10 |
| Net Btu/gal | 57 578 | 406 |
| Fuel flow rate, gph | 37.51 | 5 320 |
| Air/fuel ratio, lb/gal | 125.06 | 0.791 |
| Air flow rate, lb/hr | 4 691 | 4 208 |
| Dry Gas, lb/gal | 125.54 | 0.748 |
| lb/hr | 4 709 | 3 979 |
| $H_2O$, lb/gal | 8.75 | 0.103 |
| lb/hr | 328 | 548 |
| DG w/FO, lb/hr | 71 012 | 70 282 |
| $H_2O$ w/FO, lb/hr | 4 896 | 5 116 |
| Air w/FO, lb/hr | 67 336 | 66 853 |
| Total heat in exit gas, million Btu/hr | 46.32 | 46.40 |

outlet gas flow, supplemental fuel requirement, and total air requirement. The total heat into the incinerator was calculated previously in the mass flow computations. By determing how much of this total heat produced is present in the exhaust gas, the exhaust gas temperature can be calculated. If the calculated exhaust gas temperature is equal to the desired exhaust gas outlet temperature the process is autogenous; so supplemental fuel is not required. If the desired temperature is lower than the actual outlet temperature additional air must be added (or additional water, if this is possible) to lower the outlet temperature to that desired. This condition is also autogenous.... supplemental fuel need not be added to the system.

For the case where the actual outlet temperature is less than the desired outlet temperature supplemental fuel must be added. The products of combustion must include the products of combustion of the supplemental fuel.

Flue gas properties are assumed to be identical to those of air. As with Table 9-1, Table 9-2 will be described on a step-by-step basis. *Reference t* is the base temperature for enthalpy or heat content of feed, supplemental fuel, and air entering the system = 60 F for this example.

*Cooling Air Wasted.* Assume the incinerator shell is cooled by a flow of 2000 scfm of air. A standard cubic foot of air weighs 0.075 lb/ft$^3$; therefore the mass air flow is 2000 ft$^3$/min × 60 min/hr × 0.075 lb/ft$^3$ = 9000 lb/hr. It is further assumed that this flow is wasted, i.e., discharged to the atmosphere.

The temperature of the air at the discharge point is assumed to be 450 F.

The heat content of air at 450 F, the cooling air discharge temperature, is 94 Btu/lb.

The total heat loss due to the wasted cooling air is the quantity of cooling air discharged multiplied by its heat content, 9000 lb/hr × 94 Btu/lb = 0.85 million Btu/hr.

Ash generated is 400 lb/hr, from the mass flow table.

The heating value of ash can be calculated. In lieu of an analysis of the heating value of the ash component, a figure of 130 Btu/lb is reasonable. The heat loss through ash discharge is therefore 400 lb ash/hr × 130 Btu/lb ash = 0.05 million Btu/hr.

**Table 9-3. Furnace Radiation Loss Estimates.**

| Furnace Rate, million Btu/hr | Radiation Loss, % of furnace rate |
|---|---|
| <10 | 3 |
| 15 | 2.75 |
| 20 | 2.50 |
| 25 | 2 |
| 30 | 1.75 |
| >35 | 1.50 |

*Radiation.* The heat lost by radiation from the incinerator shell can be approximated as a percentage of the total heat of combustion. Table 9-3 lists typical values of radiation loss. For this case, with a heat release of 42 million Btu/hr a radiation loss of 1.5% is used.

The total loss by radiation is equal to 42 million Btu/hr released × 1.5% or 0.63 million Btu/hr.

*Humidity* is water vapor within the air. The humidity component of the air, 626 lb/hr, is taken from the mass flow sheet.

When considering the heat absorbed by the moisture or water within the incinerator exhaust stream humidity has released its heat of vaporization because it is in the vapor phase. The other moisture components, moisture in the feed and moisture of combustion, enters the reaction as a liquid and the heat of vaporization is released by the reaction. To simplify these moisture calculations the heat of vaporization of humidity moisture at 60 F, 960 Btu/lb, is added to the total heat capacity of the flue gas. Therefore the correction factor is 626 lb humidity/hr × 970 Btu/lb = 0.61 million Btu/hr.

*Total Losses* of an incinerator are the sum of the heat discharged as cooling air plus the heat lost in the ash discharge plus the radiation loss. To this is added the correction for humidity. In this case the total loss is 0.85 million Btu/hr + 0.05 million Btu/hr + 0.63 million Btu/hr − 0.61 million Btu/hr = 0.92 million Btu/hr.

*Input to the system* is the heat generated from the combustible, or volatile, portion of the feed. From the mass flow table this figure is 42.00 million Btu/hr.

*Outlet heat content* is that amount of heat exiting in the flue gas. The heat left in the flue gas is the heat generated by the feed (input) less the total heat loss, 42.00 million Btu/hr − 0.92 million Btu/hr = 41.08 million Btu/hr.

*Dry Gas* is 66 303 lb/hr, from the mass flow sheet.

*$H_2O$* is 4568 lb/hr, from the mass flow sheet.

*Temperature* is the temperature of the exhaust gas where the heat content of the dry gas flow plus the heat content of the moisture flow exiting the incinerator equals the outlet million Btu/hr.

Using enthalpy figures for dry air from Gas Tables, and for moisture from Steam Tables,

| | **at 1800 F** | **at 1900 F** |
|---|---|---|
| Btu/lb dry air | 453.24 Btu/lb | 481.57 Btu/lb |
| × 66 303 lb dry air/hr = | 30.05 million Btu/hr | 31.93 million Btu/hr |
| Btu/lb moisture | 1948.02 Btu/lb | 2007.17 Btu/lb |
| × 4568 lb moisture/hr = | 8.89 million Btu/hr | 9.17 million Btu/hr |
| Total enthalpy (30.05 + 8.89 =) | 38.94 million Btu/hr | 41.10 million Btu/hr |

By interpolation, exhaust gas temperature corresponding to the outlet heat content of 41.08 million Btu/hr must be

$$1800 \text{ F} + (1900 - 1800) \times \frac{41.08 - 38.94}{41.10 - 38.94} = 1899 \text{ F}$$

*Desired Temperature* is the temperature to which it is desired to bring the products of combustion. For this example a temperature of 2000 F will be used.

At this temperature the *heat content of the wet gas stream* is that of the dry gas and that of moisture at 2000 F or 66 303 lb dry gas/hr × 510.07 Btu/lb + 4568 lb moisture/hr × 2067.42 Btu/lb = 43.26 million Btu/hr.

*Net heat to be added* to the flue gas to raise its heat content to the desired 2000 F level is the heat content of the wet gas stream minus the outlet heat content = 43.26 million Btu/hr − 41.06 million Btu/hr = 2.16 million Btu/hr.

Note that this analysis will continue based on the use of No. 2 fuel oil as supplemental fuel. Table 9-4 lists fuel oil combustion parameters. Table 9-5 likewise is a listing of combustion parameters for natural gas. The incinerator analysis of Tables 9-2 and 9-6 includes listings for the use of natural gas as supplemental fuel. The values for natural gas are calculated in similar manner to the calculations for fuel oil.

*Fuel Oil, Air Fraction* is the total air fraction required for combustion of supplementary fuel. The total air normally required for efficient combustion of gas fuel is from 1.05-1.15 and for light fuel oil is 1.10-1.30. In this case assume No. 2 fuel oil is used for supplemental heat, with a total air supply of 1.20.

*Net Btu/gal.* When fuel is combusted the products of combustion must be heated to the desired flue gas temperature. The amount of heat required to heat these combustion products must be subtracted from the total heat of combustion to obtain the effective heating value of the fuel. As the temperature to which the fuel products must be raised increases, the net heat available from the fuel decreases. From Table 9-4, for No. 2 fuel oil, with 1.2 total air, and 2000 F, a net heating value of 57 578 Btu/gal is available.

The gph *fuel oil* rate required to provide the heat required to bring the exhaust gas temperature from its actual temperature, 1899 F, to the desired temperature, 2000 F, is equal to the net million Btu/hr required divided by the net Btu/gallon available = 2.16 million Btu/hr net ÷ 57 578 Btu/gal = 37.51 gph.

*Air/fuel ratio* required for combustion of fuel oil, with 1.2 total air, from Table 9-4, is 125.06 lb/gal of fuel oil.

The combustion air flow rate is the air/fuel ratio multiplied by the fuel flow rate, = 125.06 lb air/gal × 37.51 gph fuel oil = 4691 lb air/hr.

*Dry Gas:* From Table 9-4 the dry gas produced from combustion of fuel oil with 1.2 total air is 125.54 lb/gal of fuel oil.

The dry gas flow rate is 125.54 lb dry gas/gal × 37.51 gal/hr fuel oil = 4709 lb/hr.

*$H_2O$* produced from combustion of fuel oil with 1.2 total air and 0.013 humidity is 8.75 lb/gal fuel oil from Table 9.4.

**Table 9-4. No. 2 Fuel Oil, 139 703 Btu/gal, 7.6 lb/gal.**

| **Total Air:** | **1.1** | **1.2** | **1.3** |
|---|---|---|---|
| lb air/gal | 114.640 | 125.062 | 135.483 |
| lb dry gas/gal | 115.115 | 125.537 | 135.958 |
| lb $H_2O$/gal | 8.615 | 8.751 | 8.886 |
| **Temp, F** | **Heat Available, Btu/gal** | | |
| 200 | 126 210 | 125 707 | 125 206 |
| 300 | 123 016 | 122 255 | 121 495 |
| 400 | 119 802 | 118 780 | 117 760 |
| 500 | 116 562 | 115 277 | 113 995 |
| 600 | 113 283 | 111 732 | 110 184 |
| 700 | 109 965 | 108 146 | 106 328 |
| 800 | 106 029 | 103 885 | 101 742 |
| 900 | 103 197 | 100 829 | 98 463 |
| 1000 | 99 747 | 97 099 | 99 747 |
| 1100 | 96 255 | 93 325 | 90 397 |
| 1200 | 92 721 | 89 505 | 86 291 |
| 1300 | 89 147 | 85 643 | 82 140 |
| 1400 | 85 535 | 81 738 | 77 943 |
| 1500 | 81 887 | 77 796 | 73 707 |
| 1600 | 78 205 | 73 817 | 69 431 |
| 1700 | 74 487 | 69 799 | 65 115 |
| 1800 | 70 746 | 65 757 | 60 771 |
| 1832 | 69 538 | 64 452 | 59 369 |
| 1900 | 66 971 | 61 679 | 56 389 |
| 2000 | 63 175 | 57 578 | 51 984 |
| 2100 | 59 341 | 53 445 | 59 349 |
| 2192 | 55 813 | 49 628 | 44 385 |
| 2200 | 55 507 | 49 294 | 43 084 |
| 2300 | 41 637 | 45 114 | 38 594 |
| 2400 | 47 750 | 40 916 | 34 085 |
| 2500 | 43 852 | 36 706 | 29 562 |
| 2600 | 39 914 | 32 453 | 24 995 |
| 2700 | 35 938 | 28 162 | 20 388 |

The moisture flow rate from combustion of fuel oil is 8.75 lb $H_2O$/gal fuel oil × 37.51 gal fuel oil/hr = 328 lb/hr.

*DG w/FO* is the total quantity of dry gas exiting the system. It is equal to the dry gas produced from combustion of the waste plus the dry gas produced from fuel combustion 66 303 lb/hr + 4709 lb/hr = 71 012 lb dry gas/hr.

*$H_2O$ w/FO* is the total quantity of moisture exiting the system, that calculated in the mass flow sheet plus the contribution from combustion of supplementary fuel, 4568 lb/hr + 328 lb/hr = 4896 lb/hr.

*Air w/FO* is the total amount of air entering the incinerator, calculated from the mass flow sheet, plus that needed for supplemental fuel combustion, 62 645 lb/hr + 4691 lb/hr = 67 336 lb/hr.

**Table 9-5. Natural Gas, 1000 Btu/standard cubic foot, 0.050 lb/standard cubic foot.**

| **Total Air:** | 1.05 | 1.10 | 1.15 |
|---|---|---|---|
| lb air/scf | 0.755 | 0.791 | 0.827 |
| lb dry gas/scf | 0.712 | 0.748 | 0.784 |
| lb $H_2O$/scf | 0.103 | 0.103 | 0.104 |
| **Temp., F** | **Heat Available, Btu/scf** | | |
| 200 | 861 | 860 | 857 |
| 300 | 839 | 837 | 834 |
| 400 | 817 | 814 | 810 |
| 500 | 794 | 790 | 785 |
| 600 | 772 | 767 | 761 |
| 700 | 749 | 743 | 736 |
| 800 | 722 | 715 | 707 |
| 900 | 702 | 694 | 685 |
| 1000 | 678 | 670 | 660 |
| 1100 | 654 | 644 | 633 |
| 1200 | 629 | 619 | 607 |
| 1300 | 605 | 593 | 580 |
| 1400 | 579 | 567 | 553 |
| 1500 | 554 | 541 | 526 |
| 1600 | 529 | 514 | 498 |
| 1700 | 503 | 487 | 470 |
| 1800 | 477 | 460 | 442 |
| 1832 | 468 | 451 | 433 |
| 1900 | 450 | 433 | 414 |
| 2000 | 424 | 406 | 385 |
| 2100 | 397 | 378 | 356 |
| 2192 | 372 | 352 | 329 |
| 2200 | 370 | 350 | 327 |
| 2300 | 343 | 322 | 298 |
| 2400 | 316 | 294 | 269 |
| 2500 | 289 | 265 | 239 |
| 2600 | 261 | 237 | 210 |
| 2700 | 233 | 208 | 179 |

*Total heat in exit flue gas* from the incinerator, is the sum of the heat content of the gas prior to adding supplemental fuel plus the heat addition of the supplemental fuel. The supplemental fuel adds 37.51 gal/hr × 139 703 Btu/gal = 5.24 million Btu/hr to the flue gas. Therefore, the flue gas outlet contains 41.08 million Btu/hr + 5.24 million Btu/hr = 46.32 million Btu/hr.

**Table 9-6. Flue Gas Discharge.**

| | Example w/Fuel Oil | Example w/Gas |
|---|---|---|
| Inlet temperature, F | 2000 | 2000 |
| Dry gas, lb/hr | 71012 | 70282 |
| Total inlet heat content, million Btu/hr | 46.32 | 46.60 |
| Btu/lb dry gas | 652 | 660 |
| Adiabatic *t*, F | 176 | 176 |
| $H_2O$ saturation, lb/lb dry gas | 0.5511 | 0.5511 |
| lb/hr | 39135 | 38732 |
| $H_2O$ inlet, lb/hr | 4896 | 5116 |
| Quench $H_2O$, lb/hr | 34239 | 33616 |
| Quench $H_2O$, gpm | 68.5 | 67 |
| Outlet temp., F | 120 | 120 |
| Raw $H_2O$ temp., F | 60 | 60 |
| Sump temp., F | 147 | 147 |
| Temp. diff., F | 87 | 87 |
| Outlet, Btu/lb dry gas | 111.65 | 111.65 |
| million Btu/hr | 7.93 | 7.85 |
| Req'd cooling, million Btu/hr | 38.39 | 38.55 |
| Cooling water reqd, lb/hr | 441264 | 443103 |
| Cooling water reqd, gpm | 883 | 886 |
| Outlet, specific volume, $ft^3$/lb dry gas | 16.515 | 16.515 |
| Outlet gas flow rate, scfm | 19546 | 19345 |
| Fan pressure differential, ″wc | 30 | 30 |
| Outlet, acfm | 21101 | 20884 |
| Outlet humidity, $H_2O$/lb dry gas | 0.08128 | 0.08128 |
| $H_2O$, lb/hr | 5772 | 5713 |
| Recir. (ideal), gpm | 68.5 | 67 |
| Recir. (actual), gpm | 548 | 538 |
| Cooling $H_2O$, gpm | 883 | 886 |

As a check on this figure, the outlet temperature will be calculated using the flue gas flow (71 012 lb/hr dry gas and 4896 lb/hr moisture):

| | |
|---|---|
| 1900 F | 44.02 million Btu/hr |
| $X$ | 46.32 million Btu/hr |
| 2000 F | 46.34 million Btu/hr |

$$X = 1900 + (2000 - 1900)\frac{46.32 - 44.02}{46.34 - 44.02} = 1999 \text{ F}$$

This interpolation of flue gas outlet temperature, 1999 F, is a good approximation of the desired gas outlet temperature, 2000 F.

## FLUE GAS DISCHARGE

To meet the rigorous air pollution codes in effect today, wet gas scrubbing equipment is often necessary. Table 9-6, the flue gas discharge table, provides a method of calculating gas flow volumes exiting a wet scrubbing system as well as scrubber flow quantities. This table can also be used when calculating volumetric flow from a dry flue gas system.

Table 9-6 entries are as follows:

*Inlet temperature* at the incinerator outlet = 2000 F, from the heat balance table. This is the inlet temperature to the flue gas processing system.

*Dry Gas* is the total flow of dry gas exiting the incinerator. From the heat balance table, this figure is 71 012 lb/hr.

Total *inlet heat content* of flue gases exiting the incinerator, in million Btu/hr, (from the heat balance table) is 46.32 million Btu/hr.

The heat is calculated in terms of the dry gas component of the flue gas, as Btu/lb dry gas, (from the entries for total heat and dry gas flow, the heat) = 46.32 million Btu/hr ÷ 71 012 lb/hr = 652 Btu/lb DG.

*Adiabatic t.* When one pound of water evaporates, it absorbs approximately 1000 Btu, without a change in temperature. This heat absorption is called the latent heat of vaporization. (Latent heat as opposed to sensible heat, which is the heat required for a change in temperature without a change in phase.) For evaporated water (steam), the sensible heat is approximately 0.5 Btu per pound of steam for every rise of one degree Fahrenheit.

The adiabatic temperature of the flue gas is the quench temperature. Quenching of a gas is defined as use of the latent heat of water (or other liquid) to decrease the gas temperature. This process does not involve the addition or removal of heat, only the use of the heat of vaporization of the quench liquid, i.e., water. The term adiabatic defines a process where heat is neither added nor removed from a system.

The properties of dry flue gas are considered the same as properties of dry air. Note that a maximum amount of moisture can be held in dry air at a particular temperature. Table 9-7 lists saturation moisture quantities, volume and enthalpy at saturation, as a function of temperature. The temperature at which the enthalpy of the saturated dry flue gas (dry air) is equal to the enthalpy calculated above, Btu/lb dry gas, is the adiabatic temperature of the system. There has been no transfer of heat from the system, only the conversion of latent heat in the quench water to sensible heat in the dry flue gas.

In this example the adiabatic temperature is found in Table 9-7 as that temperature where the dry flue gas (saturated mixture) will have an enthalpy of approximately 652 Btu/lb, 176 F.

*$H_2O$ Saturation.* The quenched flue gas, at the adiabatic temperature (176 F in this example) will contain an amount of moisture equal to the maximum amount of moisture that it can hold, saturation. From Table 9-7 the saturation moisture, in lb $H_2O$/lb dry gas (air) is read opposite the adiabatic temperature. In this case, for 176 F adiabatic temperature the saturation moisture is 0.5511 lb $H_2O$/lb dry gas.

The moisture flow is the saturation moisture multiplied by the dry gas flow, 0.5511 lb $H_2O$/lb dry gas $\times$ 71 012 lb dry gas/hr $=$ 39 135 lb $H_2O$/hr.

*$H_2O$ Inlet.* The moisture component of the flue gas exiting the incinerator is inserted here from the heat balance table: 4896 lb/hr.

*Quench $H_2O$ lb/hr* is the moisture required for quenching the incoming flue gas to its adiabatic temperature. This is equal to the saturated moisture content of the flue gas less the moisture initially carried into the system with the flue gas. This figure is equal to $H_2O$ saturation (39 135 lb/hr) less $H_2O$ inlet (4896 lb/hr), which in this example is equal to 34 239 lb/hr.

*Quench $H_2O$ gpm.* The conversion factor from lb/hr of water to gpm (8.34 lb/gal $\times$ 60 min/hr) is 500. The quench water required is equal to 34 239 lb/hr $\div$ 500 $=$ 68.5 gpm.

*Outlet Temp.* The outlet temperature is that temperature entering the low (negative) pressure side of the ID fan, or, with no ID fan, the temperature within the stack. This temperature is normally selected in the range of 120-160 F. The lower this temperature, the smaller the size of the outlet plume and the lower the volumetric flow of flue gas. For this example 120 F was chosen as the outlet temperature. As can be seen below, a lower outlet temperature would require additional amounts of cooling water.

*Raw $H_2O$ Temp.* This entry is the temperature of the water available for cooling the flue gas from the adiabatic temperature to the outlet temperature. In this example a raw water temperature of 60 F was chosen.

*Sump Temp.* Normally a quantity of water in excess of that calculated for quenching and for cooling the flue gas is provided for particulate removal. The excess water is generally collected in a sump where a quiescent period is allowed to permit larger particles within the spent water to settle to the sump floor, eventually to be drained. The temperature of the water in the sump must be ascertained in order to determine the effective cooling rate of the water flow. The sump temperature is a practical impossiblity to forecast accurately, but an empirical relationship has been established. The sump temperature is assumed equal to the adiabatic temperature divided by 1.2. In this example the sump temperature is 176 ÷ 1.2 = 147 F.

*Temp. Diff.* The temperature differential of note is the difference in temperature between the sump temperature (water exiting the cooling tower) and the raw water entering the cooling tower (or scrubber). In this example the temperature differential = 147 F − 60 F = 87 F.

*Outlet.* The gas exiting the scrubber system is designed to be at the outlet temperature, 120 F in this example, saturated with moisture. The outlet enthalpy is inserted from Table 9-7 for the outlet temperature chosen: 111.65 Btu/lb DG.

The total heat in the outlet flue gas is its enthalpy multiplied by its flow i.e., 111.65 Btu/lb dry gas × 71 012 lb dry gas/hr = 7.93 million Btu/hr.

*Req'd Cooling.* As noted previously the flue gas is initially quenched, without heat addition or removal, to its adiabatic temperature. To reduce the adiabatic temperature of the flue gas to the desired outlet temperature a supply of cooling water is required. This cooling water must remove the heat content at adiabatic conditions relative to the heat content at outlet conditions. The required cooling is therefore the total inlet heat content minus the outlet dry gas million Btu/hr = 46.32 − 7.93 = 38.39 million Btu/hr required cooling, i.e., with removal of 38.39 million Btu/hr from the saturated flue gas stream, the flue gas temperature will fall from 176 F to 120 F.

*Cooling water required* to achieve the desired cooling effect. With $Q = WC\Delta t$ or $W = Q/C\Delta t$ and $W$ the lb/hr cooling water, $Q$ the cooling load in Btu/hr, $C$ the specific heat of water (1 Btu/lb °F), and $\Delta t$ the temperature difference of the cooling water across the flue gas stream (sump water temperature less raw water temperature) the required cooling water flow can be calculated. For this example:

$$\begin{aligned} W &= 38.39 \text{ million Btu/hr} \div 1 \text{ Btu/lb °F} \div 87 \text{ F} \\ &= 441\ 264 \text{ lb/hr, or } 441\ 264 \div 500 = 883 \text{ gpm.} \end{aligned}$$

*Outlet gas volume flow rate.* The specific volume, $ft^3$ mixture per lb dry gas (air), is found in Table 9-7. For this example, with an outlet of 120 F, the specific volume is 16.515 $ft^3$/lb dry gas. The volumetric flow is equal to the specific volume multiplied by the dry gas flow divided by 60 min/hr, i.e., 16.515 $ft^3$/lb dry gas $\times$ 71 012 lb/hr $\div$ 60 min/hr = 19 546 $ft^3$/min.

*Fan Pressure Differential.* Induced draft fans used to clean the incinerator of gases may require a relatively high pressure. The actual differential pressure across the fan, inserted here, will be used to modify the volumetric flow value. For this example the fan pressure is 30″wc.

*Outlet acfm.* The volumetric flow immediately prior to entering the ID fan will experience an expansion because of the fan suction. The volumetric flow correction is as follows: multiply the cfm determined above by the ratio 407 $\div$ (407-*p*), where *p* is the pressure across the fan. The figure 407 is atmospheric pressure (14.7 psia) expressed in inches of water column (14.7 psia $\div$ 62.4 lb $H_2O/ft^3$ $\times$ 1728 in.$^3$/ft = 407″wc). In this example the corrected, or actual, volumetric flow entering the ID fan is [407 $\div$ (407 $-$ 30)] $\times$ 19 546 $ft^3$/min = 21 101 $ft^3$/min actual flow (acfm).

*Outlet humidity.* From Table 9-7, read the saturation humidity, lb $H_2O$/lb dry gas (dry air) corresponding to the outlet temperature. For this example, with an outlet temperature of 120 F, the humidity is 0.081 28 lb $H_2O$/lb dry gas.

The total moisture exiting the stack is the saturation humidity multiplied by the dry gas flow. For this case, 0.081 28 lb $H_2O$/lb dry gas $\times$ 71 012 lb dry gas/hr = 5772 lb $H_2O$/hr.

*Recirc. (Ideal).* The recirculation flow is that amount of flow required for quenching, as calculated above (68.5 gpm for this example). The temperature of this water flow is not critical. This flow is used adiabatically, where the latent heat (not a temperature change) in the water flow reduces the temperature of the flue gas. It is termed a recirculation flow because spent scrubber water from the scrubber sump can be recirculated to the venturi for use at 140 F or greater, instead of a cooler flow of water.

*Recirc. (Actual).* In practice the ideal flow is inadequate to fully clean the gas stream of particulate. Ideal quenching requires intimate contact of each molecule of water with gas, instantaneous evaporation, and instantaneous heat transfer between the moisture and the gas, none of which occurs. To compensate for actual versus ideal conditions an empirical factor is used. In this case this factor is 8. Therefore, for actual recirculation flow use the ideal flow multiplied by eight, 68.5 gpm $\times$ 8 = 548 gpm.

*Cooling $H_2O$* = the flow of cooling water, calculated above = 883 gpm.

## COMPUTER PROGRAM

The mass flow, heat balance, and flue gas discharge analyses presented in this chapter have been developed into a computer program written in the basic language and menu driven. The program accepts up to four concurrent waste feed streams. It considers individual gas components and has waste heat boiler and wet scrubber systems options. The program displays and prints out more comprehensive information than is immediately available from the analysis sheets in this chapter. This program and related programs are available from Incinerator Consultants Incorporated, 11204 Longwood Grove Drive, Reston, Virginia 22094 (703/437-1790).

**Table 9-7. Properties of Dry Air, Water Vapor, and Saturated Air-Water Vapor Mixtures.** Temperature Range = 80 to 211 F; Pressure = 29.921″ Hg. *(Source* = Zimmerman and Lavine, Psychrometric Tables and Charts.)

| | Saturation pressure | | Saturation humidity, weight of water vapor/lb dry air | | Saturation moisture content, weight of water vapor/cf saturated mixture | |
|---|---|---|---|---|---|---|
| **Temp., F** | **psi** | **″Hg** | **lb/lb** | **grains/lb** | **pounds per cf** | **grains per cf** |
| 80 | 0.50689 | 1.0320 | 0.02231 | 156.1 | 0.001583 | 11.08 |
| 81 | .52370 | 1.0663 | .02308 | 161.5 | .001645 | 11.51 |
| 82 | .54099 | 1.1015 | .02387 | 167.1 | .001684 | 11.79 |
| 83 | .55878 | 1.1377 | .02468 | 172.8 | .001736 | 12.15 |
| 84 | .57707 | 1.1749 | .02552 | 178.7 | .001789 | 12.53 |
| 85 | 0.59588 | 1.2132 | 0.02639 | 184.7 | 0.001844 | 12.91 |
| 86 | .61522 | 1.2526 | .02728 | 191.0 | .001901 | 13.30 |
| 87 | .63510 | 1.2931 | .02821 | 197.4 | .001958 | 13.71 |
| 88 | .65555 | 1.3347 | .02916 | 204.1 | .002018 | 14.12 |
| 89 | .67656 | 1.3775 | .03014 | 210.9 | .002079 | 14.55 |
| 90 | 0.69816 | 1.4215 | 0.03115 | 218.0 | 0.002141 | 14.99 |
| 91 | .72036 | 1.4667 | .03219 | 225.3 | .002205 | 15.44 |
| 92 | .74316 | 1.5131 | .03326 | 232.8 | .002271 | 15.90 |
| 93 | .76659 | 1.5608 | .03437 | 240.6 | .002338 | 16.37 |
| 94 | .79065 | 1.6098 | .03551 | 248.5 | .002407 | 16.85 |
| 95 | 0.81537 | 1.6601 | 0.03668 | 256.8 | 0.002478 | 17.34 |
| 96 | .84074 | 1.7118 | .03789 | 265.3 | .002550 | 17.85 |
| 97 | .86681 | 1.7648 | .03914 | 274.0 | .002624 | 18.37 |
| 98 | .89358 | 1.8193 | .04043 | 283.0 | .002701 | 18.91 |
| 99 | .02105 | 1.8753 | .04175 | 292.3 | .002779 | 19.45 |
| 100 | 0.94926 | 1.9327 | 0.04312 | 301.8 | 0.002859 | 20.01 |
| 101 | 0.97821 | 1.9916 | .04453 | 311.7 | .002941 | 20.58 |
| 102 | 1.00792 | 2.0521 | .04498 | 321.9 | .003025 | 21.17 |
| 103 | 1.03842 | 2.1142 | .04748 | 332.3 | .003111 | 21.77 |
| 104 | 1.06965 | 2.1788 | .04902 | 343.1 | .003198 | 22.39 |
| 105 | 1.1018 | 2.2432 | 0.05061 | 354.3 | 0.003289 | 23.02 |
| 106 | 1.1347 | 2.3103 | .05225 | 365.7 | .003381 | 23.67 |
| 107 | 1.1685 | 2.3790 | .05394 | 377.6 | .003476 | 24.33 |
| 108 | 1.2031 | 2.4495 | .05568 | 389.7 | .003572 | 25.00 |
| 109 | 1.2386 | 2.5218 | .05747 | 402.3 | .003671 | 25.70 |
| 110 | 1.2750 | 2.5959 | 0.05932 | 415.3 | 0.003772 | 26.41 |
| 111 | 1.3123 | 2.6719 | .06123 | 428.6 | .003876 | 27.13 |
| 112 | 1.3506 | 2.7497 | .06319 | 442.4 | .003982 | 27.88 |
| 113 | 1.3897 | 2.8295 | .06522 | 456.5 | .004090 | 28.63 |
| 114 | 1.4300 | 2.9114 | .06731 | 471.2 | .004202 | 29.41 |

* See footnote on page 118.

| Saturation density, weight of air plus water vapor/cf saturated mixture | | Volume | | | Enthalpy | | | |
|---|---|---|---|---|---|---|---|---|
| pounds per cf | grains per cf | dry air, cf/lb | water vapor, cf/lb* | saturated mixture, dry air, cf/lb | dry air, Btu/lb | water vapor, Btu/lb | saturated mixture, dry air, Btu/lb | Temp., F |
| 0.07257 | 508.0 | 13.601 | 21.844 | 14.088 | 11.533 | 1096.2 | 35.985 | 80 |
| .07240 | 506.8 | 13.626 | 21.884 | 14.131 | 11.773 | 1096.6 | 37.077 | 81 |
| .07223 | 505.6 | 13.652 | 21.923 | 14.175 | 12.013 | 1097.1 | 38.197 | 82 |
| .07206 | 504.4 | 13.677 | 21.964 | 14.219 | 12.254 | 1097.5 | 39.342 | 83 |
| .07190 | 503.3 | 13.702 | 22.003 | 14.264 | 12.494 | 1097.9 | 40.515 | 84 |
| 0.07173 | 502.1 | 13.728 | 22.043 | 14.309 | 12.735 | 1098.3 | 41.718 | 85 |
| .07156 | 500.9 | 13.753 | 22.082 | 14.355 | 12.975 | 1098.7 | 42.952 | 86 |
| .07139 | 499.8 | 13.778 | 22.122 | 14.402 | 13.217 | 1099.1 | 44.216 | 87 |
| .07123 | 498.6 | 13.803 | 22.161 | 14.449 | 13.457 | 1099.6 | 45.515 | 88 |
| .07106 | 497.4 | 13.829 | 22.200 | 14.497 | 13.696 | 1100.0 | 46.845 | 89 |
| 0.07088 | 496.2 | 13.854 | 22.240 | 14.547 | 13.938 | 1100.5 | 48.212 | 90 |
| .07071 | 495.0 | 13.879 | 22.279 | 14.597 | 14.177 | 1100.9 | 49.612 | 91 |
| .07054 | 493.8 | 13.904 | 22.319 | 14.647 | 14.418 | 1101.4 | 51.050 | 92 |
| .07037 | 492.6 | 13.930 | 22.358 | 14.699 | 14.658 | 1101.8 | 52.522 | 93 |
| .07020 | 491.4 | 13.955 | 22.398 | 14.751 | 14.899 | 1102.3 | 54.037 | 94 |
| 0.07003 | 490.2 | 13.980 | 22.437 | 14.804 | 15.139 | 1102.7 | 55.586 | 95 |
| .06985 | 489.0 | 14.006 | 22.476 | 14.854 | 15.380 | 1103.1 | 57.179 | 96 |
| .06968 | 487.8 | 14.031 | 22.515 | 14.913 | 15.620 | 1103.5 | 58.810 | 97 |
| .06951 | 486.6 | 14.056 | 22.555 | 14.968 | 15.860 | 1103.9 | 60.486 | 98 |
| .06933 | 485.3 | 14.801 | 22.594 | 15.025 | 16.101 | 1104.3 | 62.209 | 99 |
| 0.06916 | 484.1 | 14.107 | 22.633 | 15.083 | 16.341 | 1104.8 | 63.980 | 100 |
| .06898 | 482.9 | 14.132 | 22.673 | 15.142 | 16.582 | 1105.2 | 65.794 | 101 |
| .06880 | 481.6 | 14.157 | 22.712 | 15.202 | 16.822 | 1105.6 | 67.657 | 102 |
| .06863 | 480.4 | 14.182 | 22.751 | 15.263 | 17.063 | 1106.1 | 69.577 | 103 |
| .06845 | 479.2 | 14.208 | 22.790 | 15.325 | 17.304 | 1106.5 | 71.541 | 104 |
| 0.06827 | 477.9 | 14.233 | 22.829 | 15.389 | 17.544 | 1106.9 | 73.563 | 105 |
| .06809 | 476.7 | 14.258 | 22.868 | 15.453 | 17.785 | 1107.3 | 75.639 | 106 |
| .06791 | 475.4 | 14.283 | 22.907 | 15.519 | 18.025 | 1107.7 | 77.771 | 107 |
| .06773 | 474.1 | 14.309 | 22.946 | 15.587 | 18.266 | 1108.1 | 79.961 | 108 |
| .06755 | 472.8 | 14.334 | 22.985 | 15.655 | 18.506 | 1108.5 | 82.215 | 109 |
| 0.06737 | 470.6 | 14.359 | 23.024 | 15.725 | 18.747 | 1109.0 | 84.535 | 110 |
| .06718 | 470.3 | 14.384 | 23.063 | 15.796 | 18.987 | 1109.4 | 86.915 | 111 |
| .06700 | 469.0 | 14.410 | 23.101 | 15.869 | 19.228 | 1109.8 | 89.360 | 112 |
| .06681 | 467.7 | 14.435 | 23.140 | 15.944 | 19.469 | 1110.2 | 91.873 | 113 |
| .06662 | 466.4 | 14.460 | 23.178 | 16.020 | 19.709 | 1110.6 | 94.461 | 114 |

**Table 9-7. Properties of Dry Air, Water Vapor, and Saturated Air-Water Vapor Mixtures** *(continued)*

| Temp., F | Saturation pressure | | Saturation humidity, weight of water vapor/lb dry air | | Saturation moisture content, weight of water vapor/cf saturated mixture | |
|---|---|---|---|---|---|---|
| | psi | "Hg | lb/lb | grains/lb | pounds per cf | grains per cf |
| 115 | 1.4711 | 2.9952 | 0.06946 | 486.2 | 0.004315 | 30.20 |
| 116 | 1.5133 | 3.0811 | .07168 | 501.8 | .004431 | 31.02 |
| 117 | 1.5566 | 3.1691 | .07397 | 517.8 | .004550 | 31.85 |
| 118 | 1.6008 | 3.2593 | .07633 | 534.3 | .004671 | 32.69 |
| 119 | 1.6462 | 3.3517 | .07877 | 551.4 | .004795 | 33.57 |
| 120 | 1.6927 | 3.4463 | 0.08128 | 569.0 | 0.004922 | 34.45 |
| 121 | 1.7403 | 3.5432 | .08388 | 587.1 | .005052 | 35.36 |
| 122 | 1.7890 | 3.6424 | .08655 | 605.9 | .005184 | 36.29 |
| 123 | 1.8389 | 3.7440 | .08931 | 625.2 | .005320 | 37.24 |
| 124 | 1.8900 | 3.8480 | .09216 | 645.1 | .005458 | 38.21 |
| 125 | 1.9423 | 3.9544 | 0.09511 | 665.8 | 0.005600 | 39.20 |
| 126 | 1.9958 | 4.0634 | .09815 | 687.1 | .005745 | 40.22 |
| 127 | 2.0506 | 4.1749 | .10129 | 709.0 | .005893 | 41.25 |
| 128 | 2.1066 | 4.2891 | .10453 | 731.7 | .006045 | 42.33 |
| 129 | 2.1640 | 4.4059 | .10788 | 755.2 | .006199 | 43.40 |
| 130 | 2.2227 | 4.5255 | 0.1113 | 779.4 | 0.006357 | 44.50 |
| 131 | 2.2828 | 4.6479 | .1149 | 804.4 | .006519 | 45.63 |
| 132 | 2.3442 | 4.7729 | .1186 | 830.2 | .006683 | 46.78 |
| 133 | 2.4072 | 4.9010 | .1224 | 856.9 | .006851 | 47.96 |
| 134 | 2.4715 | 5.0320 | .1264 | 884.5 | .007024 | 49.16 |
| 135 | 2.5373 | 5.1659 | 0.1304 | 913.0 | 0.007199 | 50.39 |
| 136 | 2.6045 | 5.3028 | .1346 | 942.5 | .007377 | 51.64 |
| 137 | 2.6733 | 5.4429 | .1390 | 973.1 | .007561 | 52.92 |
| 138 | 2.7436 | 5.5861 | .1435 | 1004.5 | .007747 | 54.23 |
| 139 | 2.8155 | 5.7324 | .1482 | 1037.3 | .007937 | 55.56 |
| 140 | 2.8890 | 5.8821 | 0.1530 | 1071 | 0.008131 | 56.92 |
| 141 | 2.9641 | 6.0349 | .1580 | 1106 | .008329 | 58.31 |
| 142 | 3.0409 | 6.1912 | .1632 | 1142 | .008532 | 59.72 |
| 143 | 3.1193 | 6.3509 | .1685 | 1180 | .008738 | 61.17 |
| 144 | 3.1915 | 6.5141 | .1741 | 1219 | .008949 | 62.64 |
| 145 | 3.2814 | 6.6809 | 0.1798 | 1259 | 0.009163 | 64.14 |
| 146 | 3.3651 | 6.8512 | .1858 | 1301 | .009383 | 65.68 |
| 147 | 3.4506 | 7.0253 | .1920 | 1344 | .009607 | 67.25 |
| 148 | 3.5379 | 7.2032 | .1984 | 1389 | .009835 | 68.84 |
| 149 | 3.6271 | 7.3847 | .2051 | 1436 | .010066 | 70.46 |

* See footnote on page 118.

| Saturation density, weight of air plus water vapor/cf saturated mixture | | Volume | | | Enthalpy | | | |
|---|---|---|---|---|---|---|---|---|
| pounds per cf | grains per cf | dry air, cf/lb | water vapor, cf/lb* | saturated mixture, dry air, cf/lb | dry air, Btu/lb | water vapor, Btu/lb | saturated mixture, dry air, Btu/lb | Temp., F |
| 0.06643 | 465.0 | 14.486 | 23.218 | 16.098 | 19.950 | 1111.1 | 97.128 | 115 |
| .06624 | 463.7 | 14.516 | 23.256 | 16.178 | 20.190 | 1111.5 | 99.866 | 116 |
| .06605 | 462.4 | 14.536 | 23.295 | 16.259 | 20.431 | 1112.0 | 102.688 | 117 |
| .06586 | 461.0 | 14.561 | 23.333 | 16.343 | 20.672 | 1112.4 | 105.586 | 118 |
| .06567 | 459.7 | 14.587 | 23.372 | 16.428 | 20.912 | 1112.8 | 108.571 | 119 |
| 0.06547 | 458.3 | 14.612 | 23.408 | 16.515 | 21.153 | 1113.3 | 111.65 | 120 |
| .06528 | 457.0 | 14.637 | 23.448 | 16.603 | 21.394 | 1113.7 | 114.81 | 121 |
| .06508 | 455.6 | 14.662 | 23.487 | 16.695 | 21.634 | 1114.2 | 118.07 | 122 |
| .06488 | 545.2 | 14.688 | 23.526 | 16.789 | 21.875 | 1114.6 | 121.42 | 123 |
| .06468 | 452.8 | 14.713 | 23.565 | 16.885 | 22.116 | 1115.0 | 124.88 | 124 |
| 0.06448 | 451.4 | 14.738 | 23.602 | 16.983 | 22.356 | 1115.4 | 128.44 | 125 |
| .06428 | 450.0 | 14.763 | 23.641 | 17.084 | 22.597 | 1115.8 | 132.11 | 126 |
| .06408 | 448.5 | 14.789 | 23.679 | 17.187 | 22.838 | 1116.3 | 135.91 | 127 |
| .06387 | 447.1 | 14.814 | 23.718 | 17.293 | 23.079 | 1116.7 | 139.81 | 128 |
| .06366 | 445.6 | 14.839 | 23.756 | 17.402 | 23.319 | 1117.1 | 143.83 | 129 |
| 0.06345 | 441.2 | 14.864 | 23.794 | 17.514 | 23.560 | 1117.6 | 147.99 | 130 |
| .06325 | 442.7 | 14.890 | 23.833 | 17.628 | 23.801 | 1118.0 | 152.27 | 131 |
| .06303 | 441.2 | 14.915 | 23.871 | 17.746 | 24.041 | 1118.4 | 156.64 | 132 |
| .06282 | 439.7 | 14.940 | 23.910 | 17.867 | 24.282 | 1118.8 | 161.23 | 133 |
| .06261 | 438.2 | 14.965 | 23.948 | 17.991 | 24.523 | 1119.2 | 165.95 | 134 |
| 0.06239 | 436.7 | 14.991 | 23.987 | 18.119 | 24.764 | 1119.6 | 170.79 | 135 |
| .06217 | 435.2 | 15.016 | 24.025 | 18.251 | 25.005 | 1120.1 | 175.81 | 136 |
| .06195 | 433.7 | 15.041 | 24.063 | 18.386 | 25.245 | 1120.5 | 181.01 | 137 |
| .06173 | 432.1 | 15.067 | 24.103 | 18.525 | 25.486 | 1120.9 | 186.35 | 138 |
| .06150 | 430.5 | 15.092 | 24.139 | 18.669 | 25.727 | 1121.3 | 191.88 | 139 |
| 0.06128 | 428.9 | 15.117 | 24.178 | 18.816 | 25.968 | 1121.7 | 197.59 | 140 |
| .06105 | 427.3 | 15.142 | 24.215 | 18.969 | 26.209 | 1122.1 | 203.50 | 141 |
| .06082 | 425.7 | 15.168 | 24.253 | 19.126 | 26.449 | 1122.5 | 209.62 | 142 |
| .06058 | 424.1 | 15.193 | 24.290 | 19.288 | 26.690 | 1122.9 | 215.94 | 143 |
| .06035 | 422.5 | 15.218 | 24.329 | 19.454 | 26.931 | 1123.3 | 222.49 | 144 |
| 0.06012 | 420.8 | 15.243 | 24.367 | 19.626 | 27.172 | 1123.7 | 229.26 | 145 |
| .05988 | 419.1 | 15.269 | 24.405 | 19.804 | 47.413 | 1124.1 | 236.29 | 146 |
| .05964 | 417.5 | 15.294 | 24.443 | 19.987 | 27.654 | 1124.6 | 243.59 | 147 |
| .05940 | 415.8 | 15.320 | 24.481 | 20.176 | 27.895 | 1125.0 | 251.12 | 148 |
| .05915 | 414.0 | 15.344 | 24.518 | 20.374 | 28.135 | 1125.4 | 258.94 | 149 |

**Table 9-7. Properties of Dry Air, Water Vapor, and Saturated Air-Water Vapor Mixtures** *(continued)*

| | Saturation pressure | | Saturation humidity, weight of water vapor/lb dry air | | Saturation moisture content, weight of water vapor/cf saturated mixture | |
|---|---|---|---|---|---|---|
| Temp., F | psi | ″Hg | lb/lb | grains/lb | pounds per cf | grains per cf |
| 150 | 3.7182 | 7.5703 | 0.2120 | 1484 | 0.01030 | 72.13 |
| 151 | 3.8113 | 7.7597 | .2192 | 1534 | .01055 | 73.82 |
| 152 | 3.9063 | 7.9531 | .2267 | 1587 | .01079 | 75.54 |
| 153 | 4.0033 | 8.1506 | .2344 | 1641 | .01104 | 77.30 |
| 154 | 4.1023 | 8.3523 | .2425 | 1698 | .01130 | 79.09 |
| 155 | 4.2034 | 8.5581 | 0.2509 | 1756 | 0.01156 | 80.92 |
| 156 | 4.3066 | 8.7682 | .2596 | 1817 | .01183 | 82.78 |
| 157 | 4.4120 | 8.9828 | .2688 | 1881 | .01210 | 84.69 |
| 158 | 4.5195 | 9.2016 | .2782 | 1948 | .01237 | 86.61 |
| 159 | 4.6292 | 9.4251 | .2881 | 2005 | .01265 | 88.58 |
| 160 | 4.7412 | 9.6531 | 0.2985 | 2089 | 0.01294 | 90.59 |
| 161 | 4.8554 | 9.8856 | .3092 | 2165 | .01323 | 92.63 |
| 162 | 4.9720 | 10.1231 | .3205 | 2244 | .01353 | 94.72 |
| 163 | 5.0909 | 10.3652 | .3323 | 2326 | .01384 | 96.85 |
| 164 | 5.2122 | 10.6162 | .3446 | 2412 | .01414 | 99.00 |
| 165 | 5.3358 | 10.864 | 0.3575 | 2502 | 0.01446 | 101.2 |
| 166 | 5.4621 | 11.121 | .3710 | 2597 | .01478 | 103.5 |
| 167 | 5.5908 | 11.383 | .3851 | 2696 | .01511 | 105.8 |
| 168 | 5.7220 | 11.650 | .4000 | 2800 | .01543 | 108.1 |
| 169 | 5.8558 | 11.922 | .4156 | 2909 | .01577 | 110.4 |
| 170 | 5.9923 | 12.200 | 0.4320 | 3024 | 0.01612 | 112.8 |
| 171 | 6.1314 | 12.484 | .4493 | 3145 | .01647 | 115.3 |
| 172 | 6.2733 | 12.772 | .4675 | 3273 | .01682 | 117.8 |
| 173 | 6.4179 | 13.067 | .4807 | 3407 | .01719 | 120.3 |
| 174 | 6.5653 | 13.367 | .5070 | 3549 | .01756 | 122.9 |
| 175 | 6.7156 | 13.673 | 0.5284 | 3699 | 0.01793 | 125.5 |
| 176 | 6.8687 | 13.985 | .5511 | 3858 | .01832 | 128.2 |
| 177 | 7.0247 | 14.302 | .5752 | 4026 | .01871 | 130.9 |
| 178 | 7.1838 | 14.626 | .6008 | 4205 | .01910 | 133.7 |
| 179 | 7.3458 | 14.956 | .6279 | 4395 | .01950 | 136.5 |
| 180 | 7.5109 | 15.292 | 0.6569 | 4598 | 0.01992 | 139.4 |
| 181 | 7.6791 | 15.635 | .6878 | 4815 | .02033 | 142.3 |
| 182 | 7.8504 | 15.983 | .7209 | 5046 | .02076 | 145.3 |
| 183 | 8.0247 | 16.339 | .7563 | 5294 | .02119 | 148.3 |
| 184 | 8.2027 | 16.701 | .7943 | 5560 | .02163 | 151.4 |

* See footnote on page 118.

| Saturation density, weight of air plus water vapor/cf saturated mixture | | Volume | | | Enthalpy | | | |
|---|---|---|---|---|---|---|---|---|
| pounds per cf | grains per cf | dry air, cf/lb | water vapor, cf/lb* | saturated mixture, dry air, cf/lb | dry air, Btu/lb | water vapor, Btu/lb | saturated mixture, dry air, Btu/lb | Temp., F |
| 0.05890 | 412.3 | 15.370 | 24.555 | 20.576 | 28.376 | 1125.8 | 267.06 | 150 |
| .05866 | 410.6 | 15.395 | 24.593 | 20.786 | 28.617 | 1126.2 | 275.48 | 151 |
| .05840 | 408.8 | 15.420 | 24.630 | 21.004 | 28.858 | 1126.6 | 284.22 | 152 |
| .05815 | 407.0 | 15.445 | 24.668 | 21.229 | 29.099 | 1127.0 | 293.30 | 153 |
| .05789 | 405.3 | 15.471 | 24.705 | 21.462 | 29.340 | 1127.4 | 302.73 | 154 |
| 0.05763 | 403.4 | 15.496 | 24.742 | 21.704 | 29.581 | 1127.8 | 312.55 | 155 |
| .05737 | 401.6 | 15.521 | 24.780 | 21.955 | 29.822 | 1128.2 | 322.75 | 156 |
| .05711 | 399.8 | 15.546 | 24.818 | 22.216 | 30.063 | 1128.7 | 333.41 | 157 |
| .05684 | 397.9 | 15.572 | 24.855 | 22.487 | 30.304 | 1129.1 | 344.46 | 158 |
| .05657 | 396.0 | 15.597 | 24.892 | 22.769 | 30.545 | 1129.5 | 356.00 | 159 |
| 0.05630 | 394.1 | 15.622 | 24.929 | 23.063 | 30.786 | 1129.9 | 368.13 | 160 |
| .05603 | 392.2 | 15.647 | 24.966 | 23.368 | 31.027 | 1130.3 | 380.56 | 161 |
| .05575 | 390.3 | 15.673 | 25.003 | 23.685 | 31.268 | 1130.7 | 393.67 | 162 |
| .05547 | 388.3 | 15.698 | 25.040 | 24.017 | 31.509 | 1131.1 | 407.35 | 163 |
| .05519 | 386.3 | 15.723 | 25.077 | 24.365 | 31.750 | 1131.5 | 421.66 | 164 |
| 0.05490 | 384.3 | 15.748 | 25.114 | 24.725 | 31.991 | 1131.9 | 436.61 | 165 |
| .05462 | 382.3 | 15.774 | 25.151 | 25.102 | 32.232 | 1132.3 | 452.30 | 166 |
| .05434 | 380.3 | 15.799 | 25.188 | 25.492 | 32.473 | 1132.7 | 468.72 | 167 |
| .05402 | 378.2 | 15.824 | 25.225 | 25.914 | 32.714 | 1133.1 | 485.95 | 168 |
| .05373 | 376.1 | 15.849 | 25.261 | 26.347 | 32.955 | 1133.5 | 504.05 | 169 |
| 0.05343 | 374.0 | 15.875 | 25.298 | 26.804 | 33.196 | 1133.0 | 523.06 | 170 |
| .05314 | 372.0 | 15.900 | 25.335 | 27.272 | 33.437 | 1134.3 | 543.08 | 171 |
| .05281 | 369.7 | 15.925 | 25.372 | 27.787 | 33.678 | 1134.7 | 564.15 | 172 |
| .05251 | 367.5 | 15.950 | 25.408 | 28.315 | 33.919 | 1135.1 | 586.38 | 173 |
| .05219 | 365.3 | 15.976 | 25.445 | 28.876 | 34.161 | 1135.5 | 609.84 | 174 |
| 0.05187 | 363.1 | 16.001 | 25.481 | 29.465 | 34.402 | 1135.9 | 634.63 | 175 |
| .05155 | 360.9 | 16.026 | 25.518 | 30.089 | 34.643 | 1136.3 | 660.88 | 176 |
| .05123 | 358.6 | 16.051 | 25.554 | 30.749 | 34.884 | 1136.7 | 688.69 | 177 |
| .05090 | 356.3 | 16.077 | 25.591 | 31.449 | 35.125 | 1137.1 | 718.25 | 178 |
| .05057 | 356.1 | 16.102 | 25.627 | 32.193 | 35.366 | 1137.5 | 749.63 | 179 |
| 0.05023 | 351.6 | 16.127 | 25.664 | 32.984 | 35.607 | 1137.9 | 783.08 | 180 |
| .04989 | 349.2 | 16.152 | 25.700 | 33.829 | 35.848 | 1138.3 | 818.78 | 181 |
| .04955 | 346.8 | 16.178 | 25.736 | 34.731 | 36.090 | 1138.7 | 856.97 | 182 |
| .04920 | 344.4 | 16.203 | 25.772 | 35.694 | 36.331 | 1139.1 | 897.79 | 183 |
| .04885 | 342.0 | 16.228 | 25.809 | 36.728 | 36.572 | 1139.5 | 941.68 | 184 |

**Table 9-7. Properties of Dry Air, Water Vapor, and Saturated Air-Water Vapor Mixtures** *(concluded)*

| | Saturation pressure | | Saturation humidity, weight of water vapor/lb dry air | | Saturation moisture content, weight of water vapor/cf saturated mixture | |
|---|---|---|---|---|---|---|
| Temp., F | psi | "Hg | lb/lb | grains/lb | pounds per cf | grains per cf |
| 185 | 8.3836 | 17.069 | 0.8352 | 5847 | 0.02207 | 154.5 |
| 186 | 8.5678 | 17.444 | .8794 | 6156 | .02253 | 157.7 |
| 187 | 8.7554 | 17.826 | .9271 | 6490 | .02299 | 160.9 |
| 188 | 8.9465 | 18.215 | .9790 | 6853 | .02346 | 164.2 |
| 189 | 9.1411 | 18.611 | 1.0355 | 7249 | .02393 | 167.5 |
| 190 | 9.3392 | 19.015 | 1.097 | 7681 | 0.02442 | 170.9 |
| 191 | 9.5409 | 19.425 | 1.165 | 8155 | .02491 | 174.4 |
| 192 | 9.7463 | 19.844 | 1.240 | 8679 | .02542 | 177.9 |
| 193 | 9.9553 | 20.269 | 1.322 | 9257 | .02592 | 181.5 |
| 194 | 10.1684 | 20.703 | 1.414 | 9900 | .02644 | 185.1 |
| 195 | 10.385 | 21.143 | 1.517 | 10629 | 0.02697 | 188.8 |
| 196 | 10.605 | 21.591 | 1.633 | 11432 | .02750 | 192.5 |
| 197 | 10.829 | 22.048 | 1.765 | 12352 | .02805 | 196.3 |
| 198 | 11.057 | 22.513 | 1.915 | 13405 | .02860 | 200.2 |
| 199 | 11.289 | 22.985 | 2.089 | 14620 | .02916 | 204.1 |
| 200 | 11.526 | 23.466 | 2.292 | 16046 | 0.02973 | 208.1 |
| 201 | 11.766 | 23.955 | 2.532 | 17725 | .03031 | 212.2 |
| 202 | 12.010 | 24.453 | 2.820 | 19739 | .03090 | 216.3 |
| 203 | 12.259 | 24.960 | 3.173 | 22212 | .03149 | 220.5 |
| 204 | 12.512 | 25.474 | 3.614 | 25301 | .03210 | 224.7 |
| 205 | 12.769 | 25.998 | 4.181 | 29269 | 0.03272 | 229.0 |
| 206 | 13.031 | 26.531 | 4.939 | 34576 | .03346 | 234.2 |
| 207 | 13.297 | 27.073 | 6.000 | 42000 | .03398 | 237.9 |
| 208 | 13.568 | 27.624 | 7.594 | 53161 | .03463 | 242.4 |
| 209 | 13.843 | 28.184 | 10.248 | 71736 | .03528 | 247.0 |
| 210 | 14.122 | 28.753 | 15.54 | 108773 | 0.03595 | 251.6 |
| 211 | 14.407 | 29.332 | 31.49 | 220451 | .03667 | 256.7 |

* Theoretical volume at 1 atm. =
(volume at saturation pressure) × (saturation pressure "Hg)/29.921.

| Saturation density, weight of air plus water vapor/cf saturated mixture | | Volume | | | Enthalpy | | | |
|---|---|---|---|---|---|---|---|---|
| pounds per cf | grains per cf | dry air, cf/lb | water vapor, cf/lb* | saturated mixture, dry air, cf/lb | dry air, Btu/lb | water vapor, Btu/lb | saturated mixture, dry air, Btu/lb | Temp., F |
| 0.04850 | 339.5 | 16.253 | 25.844 | 37.839 | 36.813 | 1139.9 | 988.8 | 185 |
| .04814 | 337.0 | 16.279 | 25.881 | 39.037 | 37.054 | 1140.3 | 1039.79 | 186 |
| .04778 | 334.5 | 16.304 | 25.916 | 40.332 | 37.296 | 1140.7 | 1094.88 | 187 |
| .04742 | 331.9 | 16.329 | 25.953 | 41.737 | 37.537 | 1141.1 | 1154.70 | 188 |
| .04705 | 329.3 | 16.354 | 25.988 | 43.265 | 37.778 | 1141.5 | 1219.80 | 189 |
| 0.04667 | 326.7 | 16.380 | 26.024 | 44.935 | 38.019 | 1141.9 | 1291.0 | 190 |
| .04630 | 324.1 | 16.405 | 26.060 | 46.764 | 38.261 | 1142.3 | 1369.5 | 191 |
| .04592 | 321.4 | 16.430 | 26.093 | 48.780 | 38.502 | 1142.7 | 1455.1 | 192 |
| .04553 | 318.7 | 16.455 | 26.131 | 51.011 | 38.743 | 1143.1 | 1550.4 | 193 |
| .04514 | 316.0 | 16.481 | 26.167 | 53.488 | 38.985 | 1143.5 | 1656.2 | 194 |
| 0.04474 | 313.2 | 16.506 | 26.202 | 56.265 | 39.226 | 1143.9 | 1775.0 | 195 |
| .04434 | 310.4 | 16.530 | 26.238 | 59.381 | 39.467 | 1144.3 | 1908.3 | 196 |
| .04394 | 307.6 | 16.556 | 26.273 | 62.918 | 39.709 | 1144.7 | 2059.6 | 197 |
| .04353 | 304.7 | 16.582 | 26.309 | 66.963 | 39.950 | 1145.1 | 2232.8 | 198 |
| .04312 | 301.8 | 16.607 | 26.344 | 71.630 | 40.191 | 1145.5 | 2432.7 | 199 |
| 0.04270 | 298.9 | 16.632 | 26.380 | 77.102 | 40.433 | 1145.9 | 2667.2 | 200 |
| .04228 | 296.0 | 16.657 | 26.415 | 83.543 | 40.674 | 1146.2 | 2943.0 | 201 |
| .04185 | 293.0 | 16.683 | 26.450 | 91.270 | 40.916 | 1146.6 | 3274.2 | 202 |
| .04142 | 290.0 | 16.708 | 26.486 | 100.750 | 41.158 | 1146.0 | 3679.6 | 203 |
| .04098 | 286.9 | 16.733 | 26.521 | 112.590 | 41.399 | 1147.4 | 4188.6 | 204 |
| 0.04054 | 283.8 | 16.758 | 26.556 | 127.80 | 41.640 | 1147.7 | 4840.5 | 205 |
| .04024 | 281.7 | 16.784 | 26.591 | 147.60 | 41.882 | 1148.1 | 5712.8 | 206 |
| .03965 | 277.5 | 16.809 | 26.626 | 176.56 | 42.124 | 1148.4 | 6932.3 | 207 |
| .03919 | 274.3 | 16.834 | 26.661 | 219.30 | 42.366 | 1148.8 | 8766.8 | 208 |
| .03873 | 271.1 | 16.859 | 26.696 | 290.44 | 42.608 | 1149.2 | 11820.3 | 209 |
| 0.03826 | 267.8 | 16.885 | 26.731 | 432.25 | 42.849 | 1149.6 | 17906 | 210 |
| .03779 | 264.5 | 16.910 | 26.765 | 859.82 | 43.090 | 1150.0 | 36260 | 211 |

## REFERENCES

1. **Brunner, Calvin R.:** "Handbook of Hazardous Waste Incineration", published by Tab Books, Inc., Blue Ridge Summit, PA, 1989.
2. **Brunner, Calvin R.:** "Incineration Systems -- Selection and Design", Second Printing published by Incinerator Consultants Inc., 11204 Longwood Grove Dr., Reston, VA 22094; 1988.
3. **Castaldini, C. et al:** "Disposal of Hazardous Wastes in Industrial Boilers and Furnaces", published by Noyes Publications, Park Ridge, NJ, 1986.
4. **Cudahy, J.J. and Wass, W.E.:** "New Technique for Estimating Heat of Combustion Needed for Waste Disposal"; POLLUTION ENGINEERING, Febr., 1979, pages 45-46.
5. **General Electric Co.:** "Solid Waste Management Technology Assessment", Van Nostrand Reinhold Co., New York, NY, 1975.
6. **Gill, Brian J., Candidate, School of Engineering, University of Dayton:** "Incineration in the 90s".
7. **Governmental Institutes, Inc.:** "Book of Lists for Regulated Hazardous Substances", published by Governmental Institutes, Inc., Rockville, MD 20850, 1991.
8. **Hay, Nelson E. (ed.):** "Natural Gas Applications for Air Pollution Control", distributed by Prentice Hall, Inc., Englewood Cliffs, NJ 07632, 1987.
9. **Joy Mfg. Co.:** "Controlled Air Incineration", published by Joy Mfg. Co., Ecolaire EPC Division, Charlotte, NC, 1981.
10. **Lee, C.C.:** "Environmental Engineering Dictionary", published by Government Institutes, Inc., Rockville, MD 20850, 1989.
11. **Lide, D.R. (ed.):** "Handbook of Chemistry and Physics", 73rd ed. published by CRC Press, Inc., Boca Raton, Florida, 1992.
12. **Liptak, B.G. (ed.):** "Environmental Engineer's Handbook", 2 volumes, Chilton Book Company, Radnor, PA, 1974.
13. **Makansi, Jason:** "Mill Seeks Advanced Boiler Technology to Stay Profitable", POWER magazine, Febr, 1991, pp 66-68.
14. **Niessen, Walter R.:** "Combustion and Incineration Processes", published by Marcel Dekker, Inc., New York, NY 10016, 1978.
15. **Reed, Richard J. (ed.):** "Combustion Handbook" (2 vols, 3rd ed.) published by North American Mfg. Co., Cleveland, OH 44105, 1986.

*(continued)*

**REFERENCES** *(concluded)*

16. **Reynolds, J.P. et al:** "Hazardous Waste Incineration Calculations", published by John Wiley & Sons, Inc, New York, NY, 1991.

17. **The National Environmental Journal,** (a) Nov. 1991 issue, page 28, "Incineration of Municipal and Hazardous Waste", by G. Geiger; (b) July 1992 issue, page 46, "Solving Air Emissions Problems from Metal Cleaning Operations" by R. Zaman.

18. **Theodore, Louis, and Reynolds, Joseph:** "Introduction to Hazardous Waste Incineration", published by John Wiley & Sons, New York, NY, 1987.

19. **Tillman D. A. et al:** "Incineration of Municipal and Hazardous Solid Wastes", published by Academic Press, Inc., San Diego, CA 92101, 1989.

20. **U.S.E.P.A.:** "Code of Federal Regulations 40, Protection of Environment", distributed by U.S. Government Printing Office, Washington, DC 20402-9328, 1991.

21. **U.S.E.P.A.:** "Technical Implementation Document for EPA's Boiler and Industrial Furnace Regulations", March 1992, PB92-154947, distributed by National Technical Information Service, telephone 800-336-4700.

## APPENDIX CONTENTS

## Appendix A1. Incineration abbreviations* and acronyms.

| | |
|---|---|
| acfm* | actual cubic feet per minute |
| AGRE | Acid Gas Removal Efficiency, % AGRE = (ASC − AST) (100)/ASC |
| ANPR | advanced notice of proposed rulemaking |
| APC | Air Pollution Control |
| APCD | Air Pollution Control District |
| AQCR | Air Quality Control Region |
| AQMP | air quality maintenance plan |
| ARB | Air Resources Board |
| ARIP | accidental release information program |
| AS | area source |
| ASC | Acid gases exiting the SCC, volume concentration |
| AST | Acid Stack Gases, volume concentration |
| ATERIS | Air Toxics Exposure and Risk Information System |
| BACT | Best Available Control Technology |
| CAA | Clean Air Act (CAAA = CAA Amendments) |
| CAP | criteria air pollutant |
| CAPH | Combustion Air Preheat |
| CAS | Chemical Abstracts Service (of American Chemical Society) |
| CE | Combustion Efficiency, %CE = ($CO_2$) (100)/($CO_2$ + CO) where $CO_2$ and CO are stack volume concentrations (dgb) [per TSCA only -- see DRE for RCRA] |
| cem | continuous emission monitoring (CEMS = ...System) |
| CERCLA | Comprehensive Environmental Response, Compensation, and Liabilities Act = "Superfund" (See also SARA) |
| CFCs | ChloroFluoroCarbons |
| CFR | Code of Federal Regulations |
| CGVI | Combustion Gas Velocity Indication |
| COH | Coefficient Of Haze |
| $c_p$ | specific heat at constant pressure |
| CTG | Control Techniques Guideline (see RACT) |
| CWA | Clean Water Act |
| da | dry air |
| delta p | pressure difference or pressure drop |
| dgb | dry gas basis |
| DOD,E,J,L,T | Dept.of Defense,..Energy, Justice, Labor, Transportation |
| DOP | DiOctylPhthalate -- for testing HEPA |
| dp | dew point |
| DRE | Destruction and Removal Efficiency, per p 588, 40CFR#264 = (100%) $(W_{in} - W_{out})/(W_{in})$ |
| dscf* | dry standard cubic feet (at 29.92"Hg and 60 F) |
| EFSP | equivalent fan static pressure |
| ehs | extremely hazardous substance |
| ei | emissions inventory |
| eis | environmental impact statement |
| EPA | Environmental Protection Agency, US unless stated otherwise |
| FEMA | Federal Emergency Management Agency |
| FFIC | fuel flow indicating controller |
| FIM | Federal (US) Implementation Plan |

*(continued)*

* See also Abbreviations of Units in Appendix A7 and Periodic Table of the Elements in Appendix C1.

**Appendix A1.** *(continued)*

| | |
|---|---|
| FIP | Federal Implementation Plan |
| FM | Factory Mutual insurance system |
| FO | fuel oil |
| fps* | feet per second (scfs = standard cubic feet per second) |
| FR | Federal Register |
| FRP | fiberglass reinforced plastic |
| FY | ratio control device |
| | |
| GAO | General Accounting Office |
| GPO | Government Printing Office |
| gr | grains (7000 gr = 1 pound avoirdupois = 1 lb avdp) |
| | |
| hap | hazardous air pollutant; HAPPS = ...prioritization system |
| hazmat | hazardous material |
| HRS | hazard ranking system |
| HCL* | HydroChloric Acid (chemically HCl) [CAP H,CAP C, lc L] |
| HEPA | High Efficiency Particulate Air filters |
| HHV or hhv | Higher Heating Value = gross heating or calorific value |
| | |
| IFB | insulating fire brick |
| laer | lowest achievable emission rate |
| LHV or lhv | Lower Heating Value = net heating or calorific value |
| | |
| mact | maximum available control technology |
| maf | moisture- and ash-free |
| msa | metropolitan statistical area; MSA = Mine Safety Appl Co. |
| | |
| naa | non-attainment area |
| NAAQS | National Ambient Air Quality Standards |
| NaOH* | sodium hydroxide |
| NEPA | National Environmental Policy Act |
| NESHAP | National Emissions Standard for Hazardous Air Pollutants |
| NFPA | National Fire Prevention Association |
| NIOSH | National Institute for Occupational Safety and Health |
| NOD | Notice of Deficiency |
| NOx* | $NO_X$ = nitrogen oxides, specifically NO and $NO_2$ |
| NPDES | National Pollutants Discharge Elimination System |
| NPRM | Notice of Proposed Rulemaking |
| NSPS | New Source Performance Standards |
| nsr | new source review |
| NTIS | National Technical Information Service, US Dept.of Commerce |
| OGT | Off-Gas Treatment |
| OSHA | Occupational Safety and Health Administration |
| | |
| Pc | Particulate concentration, corrected to 7% $O_2$ standard, Pc = Pm (21-7)/(21-Y)<br>where Pm = Particulate concentration, measured<br>Y = measured volume % $O_2$ in the stack (dgb) |
| PCB | PolyChlorinated Biphenyl |
| PCC | Primary Combustion Chamber (most often a rotary kiln) |
| PEL | Permissible Exposure Limits (OSHA) |

*(continued)*

* See also Abbreviations of Units in Appendix A7 and Periodic Table of the Elements in Appendix C1.

**Appendix A1.** *(concluded)*

| | |
|---|---|
| pH | a chemical term used to define the acidity of a solution. A pH of 0-7 is acid; 7 is neutral; 7-14 is alkaline. |
| $pm_{10}$ | particulate matter nominally $<10$ micrometers in diameter (ambient air std.) |
| pic | products of incomplete combustion |
| PIC | pressure indicating controller |
| P&ID | piping and instrument diagram |
| PID | Proportional/Integral/Derivative control (gain/rate/reset) |
| pna | poly nuclear aromatics |
| poc | products of combustion (often assumed complete combustion) |
| POHC | Principal Organic Hazard Constituent |
| ppb | parts per billion, assumed ppbv (by volume) |
| ppm | parts per million, assumed ppmv (by volume) |
| psd | prevention of significant deterioration |
| psia* | pounds per square inch absolute (vs. psig = ...gauge) |
| RACT | Reasonably Available Control Technology; RACM = ...Measure |
| RCRA | Resource Conservation and Recovery Act |
| REL | Recommended Exposure Limits (NIOSH) |
| rh | relative humidity (%) |
| RKI | Rotary Kiln Incinerator |
| RTECS | Registry of Toxic Effects of Chemical Substances |
| SAB | Science Advisory Board |
| SARA | Superfund Amendments and Reauthorization Act (see CERCLA) |
| SBA | Small Business Administration |
| SCC | Secondary Combustion Chamber |
| SGM | Stack Gas Monitoring |
| SIP | State Implementation Plan |
| SMSA | Standard Metropolitan Statistical Area |
| SOx* | Sulfur Oxides (usually $SO_2$ and $SO_3$) |
| TCLP | Toxicity Characterization Leaching Procedure |
| THC* | Total HydroCarbons |
| TIC | temperature indicating controller |
| TLV(R) | Threshold Limit Value, trademark of ACGIH, American Conference of Governmental Industrial Hygienists |
| tpy | tons per year |
| TRI | Toxic Release Inventory |
| TRV | Thermal Relief Vent |
| TSCA | Toxic Substance Control Act |
| twa | time-weighted average |
| TY | temperature control split range device |
| USC | United States Code |
| vcm | vinyl chloride monomer |
| VOCs | Volatile Organic Compounds; VOL = ..Liquid |
| wb | wet bulb temperature |
| "wc* | inches of water column or "wg = ..gauge (pressure) |
| wesp | wet electrostatic precipitator |
| wgb | wet gas basis |
| WTE | waste to energy |
| whb | waste heat boiler |
| % XSA | percent excess air |

* See also Abbreviations of Units in Appendix A7 and Periodic Table of the Elements in Appendix C1.

**Appendix A2. Thermophysical constants in American and *Metric* units.**
*(From Reference 15.)*

**Acceleration of gravity** = 32.17 ft/sec² or *9.806 m/s²*

**Gross heat release** with 1 ft³ air + stoichiometric amount of natural gas or oil = about 100 Btu
**Gross heat release** with 1 m³ air + stoichiometric amount of natural gas or oil = about *900 kcal*

**Stefan-Boltzmann constant** = $0.1713 \times 10^{-8}$ Btu/ft² hr °R⁴ or $4.88 \times 10^{-8}$ *kcal/m² h °K⁴*

**Universal gas constant**, $\overline{MR}$, in $pv = \frac{\overline{MR}T}{M}$, (M is molar weight) = $1544 \frac{\text{ft lb mass}}{\text{°R lb-mole}}$, or $84.7 \frac{\text{m kg mass}}{\text{°K kg-mole}}$, or $8.314 \frac{\text{joule}}{\text{°K g-mole}}$

**Velocity of sound** in air at *20 C* and *760 mm Hg* = 1129 ft/sec = 769.5 mph = *344.1 m/s*

**Volume of 1 lb mole** at *15.6 C* (60 F) and 1 atmosphere = *379 ft³*
**Volume of 1 kg mole** at *15.6 C* (60 F) and 1 atmosphere = *23.7 m³*

**PROPERTIES** of water, moist air, and a typical natural gas at 1 atmosphere *(760 mm Hg* = 14.696 psia) and *15.6 C* (60 F):

| | **Water** | | **Air (moist)** | | **Natl gas** | |
|---|---|---|---|---|---|---|
| Specific weight, lb/ft³; *kg/m³* | 62.35* | *1000* | 0.0763 | *1.225* | 0.046 | *0.738* |
| Specific heat, Btu/lb °F; *cal/g °C* | 1.0 | *1.0* | 0.240 | *0.240* | 0.55 | *0.55* |
| Thermal conductivity, Btu ft/hr ft² °F; *watt/m °K* | 0.344 | *0.595* | 0.0148 | *0.0256* | 0.02 | *0.03* |
| Absolute viscosity, lb mass/ft hr; *kg/m h* | 2.72 | *4.05* | 0.0434 | *0.0646* | 0.0266 | *0.0396* |
| Kinematic viscosity, ft²/hr; *(m²/s)* $\times 10^6$ | 0.0436 | *1.126* | 0.567 | *14.68* | 0.578 | *14.93* |

* Water weighs 8.335 lb/gal.

**Appendix A3. Volume equivalents,** rounded to 4 significant figures.
*(from Reference 15).*

**Force = Mass × Acceleration**

1 lb force = the force that accelerates 1 lb mass at a rate of 32.17 ft/sec$^2$
1 poundal = " " " " 1 lb mass " " " " 1 ft/sec$^2$
1 lb force = " " " " 1 slug " " " " 1 ft/sec$^2$
1 newton = " " " " 1 kg " " " " 1 m/s$^2$ = $10^5$ dynes
1 dyne = " " " " 1 g " " " " 1 cm/s$^2$

**Pressure = Force per unit Area**

1 newton/m$^2$ = 1 pascal (Pa) (See pressure conversions in Appendix A6.)
1 pound/in.$^2$ = 1 psi = 144 psf = 144 lb/ft$^2$

**Work = Force × Distance** **[Work and heat are convertible forms of energy]**

1 ft lb = the work done in exerting a 1 lb force through a distance of 1 ft = 1.3558 joules = $\frac{1}{778.8}$ Btu
1 dyne cm = " " " " " " 1 newton " " " " " 1 cm = 1 erg
1 newton metre = " " " " " " 1 dyne " " " " " 1 m = $10^7$ erg = 1 joule = 1 watt sec

**Power = rate of doing Work** **[Heat flow rate can be expressed in the same units]**

1 newton metre/s = 1 watt = 1 joule/s = $\frac{1}{746}$ hp
1 kW = 1000 joule/s = 3413 Btu/hr = 860.4 kcal/h

**Appendix A4. An easy way to convert units** *(from Referecne 15).*

Multiply a known measurement by a conversion fraction. Make conversion fractions from the equalities in Appendix A3, A4, and A6b, putting numbers and units from each side of the = sign above and below the fraction line. You can multiply anything by 1/1 = 1 without changing its value. All you have done is change its units. If you can't cancel out units (as you would cancel fractions in ½ × ⅔ × ¾ = ¼), turn the conversion fraction upside down. For example:

$$3.5''\text{wc} \times \frac{25.4\ \text{mm}}{1''} = 88.9\ \text{mm}\ H_2O.$$

$$10\ 000\ \text{kW} \times \frac{3413\ \text{Btu/hr}}{1\ \text{kW}} \times \frac{1\ \text{kcal}}{3.968\ \text{Btu}} \times \frac{1\ \text{hr}}{3600\ \text{sec}} = 2389\ \text{kcal/s}.$$

This illustrates the use of a known intermediate unit (Btu) to convert between units for which no factor is readily available.

---

† *Small pressures are often expressed in* **wc** for water column, which is synonymous with **wg** for water gauge or **$H_2O$**. Larger pressures are often measured with a mercury column and expressed as "Hg or mm Hg.)

**Appendix A5. Volume equivalents,** rounded to 4 significant figures *(from Reference 15).*

| | m$^3$ | in.$^3$ | L | USgal | Br gal | ft$^3$ | bbl |
|---|---|---|---|---|---|---|---|
| **1 cm$^3$ (cc)** | = 0.000 001 00 m$^3$ (SI unit) | | | | | | |
| **1 in.$^3$** | = 0.000 016 39 m$^3$ = | 1 in.$^3$ | | | | | |
| **1 litre** | = 0.001 000 m$^3$ = | 61.02 in.$^3$ = | 1 L (dm$^3$) | | | | |
| **1 USgal** | = 0.003 785 m$^3$ = | 231.0 in.$^3$ = | 3.785 L = | 1 USgal | | | |
| **1 Br gal** | = 0.004 546 m$^3$ = | 277.4 in.$^3$ = | 4.546 L = | 1.201 USgal = | 1 Br gal | | |
| **1 ft$^3$** | = 0.028 32 m$^3$ = | 1 728 in.$^3$ = | 28.32 L = | 7.481 USgal = | 6.229 Br gal = | 1 ft$^3$ | |
| **1 bbl, oil** | = 0.159 0 m$^3$ = | 9 702 in.$^3$ = | 159.0 L = | 42.00 USgal = | 34.97 Br gal = | 5.615 ft$^3$ = | 1 bbl |
| **1 m$^3$** | = 1 m$^3$ = | 61 020 in.$^3$ = | 1000 L = | 264.2 USgal = | 220.0 Br gal = | 35.31 ft$^3$ = | 6.290 bbl |

**Appendix A6. Pressure equivalents,** rounded to 4 significant figures *(from Referecne 15).*

| | kPa | mm $H_2O$ | mm Hg | " $H_2O$ | osi | " Hg | psi | kg/cm$^2$ |
|---|---|---|---|---|---|---|---|---|
| **1 N/m$^2$** | = 0.001 kPa (kilopascal) (SI unit) | | | | | | | |
| **1 mm $H_2O$** | = 0.0098 kPa = | 1 mm $H_2O$ | | | | | | |
| **1 mm Hg** | = 0.1333 kPa = | 13.60 mm $H_2O$ = | 1 mm Hg (torr) | | | | | |
| **1" $H_2O$** | = 0.2488 kPa = | 25.40 mm $H_2O$ = | 1.866 mm Hg = | 1 " $H_2O$ | | | | |
| **1 osi** | = 0.4309 kPa = | 43.94 mm $H_2O$ = | 3.232 mm Hg = | 1.732" $H_2O$ = | 1 osi | | | |
| **1" Hg** | = 3.386 kPa = | 345.3 mm $H_2O$ = | 25.40 mm Hg = | 13.61 " $H_2O$ = | 7.858 osi = | 1 " Hg | | |
| **1 psi** | = 6.895 kPa = | 703.1 mm $H_2O$ = | 51.72 mm Hg = | 27.71 " $H_2O$ = | 16.00 osi = | 2.036" Hg = | 1 psi | |
| **1 kg/cm$^2$** | = 98.07 kPa = | 10 000 mm $H_2O$ = | 735.6 mm Hg = | 394.1 " $H_2O$ = | 227.6 osi = | 28.96 " Hg = | 14.22 psi = | 1 kg/cm$^2$ |
| **1 bar** | = 100.0 kPa = | 10 200 mm $H_2O$ = | 750.1 mm Hg = | 401.9 " $H_2O$ = | 232.1 osi = | 29.53 " Hg = | 14.50 psi = | 1.020 kg/cm$^2$ |
| **1 atm*** | = 101.3 kPa = | 10 330 mm $H_2O$ = | 760.0 mm Hg = | 407.3 " $H_2O$ = | 235.1 osi = | 29.92 " Hg = | 14.70 psi = | 1.033 kg/cm$^2$ |

* normal atmosphere = 760 torr (mm Hg, 0 C); a "technical atmosphere" = 1 kg/cm$^2$ = 736 torr

**Appendix A7. Abbreviations and symbols used in unit conversion table** (Appendix A8)

Å = Angstrom Unit
atm = atmosphere
avdp = avoirdupois

bar = (unit of pressure ~ 1 atmosphere)
bbl = barrel
bhp = brake horsepower, or boiler horsepower
BR gal = British (Imperial) gallon
Btu/lb °F = Btu per pound degree Fahrenheit

C = carbon
C or °C = Celsius (formerly Centigrade) or degrees Celsius
cal = calorie
cal/cm$^2$ = calories per square centimeter
cal/cm$^3$ = calories per cubic centimeter

cal/g = calories per gram
cc = cubic centimeter
cf or cu ft = cubic foot
cfh = cubic feet per hour
cfs = cubic feet per second

cm, cm$^2$ = centimeter, square centimeter
CO = carbon monoxide
$CO_2$ = carbon dioxide
cp = centipoise
cu ft or cf = cubic foot

dyne·cm = dyne-centimeter
dyne·s/cm$^2$ = dyne-seconds per square centimeter

F or °F = Fahrenheit or degrees Fahrenheit
ft = foot
ft, ft$^3$ = foot, cubic foot

g = gram
G or SGA = specific gravity relative to standard dry air
GJ = gigajoule = 1 000 000 000 joules
g/cm$^2$ = grams per square centimeter
g/cm$^3$ = grams per cubic centimeter

g/ft$^3$ = grams per cubic foot
gal = gallon
gpm = gallons per minute
gps = gallons per second

*(continued)*

**Appendix A7.** *(continued)*

h or hr = hour
Hg = mercury
$H_2O$ = water

in. = inch
J = Joule
J/kg·°K = Joules per kilogram·degree Kelvin
J/m$^2$·s = Joules per square meter·second
J/s = Joules per second
Joule/m$^3$ = Joules per cubic meter

K or °K = Kelvin (absolute Celsius) or degrees Kelvin
kg = kilogram
kj = kilojoule
kcal = kilocalorie
kcal/m$^3$ = kilocalories per cubic meter
kg/cm$^2$ = kilograms per square centimeter

kg/m$^3$ = kilograms per cubic meter
kg/L = kilograms per litre
km = kilometers
km/h = kilometers per hour
kPa = kiloPascal
kW = kilowatt
kW/m$^2$ = kilowatts per square meter

L = liter or litre = cubic decimetre = dm$^3$
lb/ft$^3$ = pounds per cubic foot
lb/in$^3$ = pounds per cubic inch
lb/USgal = pounds per United States gallon

m = meter or metre
m$^2$ = square meter
MJ = megajoule
mm = millimeter
mm$^2$ = square millimeter
mph = miles per hour
m/s = meters per second
MW = megawatt

N = newton
N·m = newton-meter
N·m/s = newton-meters per second

*(continued)*

**Appendix A7.** *(concluded)*

NO = nitric oxide = nitrogen monoxide
$NO_2$ = nitrogen dioxide
$NO_X$ = nitrogen oxides

$O_2$ = oxygen (molecular)
osi = ounces per square inch
oz = ounce

Pa = pascal
Pa·s = pascal-seconds
PCBs = polychlorinated biphynels
ppb = parts per billion
ppm = parts per million
$ppm_V$ = parts per million by volume
psi(g or a) = pounds per square inch (gauge or absolute)

R or °R = Rankine (absolute Fahrenheit) or degrees Rankine
rpm = revolutions per minute

s, sec = second
SGA or G = specific gravity relative to standard dry air
SGW = specific gravity relative to water at 4 C or 39.2 F
$SO_2$ = sulfur dioxide
SSU = seconds, Saybolt Universal

toe = ton of oil equivalent
USgal = United States gallon

W = watt
W·h = watt-hours
W·s = watts-seconds
wc or wg = water column or water gauge

yd = yard

$\mu g/m^3$ = microgram per cubic meter

1 $\mu$ = 1 micron = one millionth of a meter
$\mu g/m^3$ = micrograms per cubic meter = specific gravity relative to density of water

## Appendix A8. Unit equivalents *(from Reference 15)*.

**SI Metric To American**
**Metric To Metric**

**American To SI Metric**
**American To American**

AREA:

1 $mm^2$ = 0.001 55 $in.^2$ = 0.000 010 76 $ft^2$
1 $cm^2$ = 0.155 $in.^2$ = 0.001 076 $ft^2$
1 $m^2$ = 1550 $in.^2$ = 10.76 $ft^2$
= 10 000 $cm^2$ = 1 000 000 $mm^2$

1 $in.^2$ = 645.2 $mm^2$ = 6.452 $cm^2$
= 0.000 645 2 $m^2$
1 $ft^2$ = 92 903 $mm^2$ = 929.03 $cm^2$
= 0.0929 $m^2$
1 acre = 43 560 $ft^2$
1 circular mil = 0.7854 square mil
= 5.067 × $10^{-10}$ $m^2$ = 7.854 × $10^{-5}$ $in.^2$

DENSITY and SPECIFIC GRAVITY:

1 $g/cm^3$ = 0.036 13 $lb/in.^3$ = 62.43 $lb/ft^3$
= 1000 $kg/m^3$ = 1 kg/L
= 62.43 $lb/ft^3$ = 8.345 lb/USgal
1 $kg/m^3$ = 0.062 43 $lb/ft^3$
1 $\mu g/m^3$ = 4.37 × $10^{-7}$ grains/$ft^3$
= 6.24 × $10^{-11}$ $lb/ft^3$

1 $lb/in^3$ = 27.68 $g/cm^3$ = 27 680 $kg/m^3$
1 $lb/ft^3$ = 0.0160 $g/cm^3$ = 16.02 $kg/m^3$
= 0.0160 kg/L = 16.02 × $10^9$ $\mu g/m^3$
Specific gravity relative to water
(SGW) of 1.00 = 62.43 $lb/ft^3$ at 4 C or 39.2 F†
Specific gravity relative to dry air
(SGA) of 1.00 = 0.0765 $lb/ft^3$
= 1.225 $kg/m^3$
1 lb/USgal = 7.481 $lb/ft^3$ = 0.1198 kg/L = 119.8 $kg/m^3$
1 $g/ft^3$ = 35.3 × $10^6$ $\mu g/m^3$
1 lb/1000 $ft^3$ = 16 × $10^6$ $\mu g/m^3$
1 US gallon of water weighs 8.335 lb†

ENERGY, HEAT, and WORK:

1 kcal = 3.968 Btu = 1000 cal = 4187 J
= 0.004 187 MJ = 0.001 thermies
1 J = 0.000 948 Btu = 0.239 cal = 1 W·s
= 1 N·m = $10^7$ erg = $10^7$ dyne·cm
1 kJ = 0.239 kcal
1 MJ = 947.9 Btu
1 W·h = 0.854 83 kcal (Int'l Table) = 3600 joules
1 MW·h = 3.414 million Btu
= 3.600 GJ = 34.12 thermies
= 0.0886 toe
(ton of oil equivalent)
1 GJ = 0.278 MWh = 0.0246 toe
= 0.948 million Btu
1 thermie = 1000 kcal

3412 Btu = 1 kW·h
1 Btu = 0.2931 W·h
= 252.0 cal = 0.252 kcal
= 778 ft·lb
= 1055 J = 0.001 055 MJ
= 0.000 001 055 GJ
1 million Btu = 1.055 GJ
1 ft·lb = 0.1383 kg·m = 1.356 J
1 hp·hr = 1.98 × $10^6$ ft·lb
1 therm = 1.00 × $10^5$ Btu
1 bhp (boiler horsepower) = 33 475 Btu/hr
= 8439 kcal/h = 9.81 kW
1 quad = 1 quadrillion Btu = $10^{15}$ Btu
1 toe = 42 GJ

*(continued)*

† 62.35 $lb/ft^3$ at 60 F, 15.6 C.

## Appendix A8. Unit equivalents *(continued)*

**SI Metric To American**
**Metric To Metric**

HEAT CONTENT and SPECIFIC HEAT:

1 cal/g = 1.80 Btu/lb = 4187 J/kg
1 cal/cm$^3$ = 112.4 Btu/ft$^3$
1 kcal/m$^3$ = 0.1124 Btu/ft$^3$ = 4187 J/m$^3$
1 kJ/kg = 0.43 Btu/lb
1 MJ/m$^3$ = 26.8 Btu/ft$^3$ = 3.59 Btu/US gal

1 cal/g·°C = 1 Btu/lb·°F = 4187 J/kg·°K
1 kcal/kg = 0.004 187 MJ/kg = 1.8 Btu/lb

HEAT FLOW, POWER:

1 N·m/s = 1 W = 1 J/s
= 0.001 341 hp = 0.7376 ft·lb/sec
1 kcal/h = 1.162 J/s = 1.162 W
= 3.968 Btu/hr
= 0.001 Thermies/h
1 kW = 1000 J/s = 3412 Btu/hr = 1.341 hp
= 859.8 kcal/hr
1 MW = 3 412 000 Btu/hr = 3.600 GJ/h

HEAT FLUX and HEAT TRANSFER COEFFICIENT:

1 cal/cm$^2$·s = 3.687 Btu/ft$^2$·sec
= 41.87 kW/m$^2$
1 cal/cm$^2$·h = 1.082 W/ft$^2$ = 11.65 W/m$^2$
1 kW/m$^2$ = 317.2 Btu/ft$^2$·hr

1 kW/m$^2$·°C = 1 kW/m$^2$ °K
= 176.2 Btu/ft$^2$·hr·°F
= 860.4 kcal/m$^2$·hr·°C
1 J/m$^2$·s·°K = 0.1722 Btu/ft$^2$·hr·°F

LENGTH:

1 mm = 0.10 cm = 0.039 37 in.
= 0.003 281 ft
1 m = 100 cm = 1000 mm = 39.37 in.
= 3.281 ft = 1.094 yard
1 km = 0.6214 mile

**American to SI Metric**
**American To American**

HEAT CONTENT and SPECIFIC HEAT:

1 Btu/lb = 0.5556 cal/g = 0.5556 kcal/kg
= 2326 J/kg = 0.002 326 MJ/kg
1 Btu/ft$^3$ = 0.008 90 cal/cm$^3$
= 8.899 kcal/m$^3$ = 0.0373 MJ/m$^3$
1 Btu/USgal = 0.0666 kcal/L = 66.6 kcal/m$^3$

1 Btu/lb·°F = 1 cal/g·°C = 4187 J/kg·°K

HEAT FLOW, POWER:

1 Btu/hr = 0.2520 kcal/h
= 0.000 393 1 hp
= 0.2931 W = 0.2931 J/s
1 million Btu/hr = 1055 MJ/hr
1 hp = 33 000 ftlb/min = 550 ftlb/sec
= 745.7 W = 745.7 J/s
= 641.4 kcal/h = 2546 Btu/hr
1 boiler hp = 33 475 Btu/hr = 34.5 lb steam/hr from and at 212 F

HEAT FLUX and HEAT TRANSFER COEFFICIENT:

1 Btu/ft$^2$·sec = 0.2713 cal/cm$^2$·s = 11.35 kW/m$^2$
1 Btu/ft$^2$·hr = 0.003 153 kW/m$^2$
= 2.713 kcal/m$^2$·h
1 kW/ft$^2$ = 924.2 cal/cm$^2$·h

1 Btu/ft$^2$·hr·°F = 4.89 kcal/m$^2$·h·°C
= 5.67 W/m$^2$·°K

LENGTH:

1 in. = 25.40 mm = 2.540 cm = 0.025 40 m
1 ft = 304.8 mm = 30.48 cm = 0.3048 m
1 mile = 5280 ft; 1 nautical mile = 6076 ft
1 micron = 1 μm = $10^{-6}$ m = 1 micrometer
= 10 000 Å
1 Angstrom unit = 1 Å = $10^{-10}$ m = $10^{-4}$ μm

*(continued)*

## Appendix A8. Unit equivalents *(continued)*

| **SI Metric To American** **Metric To Metric** | **American To SI Metric** **American To American** |
|---|---|

POLLUTANT CONCENTRATION:

10 000 $ppm_V$ = 1.0% by volume
1 $ppm_V$ CO = 1183 $\mu g/sm^3$
1 $ppm_V$ $NO_2$ = 1948 $\mu g/sm^3$
1 $ppm_V$ NO = 1272 $\mu g/sm^3$
1 $ppm_V$ $SO_2$ = 2707 $\mu g/sm^3$

1 $ppm_V$ (by volume) = (M × 42.25) $\mu g/sm^3$
where M = molecular weight

| | Higher heating value, Btu | Stoichiometric air req'd, cf | Total poc, cf | Dry poc, cf | # NOx/million Btu equiv. to 100 ppm |
|---|---|---|---|---|---|
| Natural gas | 1000/cf | 9.44/cf | 10.47/cf | 8.52/cf | 0.1206 |
| Coke oven gas | 570/cf | 5.42/cf | 6.18/cf | 4.87/cf | 0.1533 |
| Commercial propane | 2560/cf | 23.8/cf | 25.76/cf | 21.78/cf | 0.1203 |
| #2 fuel oil | 137,100/gal | 1365/gal | 1449/gal | 1278/gal | 0.1314 |
| #6 fuel oil | 153,200/gal | 1487/gal | 1561/gal | 1412/gal | 0.1299 |

PRESSURE:

1 $N\cdot m^2$ = 0.001 kPa = 1.00 Pa
1 mm $H_2O$ = 0.0098 kPa
1 mm Hg = 0.1333 kPa = 13.60 mm $H_2O$
= 1 torr = 0.019 33 lb/in.$^2$
1 kg/cm$^2$ = 98.07 kPa = 10 000 kg/m$^2$
= 10 000 mm $H_2O$ = 394.1 in. $H_2O$*
= 735.6 mm Hg = 28.96 in. Hg
= 227.6 oz/in.$^2$ = 14.22 lb/in.$^2$
= 0.9807 bar
1 bar = 100.0 kPa = 1.020 kg/cm$^2$
= 10 200 mm $H_2O$ = 401.9 in. $H_2O$
= 750.1 mm Hg = 29.53 in. Hg
= 232.1 oz/in.$^2$ = 14.50 lb/in.$^2$
= 100 000 N/m$^2$
1 g/cm$^2$ = 0.014 22 lb/in.$^2$
= 0.2276 oz/in.$^2$
= 0.3937 in. $H_2O$

1 in. $H_2O$* = 0.2488 kPa = 25.40 mm $H_2O$
= 1.866 mm Hg = 5.198 lb/ft$^2$
= 0.002 54 kg/cm$^2$ = 2.540 g/cm$^2$
1 in. Hg = 3.386 kPa = 25.40 mm Hg
= 345.3 mm $H_2O$ = 13.61 in. $H_2O$
= 7.858 oz/in.$^2$ = 0.491 lb/in.$^2\mu$
= 25.4 torr
1 lb/in.$^2$ = 6.895 kPa = 6895 N/m$^2$
= 703.1 mm $H_2O$ = 27.71 in. $H_2O$
= 51.72 mm Hg = 2.036 in. Hg
= 16.00 oz/in.$^2$
= 0.0703 kg/cm$^2$ = 70.31 g/cm$^2$
= 0.068 97 bar
1 oz/in.$^2$ = 0.4309 kPa
= 43.94 mm $H_2O$ = 1.732 in. $H_2O$
= 3.232 mm Hg
= 0.004 39 kg/cm$^2$ = 4.394 g/cm$^2$
1 ft of head (water) = 12 000 milinch

*(continued)*

* 1″wc = 1″wg

‡ Volume equivalents are for the same temperature and pressure in both sets of units.
A standard cubic foot or scfh is measured at 60 F and 29.92″ Hg.
A standard cubic metre or sm$^3$/h is measured at 15 C and 760 mm Hg.
A normal cubic metre or nm$^3$/h is measured at 0 C and 760 mm Hg.

## Appendix A8. Unit equivalents *(continued)*

**SI Metric To American**
**Metric To Metric**

PRESSURE (cont'd)

(For rough calculations,
1 bar = 1 atm† = 1 $kg/cm^2$
= 10 m $H_2O$ = 100 kPa)

RADIOACTIVITY:

1 curie = $3.7 \times 10^{10}$ disintegrations per second
1 microcurie = $3.7 \times 10^4$ disintegrations per second

TEMPERATURE:

$C = {}^5/_9 (F - 32)$
$K = C + 273.15$

THERMAL CONDUCTIVITY:

1 W/m·°K = 0.5778 Btu·ft/$ft^2$·hr·°F
= 6.934 Btu·in./$ft^2$·hr·°F
1 cal·cm/$cm^2$·s·°C = 241.9 Btu·ft/$ft^2$·hr·°F
= 2903 Btu·in./$ft^2$·hr·°F
= 418.7 W/m·°K

THERMAL DIFFUSIVITY:

1 $m^2$/s = 38 760 $ft^2$/hr
1 $m^2$/h = 10.77 $ft^2$/hr

TORQUE (BENDING MOMENT):

1 lb (force)·inch = 0.112 98 N·m

VELOCITY:

1 cm/s = 0.3937 in./sec = 0.032 81 ft/sec
= 10.00 mm/s = 1.969 ft/min
1 m/s = 39.37 in./sec = 3.281 ft/sec
= 196.9 ft/min = 2.237 mph
= 3.600 km/h = 1.944 knot

**American To SI Metric**
**American To American**

PRESSURE (cont'd)

1 atm† = 101.3 kPa = 101 325 N/$m^2$
= 10 330 mm $H_2O$ = 407.3 in. $H_2O$
= 760.0 mm Hg = 29.92 in. Hg
= 235.1 oz/in.$^2$ = 14.70 lb/in.$^2$
= 1.033 $kg/cm^2$
= 1.013 bar

TEMPERATURE:

$F = ({}^9/_5)(K - 255.37)$
$F = ({}^9/_5\,C) + 32$
$K = ({}^5/_9)\,F + 255.37$
$R = F + 459.67$

THERMAL CONDUCTIVITY:

1 Btu·ft/$ft^2$·hr·°F = 1.730 W/m·°K
= 1.488 kcal/m·h·°K
1 Btu·in./$ft^2$·hr·°F = 0.1442 W/m·°K
1 Btu·ft/$ft^2$·hr·°F = 0.004 134 cal·cm/$cm^2$·s·°C
1 Btu·in./$ft^2$·hr·°F = 0.000 344 5 cal·cm/$cm^2$·s·°C

THERMAL DIFFUSIVITY:

1 $ft^2$/hr = 0.000 025 8 $m^2$/s = 0.0929 $m^2$/h

TORQUE (BENDING MOMENT):

1 lb·in. = 1.152 kg·cm
1 lb·ft = 0.1383 kg·m

VELOCITY:

1 in./sec = 25.4 mm/s = 0.0254 m/s
= 0.0568 mph
1 ft/sec = 304.8 mm/s = 0.3048 m/s
= 0.6818 mph
1 ft/min = 5.08 mm/s = 0.005 08 m/s
= 0.0183 km/h
1 mph = 0.4470 m/s = 1.609 km/h
= 1.467 ft/sec
1 knot = 0.5144 m/s
1 rpm = 0.1047 radians/sec

*(continued)*

† Normal atmosphere = 760 torr (mm Hg at 0 C)--not a "technical atmosphere", which is 736 torr or 1 $kg/cm^2$. Subtract about 0.5 lb/in.$^2$ for each 1000 ft above sea level.

## Appendix A8.. Unit equivalents *(continued)*

**SI Metric To American**
**Metric To Metric**

VISCOSITY, absolute, $\mu$:

0.1 Pa·s = 1 dyne·s/cm$^2$ = 360 kg/m·h
= 1 poise = 100 centipoise
= 242.1 lb mass/ft·hr
= 0.002 089 lb force·sec/ft$^2$
1 kg/m·h = 0.672 lb/hr·ft = 0.002 78 g/cm·s
= 0.000 005 81 lb force·sec/ft$^2$

$\mu$ of water† = 1.124 centipoise
= 2.72 lb mass/hr·ft
= 2.349 × 10$^{-5}$ lb·sec/ft$^2$

VISCOSITY, kinematic, $\nu$:

1 cm$^2$/s = 0.0001 m$^2$/s
= 1 stoke = 100 centistokes
= 0.001 076 ft$^2$/sec
= 3.874 ft$^2$/hr
1 m$^2$/s = 3600 m$^2$/h
= 38 736 ft$^2$/hr = 10.76 ft$^2$/sec

$\nu$ of water† = 1.130 centistokes
= 32 SSU
= 1.216 × 10$^{-5}$ ft$^2$/sec

VOLUME: ‡

1 cm$^3$ (cc) = 0.000 001 00 m$^3$
= 0.0610 in.$^3$ = 0.0338 US fluid oz
1 L (dm$^3$) = 0.0010 m$^3$ = 1000 cm$^3$
= 61.02 in.$^3$ = 0.035 31 ft$^3$
= 0.2642 USgal

1 m$^3$ = 1000 L = 1 000 000 cm$^3$
= 61 024 in.$^3$ = 35.31 ft$^3$
= 220.0 Br gal
= 6.290 bbl
= 264.2 USgal
= 1.308 yd$^3$

**American To SI Metric**
**American To American**

VISCOSITY, absolute, $\mu$:

1 lb mass/hr·ft = 0.000 008 634 lb force·sec/ft$^2$
= 0.413 centipoise
= 0.000 413 Pa·s
1 lb force·sec/ft$^2$ = 115 800 lb mass/ft·hr
= 47 880 centipoise
= 47.88 Pa·s
1 reyn = 1 lb force·sec/in.$^2$
= 6.890 × 10$^6$ centipoise

$\mu$ of air† = 0.0180 centipoise
= 0.0436 lb/hr·ft
= 3.763 × 10$^{-7}$ lb·sec/ft

VISCOSITY, kinematic, $\nu$:

1 ft$^2$/sec = 3600 ft$^2$/hr = 92 900 centistokes
= 0.0929 m$^2$/s
1 ft$^2$/h = 0.000 278 ft$^2$/sec = 25.8 centistokes
= 0.000 025 8 m$^2$/s

$\nu$ of air† = 14.69 centistokes
= 1.581 × 10$^{-4}$ ft$^2$/sec

VOLUME: ‡

1 in.$^3$ = 16.39 cm$^3$ = 0.000 016 39 m$^3$
= 0.016 39 L
1 ft$^3$ = 1728 in.$^3$ = 7.481 USgal
= 6.229 Br gal
= 28 320 cm$^3$ = 0.028 32 m$^3$ = 28.32 L
= 62.427 lb of 39.4 F (4 C) water
= 62.344 lb of 60 F (15.6 C) water

1 USgal = 3785 cm$^3$ = 0.003 785 m$^3$
= 3.785 L = 231.0 in.$^3$
= 0.8327 Br gal = 0.1337 ft$^3$
= sp gr × 8.335 lb
= 8.334 lb of 60 F (15.6 C) water
= $^1/_{42}$ barrel (oil)

*(continued)*

† At stp (60 F and 14.7 psia or 15.6 C and 760 mm Hg).
‡ See footnote on page 139.

## Appendix A8. Unit equivalents *(continued)*

**SI Metric To American**
**Metric To Metric**

VOLUME *(cont'd)*

**American To SI Metric**
**American To American**

1 Br gal = 277.4 $in.^3$
= 0.004 546 $m^3$ = 4.546 L
= 1.201 USgal

1 bbl, oil = 9702 $in.^3$ = 5.615 $ft^3$
= 0.1590 $m^3$ = 159.0 L
= 42.00 USgal
= 34.97 Br gal

VOLUME FLOW RATE:‡

1 $cm^3$ (cc)/s = $1 \times 10^{-6}$ $m^3/s$
1 L/s = $1 \times 10^{-3}$ $m^3/s$
1 $m^3/h$ = 4.403 US gpm (gal/min)
= 0.5887 $ft^3/min$

1 gpm (gal/min) = 60.0 gph (gal/hr)
= 0.016 67 gps (gal/sec)
= 0.002 23 cfs ($ft^3/sec$)
= 8.022 cfh ($ft^3/hr$)
= 0.8326 Br gpm
= 0.227 $m^3/h$ = 0.063 08 L/s
= 1.429 bbl/hr = 34.29 bbl/day

1 gph (gal/hr) = 0.001 05 L/s
= 0.000 037 1 cfs ($ft^3/sec$)

1 cfm ($ft^3/min$) = 6.18 Br gpm
= 0.000 471 $m^3/s$

1 cfs ($ft^3/sec$) = 448.8 gpm
= 22 250 Br gph

WEIGHT, FORCE, MASS:

1 $\mu g$ = $1.544 \times 10^{-5}$ grains
= $2.205 \times 10^{-9}$ lb avdp mass
1 g = 0.035 27 oz avdp mass
1 kg mass = 1000 g mass = $10^9$ $\mu g$
= 35.27 oz avdp mass
= 2.205 lb avdp mass
1 kg force = 1000 g force = 9.807 N
= 2.205 lb avdp force
1 metric ton (1 tonne) = 1000 kg = 2205 lb

1 oz avdp mass = 28.35 g = 0.028 35 kg
1 lb avdp mass = 453.6 g = 0.4536 kg
= $4.536 \times 10^8$ $\mu g$
1 lb avdp force = 0.4536 kg force
= 4.448 N
1 lb avdp = 16 ounces (oz) = 7000 grains
1 short ton = 2000 lb = 907.2 kg
1 long ton = 2240 lb = 1015.9 kg

‡ Volume equivalents are for the same temperature and pressure in both sets of units.
A standard cubic foot or scfh is measured at 60 F and 29.92″ Hg.
A standard cubic metre or $sm^3/h$ is measured at 15 C and 760 mm Hg.
A normal cubic metre or $nm^3/h$ is measured at 0 C and 760 mm Hg.

**Appendix A9a. Temperature scale conversions.** Formulas are in Appendix A8. Examples: From the bold type column, 600 Fahrenheit is the same as 315.6 Celsius or 588.7 Kelvin. 600 C = 1112.0 F. 600 K = 620.3 F.

| from F to K | from F to C | Convert F, C, or K | from C to F | from K to F | from F to K | from F to C | Convert F, C, or K | from C to F | from K to F |
|---|---|---|---|---|---|---|---|---|---|
| 244.3 | -28.9 | **-20.0** | -4.0 | — | 599.8 | 326.7 | **620.0** | 1148.0 | 656.3 |
| 249.8 | -23.3 | **-10.0** | 14.0 | — | 610.9 | 337.8 | **640.0** | 1184.0 | 692.3 |
| 255.4 | -17.8 | **0.0** | 32.0 | -459.7 | 622.0 | 348.9 | **660.0** | 1220.0 | 728.3 |
| 260.9 | -12.2 | **10.0** | 50.0 | -441.7 | 633.2 | 360.0 | **680.0** | 1256.0 | 764.3 |
| 266.5 | -6.7 | **20.0** | 68.0 | -423.7 | 644.3 | 371.1 | **700.0** | 1292.0 | 800.3 |
| 272.0 | -1.1 | **30.0** | 86.0 | -405.7 | 672.0 | 398.9 | **750.0** | 1382.0 | 890.3 |
| 277.6 | 4.4 | **40.0** | 104.0 | -387.7 | 699.8 | 426.7 | **800.0** | 1472.0 | 980.3 |
| 283.2 | 10.0 | **50.0** | 122.0 | -369.7 | 727.6 | 454.4 | **850.0** | 1562.0 | 1070.3 |
| 288.7 | 15.6 | **60.0** | 140.0 | -351.7 | 755.4 | 482.2 | **900.0** | 1652.0 | 1160.3 |
| 294.3 | 21.1 | **70.0** | 158.0 | -333.7 | 783.2 | 510.0 | **950.0** | 1742.0 | 1250.3 |
| 299.8 | 26.7 | **80.0** | 176.0 | -315.7 | 810.9 | 537.8 | **1000.0** | 1832.0 | 1340.3 |
| 305.4 | 32.2 | **90.0** | 194.0 | -297.7 | 838.7 | 565.6 | **1050.0** | 1922.0 | 1430.3 |
| 310.9 | 37.8 | **100.0** | 212.0 | -279.7 | 866.5 | 593.3 | **1100.0** | 2012.0 | 1520.3 |
| 316.5 | 43.3 | **110.0** | 230.0 | -261.7 | 894.3 | 621.1 | **1150.0** | 2102.0 | 1610.3 |
| 322.0 | 48.9 | **120.0** | 248.0 | -243.7 | 922.0 | 648.9 | **1200.0** | 2192.0 | 1700.3 |
| 327.6 | 54.4 | **130.0** | 266.0 | -225.7 | 949.8 | 676.7 | **1250.0** | 2282.0 | 1790.3 |
| 333.2 | 60.0 | **140.0** | 284.0 | -207.7 | 977.6 | 704.4 | **1300.0** | 2372.0 | 1880.3 |
| 338.7 | 65.6 | **150.0** | 302.0 | -189.7 | 1005.4 | 732.2 | **1350.0** | 2462.0 | 1970.3 |
| 344.3 | 71.1 | **160.0** | 320.0 | -171.7 | 1033.2 | 760.0 | **1400.0** | 2552.0 | 2060.3 |
| 349.8 | 76.7 | **170.0** | 338.0 | -153.7 | 1060.9 | 787.8 | **1450.0** | 2642.0 | 2150.3 |
| 355.4 | 82.2 | **180.0** | 356.0 | -135.7 | 1088.7 | 815.6 | **1500.0** | 2732.0 | 2240.3 |
| 360.9 | 87.8 | **190.0** | 374.0 | -117.7 | 1116.5 | 843.3 | **1550.0** | 2822.0 | 2330.3 |
| 366.5 | 93.3 | **200.0** | 392.0 | -99.7 | 1144.3 | 871.1 | **1600.0** | 2912.0 | 2420.3 |
| 372.0 | 98.9 | **210.0** | 410.0 | -81.7 | 1172.0 | 898.9 | **1650.0** | 3002.0 | 2510.3 |
| 373.2 | 100.0 | **212.0** | 413.6 | -78.1 | 1199.8 | 926.7 | **1700.0** | 3092.0 | 2600.3 |
| 377.6 | 104.4 | **220.0** | 428.0 | -63.7 | 1227.6 | 954.4 | **1750.0** | 3182.0 | 2690.3 |
| 383.2 | 110.0 | **230.0** | 446.0 | -45.7 | 1255.4 | 982.2 | **1800.0** | 3272.0 | 2780.3 |
| 388.7 | 115.6 | **240.0** | 464.0 | -27.7 | 1283.2 | 1010.0 | **1850.0** | 3362.0 | 2870.3 |
| 394.3 | 121.1 | **250.0** | 482.0 | -9.7 | 1310.9 | 1037.8 | **1900.0** | 3452.0 | 2960.3 |
| 399.8 | 126.7 | **260.0** | 500.0 | 8.3 | 1338.7 | 1065.6 | **1950.0** | 3542.0 | 3050.3 |
| 405.4 | 132.2 | **270.0** | 518.0 | 26.3 | 1366.5 | 1093.3 | **2000.0** | 3632.0 | 3140.3 |
| 410.9 | 137.8 | **280.0** | 536.0 | 44.3 | 1394.3 | 1121.1 | **2050.0** | 3722.0 | 3230.3 |
| 416.5 | 143.3 | **290.0** | 554.0 | 62.3 | 1422.0 | 1148.9 | **2100.0** | 3812.0 | 3320.3 |
| 422.0 | 148.9 | **300.0** | 572.0 | 80.3 | 1449.8 | 1176.7 | **2150.0** | 3902.0 | 3410.3 |
| 427.6 | 154.4 | **310.0** | 590.0 | 98.3 | 1477.6 | 1204.4 | **2200.0** | 3992.0 | 3500.3 |
| 433.2 | 160.0 | **320.0** | 608.0 | 116.3 | 1505.4 | 1232.2 | **2250.0** | 4082.0 | 3590.3 |
| 438.7 | 165.6 | **330.0** | 626.0 | 134.3 | 1533.2 | 1260.0 | **2300.0** | 4172.0 | 3680.3 |
| 444.3 | 171.1 | **340.0** | 644.0 | 152.3 | 1560.9 | 1287.8 | **2350.0** | 4262.0 | 3770.3 |
| 449.8 | 176.7 | **350.0** | 662.0 | 170.3 | 1588.7 | 1315.6 | **2400.0** | 4352.0 | 3860.3 |
| 455.4 | 182.2 | **360.0** | 680.0 | 188.3 | 1616.5 | 1343.3 | **2450.0** | 4442.0 | 3950.3 |
| 460.9 | 187.8 | **370.0** | 698.0 | 206.3 | 1644.3 | 1371.1 | **2500.0** | 4532.0 | 4040.3 |
| 466.5 | 193.3 | **380.0** | 716.0 | 224.3 | 1699.8 | 1426.7 | **2600.0** | 4712.0 | 4220.3 |
| 472.0 | 198.9 | **390.0** | 734.0 | 242.3 | 1755.4 | 1482.2 | **2700.0** | 4892.0 | 4400.3 |
| 477.6 | 204.4 | **400.0** | 752.0 | 260.3 | 1810.9 | 1537.8 | **2800.0** | 5072.0 | 4580.3 |
| 488.7 | 215.6 | **420.0** | 788.0 | 296.3 | 1866.5 | 1593.3 | **2900.0** | 5252.0 | 4760.3 |
| 499.8 | 226.7 | **440.0** | 824.0 | 332.3 | 1922.0 | 1648.9 | **3000.0** | 5432.0 | 4940.3 |
| 510.9 | 237.8 | **460.0** | 860.0 | 368.3 | 1977.6 | 1704.4 | **3100.0** | 5612.0 | 5120.3 |
| 522.0 | 248.9 | **480.0** | 896.0 | 404.3 | 2033.2 | 1760.0 | **3200.0** | 5792.0 | 5300.3 |
| 533.2 | 260.0 | **500.0** | 932.0 | 440.3 | 2088.7 | 1815.6 | **3300.0** | 5972.0 | 5480.3 |
| 544.3 | 271.1 | **520.0** | 968.0 | 476.3 | 2144.3 | 1871.1 | **3400.0** | 6152.0 | 5660.3 |
| 555.4 | 282.2 | **540.0** | 1004.0 | 512.3 | 2199.8 | 1926.7 | **3500.0** | 6332.0 | 5840.3 |
| 566.5 | 293.3 | **560.0** | 1040.0 | 548.3 | 2255.4 | 1982.2 | **3600.0** | 6512.0 | 6020.3 |
| 577.6 | 304.4 | **580.0** | 1076.0 | 584.3 | 2310.9 | 2037.8 | **3700.0** | 6692.0 | 6200.3 |
| 588.7 | 315.6 | **600.0** | 1112.0 | 620.3 | 2366.5 | 2093.3 | **3800.0** | 6872.0 | 6380.3 |

**Appendix A9b. Temperature scale alignment chart.** (The fahrenheit column is duplicated to make it easy to align a straight-edge across the equivalent columns.)

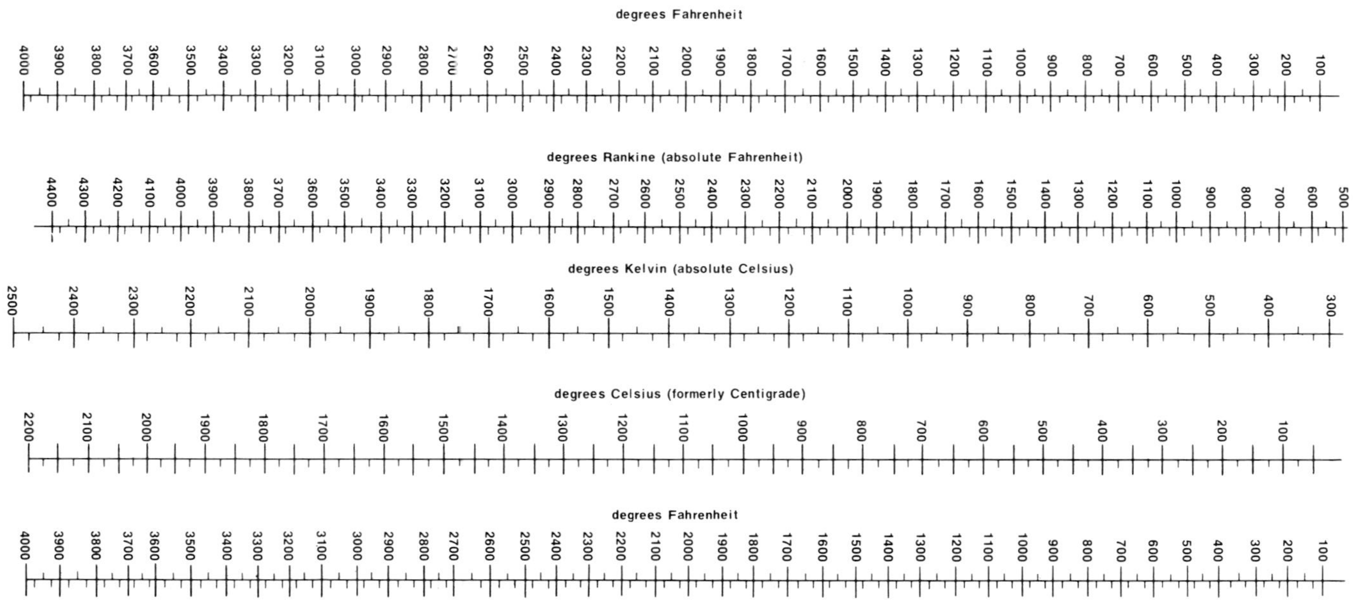

**Appendix B1. Hazardous Air Pollutants, Title III, 10-90**

(Compounds and glycol ethers include any unique chemical substance containing the named chemical as part of its infrastructure.)

| Chemical Name | Molecular Formula | CAS* Registry Number |
|---|---|---|
| Acetaldehyde | $C_2H_4O$ | 75070 |
| Acetamide | $C_2H_5NO$ | 60355 |
| Acetonitrile | $C_2H_3N$ | 75058 |
| Acetophenone | $C_8H_8O$ | 98862 |
| 2-Acetylaminofluorene | | 53963 |
| Acrolein | $C_3H_4O$ | 107028 |
| Acrylamide | $C_3H_5NO$ | 79061 |
| Acrylic acid | $C_3H_4O_2$ | 79107 |
| Acrylonitrile | $C_3H_3N$ | 107131 |
| Allyl chloride | $C_3H_5Cl$ | 107051 |
| 4-Aminobiphenyl | | 92671 |
| Aniline | $C_6H_7N$ | 62533 |
| o-Anisidine | | 90040 |
| Antimony Compounds | | |
| Arsenic Compounds (inorganic including arsine) | | |
| Asbestos | | 1332214 |
| Benzene (including benzene from gasoline) | $C_6H_6$ | 71432 |
| Benzidine | $C_{12}H_{12}N_2$ | 92875 |
| Benzotrichloride | | 98077 |
| Benzyl chloride | $C_7H_7Cl$ | 100447 |
| Beryllium Compounds | | |
| Biphenyl | | 92524 |
| Bis(2-ethylhexyl)phthalate (DEHP) | | 117817 |
| Bis(chloromethyl)ether | | 542881 |
| Bromoform | $CHBr_3$ | 75252 |
| 1,3-Butadiene | $C_4H_6$ | 106990 |
| Cadmium Compounds | | |
| Calcium cyanamide | | 156627 |
| Caprolactam | $C_6H_{11}NO$ | 105602 |
| Captan | | 133062 |
| Carbaryl | | 63252 |
| Carbon disulfide | $CS_2$ | 75150 |
| Carbon tetrachloride | $CCl_4$ | 56235 |
| Carbonyl sulfide | COS | 463581 |
| Catechol | | 120809 |

*(continued)*

* Chemical Abstracts Service of the American Chemical Society.

**Appendix B1.** *(continued)*

| Chemical Name | Molecular Formula | CAS* Registry Number |
|---|---|---|
| Chloramben | | 133904 |
| Chlordane | | 57749 |
| Chlorine | | 7782505 |
| Chloroacetic acid | $C_2H_3ClO_2$ | 79118 |
| 2-Chloroacetophenone | | 532274 |
| Chlorobenzene | $C_6H_5Cl$ | 108907 |
| Chlorobenzilate | | 510156 |
| Chloroform | $CHCl_4$ | 67663 |
| Chloromethyl methyl ether | | 107302 |
| Chloroprene | | 126998 |
| Chromium Compounds | | |
| Cobalt Compounds | | |
| Coke Oven Emissions | | |
| Cresols/Cresylic acid (isomers and mixture) | | 1319773 |
| o-Cresol | $C_7H_8O$ | 95487 |
| m-Cresol | $C_7H_8O$ | 108394 |
| p-Cresol | $C_7H_8O$ | 106445 |
| Cumene | $C_9H_{12}$ | 98828 |
| Cyanide Compounds | | |
| 2,4-D, salts and esters | | 94757 |
| DDE | | 3547044 |
| Diazomethane | $CH_2N_2$ | 334883 |
| Dibenzofurans | | 132649 |
| 1,2-Dibromo-3-chloropropane | | 96128 |
| Dibutylphthalate | $C_{16}H_{22}O$ | 84742 |
| 1,4-Dichlorobenzene(p) | $C_6H_4Cl_2$ | 106467 |
| 3,3-Dichlorobenzidine | | 91941 |
| Dichloroethyl ether (Bis(2-chloroethyl) ether) | $C_4H_8Cl_2O$ | 111444 |
| 1,3-Dichloropropene | | 542756 |
| Dichlorvos | | 62737 |
| Diethanolamine$C_4H_{11}NO_2$ | | 111422 |
| N,N Diethyl analine (N,N-Dimethylaniline) | | 121697 |
| Diethyl sulfate | | 64675 |
| 3,3-Dimethoxybenzidine | | 119904 |
| Dimethyl aminoazobenzene | | 60117 |
| 3,3-Dimethyl benzidine | | 119937 |
| Dimethyl carbamoyl chloride | | 79447 |
| Dimethyl formamide(N,N-) | $C_3H_7NO$ | 68122 |
| 1,1-Dimethyl hydrazine | $C_2H_8N_2$ | 57147 |

*(continued)*

**Appendix B1.** *(continued)*

| Chemical Name | Molecular Formula | CAS* Registry Number |
|---|---|---|
| Dimethyl phthalate | $C_{10}H_{10}O_4$ | 131113 |
| Dimethyl sulfate | | 77781 |
| 4,6-Dinitro-o-cresol, and salts | | 534521 |
| 2,4-Dinitrophenol | | 51285 |
| 2,4-Dinitrotoluene | | 121142 |
| 1,4-Dioxane (1,4-Diethyleneoxide) | | 123911 |
| 1,2-Diphenylhydrazine | | 122667 |
| Epichlorohydrin (1-Chloro-2,3-epoxypropane) | $C_3H_5ClO$ | 106898 |
| 1,2-Epoxybutane | $C_4H_8O$ | 106887 |
| Ethyl acrylate | $C_5H_8O_2$ | 140885 |
| Ethyl benzene | $C_8H_{10}$ | 100414 |
| Ethyl carbmate (Urethane) | | 51796 |
| Ethyl chloride (Chloroethane) | $C_2H_5Cl$ | 75003 |
| Ethylene dibromide (Dibromoethane) | | 106934 |
| Ethylene dichloride (1,2-Dichloroethane) | | 107062 |
| Ethylene gycol | $C_2H_6O_2$ | 107211 |
| Ethylene imine (Aziridine) | $C_2H_5N$ | 151564 |
| Ethylene oxide | $C_2H_4O$ | 75218 |
| Ethylene thiourea | | 96457 |
| Ethylidene dichloride (1,1-Dichloroethane) | $C_2H_4Cl_2$ | 75343 |
| Formaldehyde | $CH_2O$ | 50000 |
| Glycol ethers | | |
| Heptachlor | | 76448 |
| Hexachlorobenzene | $C_6Cl_6$ | 118741 |
| Hexachlorobutadiene | $C_4Cl_4$ | 87683 |
| Hexachlorocyclopentadiene | $C_5Cl_6$ | 77474 |
| Hexachloroethane | $C_2Cl_6$ | 67721 |
| Hexamethylene-1,6-diisocyanate | | 822060 |
| Hexamethylphosphoramide | | 680319 |
| Hexane | $C_6H_{14}$ | 110543 |
| Hydrazine | $H_4N_2$ | 302012 |
| Hydrochloric acid | HCl | 7647010 |
| Hydrogen fluoride (Hydrofluoric acid) | HF | 7664393 |
| Hydroquinone | $C_6H_6O_2$ | 123319 |
| Isophorone | $C_9H_NO$ | 78591 |
| Lead Compounds | | |
| Lindane (all isomers) | | 58899 |
| Maleic anhydride | $C_4H_2O_3$ | 108316 |
| Manganese Compounds | | |

*(continued)*

**Appendix B1.** *(continued)*

| Chemical Name | Molecular Formula | CAS* Registry Number |
|---|---|---|
| Mercury Compounds | | |
| Methanol | $CH_4O$ | 67561 |
| Methoxychlor | | 72435 |
| Methyl bromide (Bromomethane) | $CH_3Br$ | 74839 |
| Methyl chloride (Chloromethane) | $CH_3Cl$ | 74873 |
| Methyl chloroform (1,1,1-Trichloroethane) | $C_2H_3Cl_3$ | 71556 |
| Methyl ethyl ketone (2-Butanone) | $C_4H_8O$ | 78933 |
| Methyl hydrazine | $CH_6N_2$ | 60344 |
| Methyl iodide (Iodomethane) | $CH_3I$ | 74884 |
| Methyl isobutyl ketone (Hexone) | | 108101 |
| Methyl isocyanate | $C_2H_3NO$ | 624839 |
| Methyl methacrylate | $C_5H_8O_2$ | 80626 |
| Methyl tert butyl ether | | 1634044 |
| 4,4-Methylene bis (2-chloroaniline) | | 101144 |
| Methylene chloride (Dichloromethane) | | 75092 |
| Methylene diphenyl diisochyanate (MDI) | | 101688 |
| 4,4-Methylenedianiline | | 101779 |
| Mineral fibers, fine | | |
| Naphthalene | $C_{10}H_8$ | 91203 |
| Nickel Compounds | | |
| Nitrobenzene | $C_6H_5NO_2$ | 98953 |
| 4-Nitrobiphenyl | | 92933 |
| 4-Nitrophenol | | 100027 |
| 2-Nitropropane | $C_3H_7NO_2$ | 79469 |
| N-Nitroso-N-methylurea | | 684935 |
| N-Nitrosodimcthylamine | | 62759 |
| N-Nitrosomorpholine | | 59892 |
| Parathion | | 56382 |
| Pentachloronitrobenzene (Quintobenzene) | | 82688 |
| Pentachlorophenol | | 87865 |
| Phenol | $C_6H_6O$ | 108952 |
| p-Phenylenediamine | $C_6H_8N_2$ | 106503 |
| Phosgene | $CCl_2O$ | 75445 |
| Phosphine | $H_3P$ | 7803512 |
| Phosphorus | P | 7723140 |
| Phthalic anhydride | $C_8H_4O_3$ | 85449 |
| Polychlorinated biphenyls (Aroclors) | | 1336363 |
| Polycylic Organic Matter 1,3-Propane sultone | | 1120714 |
| beta-Propiolactone | $C_3H_4O_2$ | 57578 |

*(continued)*

**Appendix B1.** *(concluded)*

| Chemical Name | Molecular Formula | CAS* Registry Number |
|---|---|---|
| Propionaldehyde | $C_3H_6O$ | 123386 |
| Propoxur (Baygon) | | 114261 |
| Propylene dichloride (1,2-Dichloropropane) | $C_2H_6Cl_2$ | 78875 |
| Propylene oxide (1,2-) | $C_3H_6O$ | 75569 |
| 1,2-Propylenimine (2-Methyl aziridine) | $C_3H_7N$ | 75558 |
| Quinoline | $C_9H_7N$ | 91225 |
| Quinone | | 106514 |
| Radionuclides (including radon) | | |
| Selenium Compounds | | |
| Styrene | $C_8H_8$ | 100425 |
| Styrene oxide | | 96093 |
| 2,3,7,8-Tetrachlorodibenzo-p-dioxin | | 1746016 |
| 1,1,2,2-Tetrachloroethane | $C_2H_2Cl_4$ | 79345 |
| Tetrachloroethylene (Perchloroethylene) | $C_2Cl_4$ | 127184 |
| Titanium tetrachloride | | 7550450 |
| Toulene | $C_7H_8$ | 108883 |
| 2,4-Toluene diamine | $C_7H_{10}N_2$ | 95807 |
| 2,4-Toluene diisocyanate | $C_9H_6N_2O_2$ | 584849 |
| o-Toluidine | $C_7H_9N$ | 95534 |
| Toxaphene (chlorinated camphene) | | 8001352 |
| 1,2,4-Trichlorobenzene | $C_6H_3Cl_3$ | 120821 |
| 1,1,2-Trichloroethane | $C_2H_3Cl_3$ | 79005 |
| Trichloroethylene | $C_2HCl_3$ | 79016 |
| 2,4,5-Trichlorophenol | | 95954 |
| 2,4,6-Trichlorophenol | $C_6H_3Cl_3O$ | 88062 |
| Triethylamine | $C_6H_{15}N$ | 121448 |
| Trifluralin | | 1582098 |
| 2,2,4-Trimethylpentane | $C_8H_{18}$ | 540841 |
| Vinyl acetate | $C_4H_6O_2$ | 108054 |
| Vinyl bromide | $C_2H_3Br$ | 593602 |
| Vinyl chloride | $C_2H_3Cl$ | 75014 |
| Vinylidene chloride (1,1-Dichloroethylene) | | 75354 |
| Xylenes (isomers and mixtures) | | 1330207 |
| o-Xylenes | $C_8H_{10}$ | 95476 |
| m-Xylenes | $C_8H_{10}$ | 108383 |
| p-Xylenes | $C_8H_{10}$ | 106423 |

**Appendix B2. Hazardous and toxic wastes** *(from 40CFR, #261.10-40 Ref. 20).*

**Subpart B - Criteria for identifying the Characteristics of Hazardous Waste and for Listing Hazardous Waste.**

**§ 261.10 Criteria for identifying the characteristics of hazardous waste.**

(a) The Administrator shall identify and define a characteristic of hazardous waste in subpart C only upon determining that:

(1) A solid waste that exhibits the characteristic may;

(i) Cause, or significantly contribute to, an increase in mortality or an increase in serious irreversible, or incapacitating reversible, illness: or

(ii) Pose a substantial present or potential hazard to human health or the environment when it is improperly treated, stored, transported, disposed of or otherwise managed; and

(2) The characteristic can be:

(i) Measured by an available standardized test method which is reasonably within the capability of generators of solid waste; or

(ii) Reasonably detected by generators of solid waste through their knowledge of their waste.

**§ 261.11 Criteria for listing hazardous waste.**

(a) The Administrator shall list a solid waste as a hazardous waste only upon determining that the solid waste meets one of the following criteria:

(1) It exhibits any of the characteristics of hazardous waste identified in subpart C.

(2) It has been found to be fatal to humans in low doses or, in the absence of data on human toxicity, it has been shown in studies to have an oral LD 50 toxicity (rat) of less than 50 milligrams per kilogram, an inhalation LC 50 toxicity (rat) of less than 2 milligrams per liter, or a dermal LD 50 toxicity (rabbit) of less than 200 milligrams per kilogram or is otherwise capable of causing or significantly contributing to an increase in serious irreversible, or incapacitating reversible, illness. (Waste listed in accordance with these criteria will be designated Acute Hazardous Waste.)

(3) It contains any of the toxic constituents listed in appendix VIII and, after considering the following factors, the Administrator concludes that the waste is capable of posing a substantial present or potential hazard to human health or the environment when improperly treated, stored, transported or disposed of, or otherwise managed:

(i) The nature of the toxicity presented by the constituent.

**Appendix B2. Hazardous and toxic wastes** *(continued)*

(ii) The concentration of the constituent in the waste.

(iii) The potential of the constituent or any toxic degredation product of the constituent to migrate from the waste into the environment under the types of improper management considered in paragraph (a)(3)(vii) of this section.

(iv) The persistence of the constituent or any toxic degradation product of the constituent.

(v) The potential for the constituent or any toxic degradation product of the constituent to degrade into non-harmful constituents and the rate of degradation.

(vi) The degree to which the constituent or any degradation product of the constituent bioaccumulates in ecosystems.

(vii) The plausible types of improper management to which the waste could be subjected.

(viii) The quantities of the waste generated at individual generation sites or on a regional or national basis.

(ix) The nature and severity of the human health and environmental damage that has occurred as a result of the improper management of wastes containing the constituent.

(x) Action taken by other governmental agencies or regulatory programs based on the health or environmental hazard posed by the waste or waste constituent.

(xi) Such other factors as may be appropriate.

Substances will be listed on appendix VIII only if they have been shown in scientific studies to have toxic, carcinogenic, mutagenic or teratogenic effects on humans or other life forms.

(Wastes listed in accordance with these criteria will be designated Toxic wastes.)

(b) The Administrator may list classes or types of solid waste as hazardous waste if he has reason to believe that individual wastes, within the class or type of waste, typically or frequently are hazardous under the definition of hazardous waste found in sectin 1004(5) of the Act.

(c) The Administrator will use the criteria for listing specified in this section to establish the exclusion limits referred to in § 261.5(c).

[45 FR 33119 May 19, 1980, as amended at 55 FR 18726, May 4,1990]

**Appendix B2. Hazardous and toxic wastes** *(continued)*

## Subpart C - Characteristics of Hazardous Waste

### § 261.20 General

(a) A solid waste, as defined in § 261.2, which is not excluded from regulation as a hazardous waste under § 261.4(b), is a hazardous waste if it exhibits any of the characteristics identified in this subpart.

[*Comment:* § 262.11 of this chapter sets forth the generator's responsibility to determine whether his waste exhibits one or more of the characteristics identified in this subpart]

(b) A hazardous waste which is identified by a characteristic in this subpart is assigned every EPA Hazardous Waste Number that is applicable as set forth in this subpart. This number must be used in complying with the notification requirements of section 3010 of the Act and all applicable recordkeeping and reporting requirements under parts 262 through 265, 268, and 270 of this chapter.

(c) For purposes of this subpart, the Administrator will consider a sample obtained using any of the applicable sampling methods specified in appendix I to be a respresentative sample within the meaning of part 260 of this chapter.

[*Comment:* Since the appendix I sampling methods are not being formally adopted by the Administrator, a person who desires to employ an alternative sampling method is not required to demonstrate the equivalency of his method under the procedures set forth in §§ 260.20 and 260.21.]

[45 FR 33119, May 19, 1980, as amended at 51 FR 40636, Nov. 7, 1986; 55 FR 22684, June 1, 1990; 56 FR 3876, Jan. 31, 1991]

### § 261.21 Characteristics of ignitability.

(a) A solid waste exhibits the characteristic of ignitability if a respresentative sample of the waste has any of the following properties:

(1) It is a ligquid, other than an aqueous solution containing less than 24 percent alcohol by volume and has flash point less than 60 C (140 F), as determined by a Pensky-Martens Closed Cup Tester, using the test method specified in ASTM Standard D-93-79 or D-93-80 (incorporated by reference, see § 260.11), or a Setaflash Closed Cup Tester, using the test method specified in ASTM Standard D-3278-79 (incorporated by reference, see § 260.11), or as determined by an equivalent test method approved by the Administrator under procedures set forth in §§ 260.20 and 260.21.

**Appendix B2. Hazardous and toxic wastes** *(continued)*

(2) It is not a liquid and is capable, under standard temperature and pressure, of causing fire through friction, absoprtion of moisture or spontaneous chemical changes and, when ignited, burns so vigorously and persistently that it creates a hazard.

(3) It is an ignitable compressed gas as defined in 49 CFR 173.300 and as determined by the test methods describedin that regulation or equivalent test methods approved by the Adminstrator under §§ 260.20 and 260.21.

(4) It is an oxidizer as defined in 49 CFR 173.151.

(b) A solid waste that exhibits the characteristic of ignitability has the EPA Hazardous Waste Number of D001.

[45 FR 33119, May 19, 1980, as amended at 46 FR 35247, July 7, 1981; 55 FR 22684, June 1, 1990]

**§ 261.22 Characteristic of corrosivity.**

(a) A solid waste exhibits the characteristic of corrosivity if a representative sample of the waste has either of the following properties:

(1) It is aqueous and has a pH less than or equal to 2 or greater than or equal to 12.5, as determined by a pH meter using either an EPA test method or an equivalent test method approved by the Administrator under the procedures set forth in §§ 260.20 and 260.21. The EPA test method for pH is specified as Method 5.2 in "Test Methods for the Evaluation of Solid Waste, Physical/Chemical Methods" (incorporated by reference, see § 260.11).

(2) It is a liquid and corrodes steel (SAE 1020) at a rate greater than 6.35 mm (0.250 inch) per year at a test temperature of 55 C (130 F) as determined by the test method specified in NACE (National Association of Corrosion Engineers) Standard TM-01-69 as standardized in "Test Methods for the Evaluation of Solid Waste, Physical/Chemical Methods" (incorporated by reference, see § 260.11) or an equivalent test method approved by the Administrator under the procedures set forth in §§ 260.20 and 260.21.

(b) A solid waste that exhibits the characteristic of corrosivity has the EPA Hazardous Waste Number of D002.

[45 FR 33119, May 19. 1980, as amended at 46 FR 35247, July 7, 1981; 55 FR 22684, June 1, 1990]

**Appendix B2. Hazardous and toxic wastes** *(continued)*

### § 261.23 Characteristics of reactivity.

(a) A solid waste exhibits the characteristic of reactivity if a representative sample of the waste has any of the following properties:

(1) It is normally unstable and readily undergoes violent change without detonating.

(2) It reacts violently with water.

(3) It forms potentially explosive mixtures with water.

(4) When mixed with water, it generates toxic gases, vapors or fumes in a quantity sufficient to present a danger to human health or the environment.

(5) It is a cyanide or sulfide bearing waste which, when exposed to pH conditions between 2 and 12.5, can generate toxic gases, vapors or fumes in a quantity sufficient to present a danger to human healt or the environment.

(6) It is capable of detonation or explosive reation if it is subjected to a strong initiating source or if heated under confinement.

(7) It is readily capable of detonation or explosive decomposition or reaction at standard temperature and pressure.

(8) It is a forbidden explosive as defined in 49 CFR 173.51, or a Class A explosive as defined in 49 CFR 173.53 or a Class B explosive as defined in 49 CFR 173.88.

(b) A solid waste that exhibits the characteristic of reactivitiy has the EPA Hazardous Waste Number of D003.

[45 FR 33119, May 19, 1980, as amended at 55 FR 22684, June 1, 1990]

### § 261.24 Toxicity characteristic.

(a) A solid waste exhibits the characteristic of toxicity if, using the test methods described in appendix II or equivalent methods approved by the Administrator under the procedures set forth in §§ 260.20 and 260.21, the extract from a representative sample of the waste contains any of the contaminants listed in table 1 at the concentration equal to or greater than the respective value given in that table. Where the waste contains less than 0.5 percent filterable solids, the waste itself, after filtering using the methodology outlined in appendix II, is considered to be, the extract for the purpose of this section.

(b) A solid waste that exhibits the characteristic of toxicity has the EPA Hazardous Waste Number specified in Table I which corresponds to the toxic contaminant causing it to be hazardous.

**Appendix B2. Toxic and hazardous wastes.** *(continued)*

**Table 1. Maximum concentration of contaminants for the toxicity characteristic.**

| EPA HW No.[1] | Contaminant | CAS No.[2] | Regulatory Level (mg/L) |
|---|---|---|---|
| D004 | Arsenic | 7440-38-2 | 5.0 |
| D005 | Barium | 7440-39-3 | 100.0 |
| D018 | Benzene | 71-43-2 | 0.5 |
| D006 | Cadmium | 7440-43-9 | 1.0 |
| D019 | Carbon tetrachloride | 56-23-5 | 0.5 |
| D020 | Chlordane | 57-74-9 | 0.03 |
| D021 | Chlorobenzene | 108-90-7 | 100.0 |
| D022 | Chloroform | 67-66-3 | 6.0 |
| D007 | Chromium | 7440-47-3 | 5.0 |
| D023 | o-Cresol | 95-48-7 | [4]200.0 |
| D024 | m-Cresol | 108-39-4 | [4]200.0 |
| D025 | p-Cresol | 106-44-5 | [4]200.0 |
| D016 | 2,4-D | 94-75-7 | 10.0 |
| D027 | 1,4-Dichlorobenzene | 106-46-7 | 7.5 |
| D028 | 1,2-Dichloroethane | 107-06-2 | 0.5 |
| D029 | 1,1-Dichlorethylene | 75-35-4 | 0.7 |
| D030 | 2,4-Dinitrotoluene | 121-14-2 | 0.13 |
| D012 | Endrin | 72-20-8 | 0.02 |
| D031 | Heptachlor (and its epoxide) | 76-44-8 | 0.008 |
| D032 | Hexachlorobenzene | 118-74-1 | [3]0.13 |
| D033 | Hexachlorobutadiene | 87-68-3 | 0.5 |
| D034 | Hexachloroethane | 67-72-1 | 3.0 |
| D008 | Lead | 7439-92-1 | 5.0 |
| D013 | Lindane | 58-89-9 | 0.4 |
| D009 | Mercury | 7439-97-6 | 0.2 |
| D014 | Methoxychlor | 72-43-5 | 10.0 |
| D035 | Methyl ethyl ketone | 78-93-3 | 200.0 |
| D036 | Nitrobenzene | 98-95-3 | 2.0 |
| D037 | Pentrachlorophenol | 87-86-5 | 100.0 |
| D038 | Pyridine | 110-86-1 | [3]5.0 |
| D010 | Selenium | 7782-49-2 | 1.0 |
| D011 | Silver | 7440-22-4 | 5.0 |
| D039 | Tetrachloroethylene | 127-18-4 | 0.7 |
| D015 | Toxaphene | 8001-35-2 | 0.5 |

See footnotes on page 153. *(continued)*

**Appendix B2. Hazardous and toxic wastes** *(continued)*

**Table 1. Maximum concentration of contaminants for the toxicity characteristic** *(concluded)*

| EPA HW No.[1] | Contaminant | CAS No.[2] | Regulatory Level (mg/L) |
|---|---|---|---|
| D040 | Trichloroethylene | 79-01-6 | 0.5 |
| D041 | 2,4,5-Trichlorophenol | 95-95-4 | 400.0 |
| D042 | 2,4,6-Trichlorophenol | 88-06-2 | 2.0 |
| D017 | 2,4,5-TP (Silvex) | 93-72-1 | 1.0 |
| D043 | Vinyl Chloride | 75-01-4 | 0.2 |

[1] Hazardous waste number.

[2] Chemical abstracts service number.

[3] Quantitation limit is greater than the calculated regulatory level. The quantitation limit therefore becomes the regulatory level.

[4] If o-, m-, and p-Cresol concentrations can not be differentiated, the total cresol (D026) concentration is used. The regulatory level of total cresol is 200 mg/l.

[55 FR 11862, Mar. 29, 1990, as amended at 55 FR 22684, June 1, 1990; 55 FR 26987, June 29, 1990]

**Subpart D - Lists of Hazardous Wastes**

**§ 261.30 General.**

(a) A solid waste is a hazardous waste if it is listed in this subpart, unless has been excluded from this list under §§ 260.20 and 260.22.

(b) The Administrator will indicate his basis for listing the classes or types of wastes listed in this subpart by employing one or more of the following Hazard Codes:

Ignitable Waste ........................(I)
Corrosive Waste ......................(C)
Reactive Waste ........................(R)
Toxicity Characteristic Waste ...........(E)
Acute Hazardous Waste.................(H)
Toxic Waste..........................(T)

Appendix VII identifies the constituent which caused the Administrator to list the waste as a Toxicity Characteristic Waste (E) or Toxic Waste (T) in §§ 261.31 and 261.32.

(c) Each hazardous waste listed in this subpart is assigned an EPA Hazardous Waste Number which precedes the name of the waste. This number must be used in complying with the notification requirements of Section 3010 of the Act

**Appendix B2. Hazardous and toxic wastes** *(continued)*

and certain recordkeeping and reporting requirements under parts 262 through 265, 268, and part 270 of this chapter.

(d) The following hazardous wastes listed in § 261.31 and § 261.32 are subject to the exclusion limits for acutely hazardous wastes established in § 261.5: EPA Hazardous Wastes Nos. F020, F021, F022, F023, F026, and F027.

[45 FR 33119, May 19, 1980, as amended at 48 FR 14294, Apr. 1, 1983; 50 FR 2000, Jan. 14, 1985; 51 FR 40636, Nov. 7, 1986; 55 FR 11863, Mar. 29,1990]

### § 261.31 Hazardous wastes from non-specific sources.

(a) The following solid wastes are listed hazardous wastes from non-specific sources unless they are excluded under §§ 260.20 and 260.22 and listed in appendix IX.

| Industry and EPA hazardous waste No. | Hazardous waste | Hazard code |
|---|---|---|
| Generic: | | |
| F001 ........... | The following spent halogenated solvents used in degreasing: Tetrachloroethylene, trichloroethylene, methylene chloride, 1,1,1-trichloroethane, carbon tetrachloride, and chlorinated fluorocarbons; all spend solvent mixtures/blends used in degreasing containing, before use, a total of ten percent or more (by volume) of one or more of the above halogenated solvents or those solvents listed in F002, F004, and F005; and still bottoms from the recovery of these spent solvents and spend solvent mixtures. | (T) |
| F002 ........... | The following spent halogenated solvents: Tetrachloroethylene, methylene chloride, trichloroethylene, 1,1,1-trichloroethane, chlorobenzene, 1,1,2-trichloro-1,2,2,-trifluoroethane, ortho-dichlorobenzene, trichlorofluoromethane, and 1,1,2-trichloroethane; all spent solvent mixtures/blends containing, before use, a total of ten percent or more (by volume) of one or more of the above halogenated solvents or those listed in F001, F004, or F005; and still bottoms from the recovery of these spent solvents and spent solvent mixtures. | (T) |
| F003 ........... | The following spent non-halogenated solvents: Xylene, acetone, ethyl acetate, ethyl benzene, ethyl ether, methyl isobutyl ketone, n-butyl alcohol, cyclohexanone, and methanol; all spent solvent mixtures/blends containing, before use, only the above spent non-halogenated solvents; and all spent solvent mixtures/blends containing, before use, one or more of the above non-halogenated solvents, and, a total of ten percent or more (by volume) of one or more of those solvents listed in F001, F002, F004, and F005; and still bottoms from the recovery of these spent solvents and spent solvent mixtures. | (I)* |

* See footnote on page 159.

## Appendix B2. Hazardous and toxic wastes *(continued)*

| Industry and EPA hazardous waste No. | Hazardous waste | Hazard code |
|---|---|---|
| F004 | The following spent non-halogenated solvents: Cresols and cresylic acid, and nitrobenzene; all spent solvent mixtures/blends containing, before use, a total of ten percent or more (by volume) of one or more of the above non-halogenated solvents or those solvents listed in F001, F002, F005; and still bottoms from the recovery of these spent solvents and spent solvent mixtures. | (T) |
| F005 | The following spent non-halogenated solvents: Toluene, methyl ethyl ketone, carbon disulfide, isobutanol, pyridine, benzene, 2-ethoxyethanol, and 2-nitropropane; all spent solvent mixtures/blends containing, before use, a total of ten percent or more (by volume) of one or more of the above non-halogenated solvents or those solvents listed in F001, F002, or F004; and still bottoms from the recovery of these spent solvents and spent solvent mixtures. | (I,T) |
| F006 | Wastewater treatment sludges from electroplating operations except from the following processes; (1) Sulfuric acid anodizing of aluminum; (2) tin plating on carbon steel; (3) zinc plating (segregated basis) on carbon steel; (4) aluminum or zinc-aluminum plating on carbon steel: (5) cleaning/stripping associated with tin, zinc and aluminum plating on carbon steel; and (6) chemical etching and milling of aluminum. | (T) |
| F007 | Spent cyanide plating bath solutions from electroplating operations. | (R,T) |
| F008 | Plating bath residues from the bottom of plating baths from electroplating operations where cyanides are used in the process. | (R,T) |
| F009 | Spent stripping and cleaning bath solutions from electroplating operations where cyanides are used in the process. | (R,T) |
| F010 | Quenching bath residues from oil baths from metal heat treating operations where cyanides are used in the process. | (R,T) |
| F011 | Spent cyanide solutions from salt bath pot cleaning from metal heat treating operations. | (R,T) |
| F012 | Quenching waste water treatment sludges from metal heat treating operations where cyanides are used in the process. | (T) |
| F019 | Wastewater treatment sludges from the chemical conversion coating of aluminum except from zirconium phosphating in aluminum can washing when such phosphating is an exclusive conversion coating process. | (T) |

## Appendix B2. Hazardous and toxic wastes *(continued)*

| Industry and EPA hazardous waste No. | Hazardous waste | Hazard code |
|---|---|---|
| F020 ........... | Wastes (except wastewater and spent carbon from hydrogen chloride purification) from the production or manufacturing use (as a reactant, chemical intermediate, or component in a formulating process) of tri- or tetrachlorophenol, or of intermediates used to produce their pesticide derivatives. (This listing does not include wastes from the production of Hexachlorophene from highly purified 2,4,5-trichlorophenol.) | (H) |
| F021 ........... | Wastes (except wastewater and spent carbon from hydrogen chloride purification) from the production or manufacturing use (as a reactant, chemical intermediate, or component in a formulating process) of pentachlorophenol, or of intermediates used to produce its derivatives. | (H) |
| F022 ........... | Wastes (except wastewater and spent carbon from hydrogen chloride purification) from the manufacturing use (as a reactant, chemical intermediate, or component in a formulating process) of tetra-, penta-, or hexachlorobenzenes under alkaline conditions. | (H) |
| F023 ........... | Wastes (except wastewater and spent carbon from hydrogen chloride purification) from the production of materials on equipment previously used for the production or manufacturing use (as a reactant, chemical intermediate, or component in a formulating process) of tri- and tetrachlorophenois. (This listing does not include wastes from equipment used only for the production or use of Hexachlorophene from highly purified 2,4,5-trichlorophenol.) | (H) |
| F024 ........... | Process wastes, including but not limited to, distillation residues, heavy ends, tars, and reactor clean-out wastes, from the production of certain chlorinated aliphatic hydrocarbons by free radical catalyzed processes. These chlorinated aliphatic hydrocarbons are those having carbon chain lengths ranging from one to and including five, with varying amounts and positions of chlorine substitution. (This listing does not include wastewaters, wastewater treatment sludges, spent catalysts, and wastes listed in § 261.31 or § 261.32.) | (T) |
| F025 ........... | Condensed light ends, spent filters and filter aids, and spent desiccant wastes from the production of certain chlorinated aliphatic hydrocarbons, by free radical catalyzed processes. These chlorinated aliphatic hydrocarbons are those having carbon chain lengths ranging from one to and including five, with varying amounts and positions of chlorine substitution. | (T) |

## Appendix B2. Hazardous and toxic wastes *(continued)*

| Industry and EPA hazardous waste No. | Hazardous waste | Hazard code |
|---|---|---|
| F026 ........... | Wastes (except wastewater and spent carbon from hydrogen chloride purification) from the production of materials on equipment previously used for the manufacturing use (as a reactant, chemical intermediate, or component in a formulating process) of tetra-, penta-, or hexachlorobenzene under alkaline conditions. | (H) |
| F027 ........... | Discarded unused formulations containing tri-, tetra-, or pentachlorophenol or discarded unused formulations containing compounds derived from these chlorophenols. (This listing does not include formulations containing Hexachlorophene sythesized from prepurified 2,4,5-trichlorophenol as the sole component.) | (H) |
| F028 ........... | Residues resulting from the incineration or thermal treatment of soil contaminated with EPA Hazardous Waste Nos. F020, F022, F023, F026, and F027. | (T) |
| F032[1] ........... | Wastewaters, process residuals, preservative drippage, and spent formulations from wood preserving processes generated at plants that currently use or have previously used chlorophenolic formulations (except potentially cross-contaminated wastes that have had the F032 waste code deleted in accordance with § 261.35 of this chapter and where the generator does not resume or initiate use of chlorophenolic formulations). This listing does not include K100 bottom sediment sludge from the treatment of wastewater from wood preserving processes that use creosote and/or pentachlorophenol. (Note: The listing of wastewaters that have not come into contact with process contaminants is stayed administratively. The listing for plants that have previously used chlorophenolic formulations is administratively stayed whenever these wastes are covered by the F034 or F035 listings. These stays will remain in effect until further administrative action is taken.) | (T) |
| F034[1] ........... | Wastewaters, process residuals, preservative drippage, and spent formulations from wood preserving process generated at plants that use creosote formulations. This listing does not include K001 bottom sediment sludge from the treatment of wastewater from wood preserving processes that use creosote and/or pentachlorophenol. (Note: The listing of wastewaters that have not come into contact with process contaminants is stayed administratively. The stay will remain in effect until further administrative action is taken.) | (T) |

[1] See footnote on page 159.

## Appendix B2. Hazardous and toxic wastes *(continued)*

| Industry and EPA hazardous waste No. | Hazardous waste | Hazard code |
|---|---|---|
| F035[1] ........... | Wastewaters, process residuals, preservative drippage, and spent formulations from wood preserving process generated at plants that use inorganic preservatives containing arsenic or chromium. This listing does not include K001 bottom sediment sludge from the treatment of wastewater from wood preserving processes that use creosote and/or pentachlorophenol. (Note: The listing of wastewaters that have not come into contact with process contaminants is stayed administratively. The stay will remain in effect until further administrative action is taken.) | (T) |
| F037 ........... | Petroleum refinery primary oil/water/solids separation sludge: Any sludge generated from the gravitational separation of oil/water/solids during the storage or treatment of process wastewaters and oily cooling wastewaters from petroleum refineries. Such sludges include, but are not limited to, those generated in: oil/water/solids separators; tanks and impoundments; ditches and other conveyances; sumps; and stormwater units receiving dry weather flow. Sludge generated in stormwater units that do not receive dry weather flow, sludges generated from non-contact once-through cooling waters segregated for treatment from other process or oily cooling waters, sludged generated in aggressive biological treatment units as defined in § 261.31(b)(2) (including sludges generated in one or more additional units after wastewaters have been treated in aggressive biological treatment units) and K051 wastes are not included in this listing. | (T) |
| F038 ........... | Petroleum refinery secondary (emulsified) oil/water/solids separation sludge: Any sludge and/or float generated from the physical and/or chemical separation of oil/water/solids in process wastewaters and oily cooling wastewaters from petroleum refineries. Such wastes include, but are not limited to, all sludges and floats generated in: induced air flotation (IAF) units, tanks and impoundments, and all sludges generated in DAF units. Sludges generated in stormwater units that do not receive dry weather flow, sludges generated from non-contact once-through cooling waters segregated for treatment from other process or oily cooling waters, sludges and floats generated in aggressive biological treatment units as defined in § 261.31(b)(2) (including sludges and floats generated in one or more additional units after wastewaters have been treated in aggressive biological treatment units) and F037, K048, and K051 wastes are not included in this listing. | (T) |

[1] See footnote on page 159.

**Appendix B2. Hazardous and toxic wastes** *(continued)*

| Industry and EPA hazardous waste No. | Hazardous waste | Hazard code |
|---|---|---|
| F039 ........... | Leachate (liquids that have percolated through land disposed wastes) resulting from the disposal of more than one restricted waste classified as hazardous under subpart D of this part. (Leachate resulting from the disposal of one or more of the following EPA Hazardous Wastes and no other Hazardous Wastes retains its EPA Hazardous Waste Number(s): F020, F021, F022, F026, F027, and/or F028.). | (T) |

[1] The F032, F034, and F305 listings are administratively stayed with respect to the process area receiving drippage of these wastes provided persons desiring to continue operating notify EPA by August 6, 1991 of their intent to upgrade or install drip pads, and by November 6, 1991 provide evidence to EPA that they have adequate financing to pay for drip pad upgrades or installation, as provided in the administrative stay. The stay of the listings will remain in effect until February 6, 1992 for existing drip pads and until May 6, 1992 for new drip pads.

* (I,T) should be used to specify mixtures containing ignitable and toxic constituents.

(b) Listing Specific Definitions: (1) For the purposes of the F037 and F038 listings, oil/water/solids is defined as oil and/or water and/or solids.

(2) (i) For the purposes of the F037 and F038 listings, aggressive biological treatment units are defined as units which employ one of the following four treatment methods: activated sludge; trickling filter; rotating biological contactor for the continuous accelerated biological oxidation of wastewaters; or high-rate aeration. High-rate aeration is a system of surface impoundments or tanks, in which intense mechanical aeration is used to completely mix the wastes, enhance biological activity, and (A) the units employs a minimum of 6 hp per million gallons of treatment volume; and either (B) the hydraulic retention time of the unit is no longer than 5 days; or (C) the hydraulic retention time is no longer than 30 days and the units does not generate a sludge that is a hazardous waste by the Toxicity Characteristic.

(ii) Generators and treatment, storage and disposal facilities have the burden of proving that their sludges are exempt from listing as F037 and F038 wastes under this definition. Generators and treatment, storage and disposal facilities must maintain, in their operating or other onsite records, documents and data sufficient to prove that: (A) the unit is an aggressive biological treatment unit as defined in this subsection; and (B) the sludges sought to be exempted from the definitions of F037 and/or F038 were actually generated in the aggressive biological treatment unit.

(3) (i) For the purposes of the F037 listing, sludges are considered to be generated at the moment of deposition in the unit, where deposition is defined as at least a temporary cessation of lateral particle movement.

**Appendix B2. Hazardous and toxic wastes** *(continued)*

(ii) For the purposes of the F038 listing,

(A) sludges are considered to be generated at the moment of deposition in the unit, where deposition is defined as at least a temporary cessation of lateral particle movement and

(B) floats are considered to be generated at the moment they are formed in the top of the unit.

[46 FR 4617, Jan. 16, 1981]

Editorial Note: For Federal Register citations affecting § 261.31, see the List of CFR Sections Affected in the Finding Aids section of this volume.

**§ 261.32 Hazardous wastes from specific sources.**

The following solid wastes are listed hazardous wastes from specific sources unless they are excluded under §§ 260.20 and 260.22 and listed in appendix IX.

| Industry and EPA hazardous waste No. | Hazardous waste | Hazard code |
|---|---|---|
| Wood preservation: | | |
| K001 | Bottom sediment sludge from the treatment of wastewaters from wood preserving processes that use creosote and/or pentachlorophenol. | (T) |
| Inorganic pigments: | | |
| K002 | Wastewater treatment sludge from the production of chrome yellow and orange pigments. | (T) |
| K003 | Wastewater treatment sludge from the production of molybdate orange pigments. | |
| K004 | Wastewater treatment sludge from the production of zinc yellow pigments. | (T) |
| K005 | Wastewater treatment sludge from the production of chrome green pigments. | (T) |
| K006 | Wastewater treatment sludge from the production of chrome oxide green pigments (anhydrous and hydrated). | (T) |
| K007 | Wastewater treatment sludge from the production of iron blue pigments. | (T) |
| K008 | Oven residue from the production of chrome oxide green pigments. | (T) |
| Organic chemicals: | | |
| K009 | Distillation bottoms from the production of acetaldehyde from ethylene. | (T) |
| K010 | Distillation side cuts from the production of acetaldehyde from ethylene. | (T) |

## Appendix B2. Hazardous and toxic wastes *(continued)*

| Industry and EPA hazardous waste No. | Hazardous waste | Hazard code |
|---|---|---|
| K011 ........... | Bottom stream from the wastewater stripper in the production of acrylonitrile. | (R,T) |
| K013 ........... | Bottom stream from the acetonitrile column in the production of acrylonitrile. | (R,T) |
| K014 ........... | Bottoms from the acetonitrile purification column in the production of acrylonitrile. | (T) |
| K015 ........... | Still bottoms from the distillation of benzyl chloride. | (T) |
| K016 ........... | Heavy ends or distillation resides from the production of carbon tetrachloride | (T) |
| K017 ........... | Heavy ends (still bottoms) from the purification column in the production of epichlorohydrin. | (T) |
| K018 ........... | Heavy ends from the fractionation column in ethyl chloride production. | (T) |
| K019 ........... | Heavy ends from the distillation of ethylene dichloride in ethylene dichloride production. | (T) |
| K020 ........... | Heavy ends from the distillation of vinyl chloride in vinyl chloride monomer production. | (T) |
| K021 ........... | Aqueous spent antimony catalyst waste from fluoromethanes production. | (T) |
| K022 ........... | Distillation bottom tars from the production of phenol/acetone from cumene. | (T) |
| K023 ........... | Distillation light ends from the production of phthalic anhydride from napthalene. | (T) |
| K024 ........... | Distillation bottoms from the production of phthalic anhydride from napthalene. | (T) |
| K025 ........... | Distillation bottoms from the production of nitrobenzene by the nitration of benzene. | (T) |
| K026 ........... | Stripping still tails from the production of methy ethyl pyridines. | (T) |
| K027 ........... | Centrifuge and distillation residues from toluene diisocyanate production. | (R,T) |
| K028 ........... | Spent catalyst from the hydrochlorinator reactor in the production of 1,1,1-Trichloroethane. | (T) |
| K029 ........... | Waste from the product steam stripper in the production of 1,1,1-trichloroethane. | (T) |
| K030 ........... | Column bottoms or heavy ends from the combined production of trichloroethylene and perchloroethylene. | |
| K083 ........... | Distillation bottoms from aniline production. | (T) |
| K085 ........... | Distillation or fractionation column bottoms from the production of chlorobenzenes. | (T) |
| K093 ........... | Distillation light ends from the production of phthalic anhydride from ortho-xylene. | (T) |
| K094 ........... | Distillation bottoms from the production of phthalic anhydride from ortho-xylene. | (T) |
| K095 ........... | Distillation bottoms from the production of 1,1,1-trichloroethane. | (T) |
| K096 ........... | Heavy ends from the heavy ends column from the production of 1,1,1-trichloroethane. | (T) |

## Appendix B2. Hazardous and toxic wastes *(continued)*

| Industry and EPA hazardous waste No. | Hazardous waste | Hazard code |
|---|---|---|
| K103 . . . . . . . . . . . | Process residues from aniline extraction from the production of aniline. | (T) |
| K104 . . . . . . . . . . . | Combined wastewater streams generated from nitrobenzene/aniline production. | (T) |
| K105 . . . . . . . . . . . | Separated aqueous stream from the reactor product washing step in the production of chlorobenzenes. | (T) |
| K107 . . . . . . . . . . . | Column bottoms from product separation from the production of 1,1-dimethylhydrazine (UDMH) from carboxylic acid hydrazines. | (C,T) |
| K108 . . . . . . . . . . . | Condensed column overheads from product separation and condensed reactor vent gases from the production of 1,1-dimethylhydrazine (UDMH) from carboxylic acid hydrazides. | (I,T) |
| K109 . . . . . . . . . . . | Spent filter cartridges from product purification from the production of 1,1-dimethylhydrazine (UDMH) from carboxylic acid hydrazides. | (T) |
| K110 . . . . . . . . . . . | Condensed column overheads from intermediate separation from the production of 1,1-dimethylhydrazine (UDMH) from carboxylic acid hydrazides. | (T) |
| K111 . . . . . . . . . . . | Product washwaters from the production of dinitrotoluene via nitration of toluene. | (C,T) |
| K112 . . . . . . . . . . . | Reaction by-product water from the drying column in the production of toluenediamine via hydrogenation of dinitrotoluene. | (T) |
| K113 . . . . . . . . . . . | Condensed liquid light ends from the purification of toluenediamine in the production of toluenediamine via hydrogenation of dinitrotoluene. | (T) |
| K114 . . . . . . . . . . . | Vicinals from the purification of toluenediamine in the production of toluenediamine via hydrogenation of dinitrotoluene. | (T) |
| K115 . . . . . . . . . . . | Heavy ends from the purification of toluenediamine in the production of toluenediamine via hydrogenation of dinitrotoluene. | (T) |
| K116 . . . . . . . . . . . | Organic condesate from the solvent recovery column in the production of toluene diisocyanate via phosgenation of toluenediamine. | (T) |
| K117 . . . . . . . . . . . | Wastewater from the reactor vent gas scrubber in the productin of ethylene dibromide via bromination of ethene. | (T) |
| K118 . . . . . . . . . . . | Spent adsorbent solids from purification of ethylene dibromide in the production of ethylene dibromide via bromination of ethene. | (T) |
| K136 . . . . . . . . . . . | Still bottoms from the purification of ethylene dibromide in the production of ethylene dibromide via bromination of ethene. | (T) |
| Inorganic chemicals: | | |
| K071 . . . . . . . . . . . | Brine purification muds from the mercury cell process in chlorine production, where separately prepurified brine is not used. | (T) |

## Appendix B2. Hazardous and toxic wastes *(continued)*

| Industry and EPA hazardous waste No. | Hazardous waste | Hazard code |
|---|---|---|
| K073 ........... | Chlorinated hydrocarbon waste from the purification step of the diaphragm cell process using graphite anodes in chlorine production. | (T) |
| K106 ........... | Wastewater treatment sludge from the mercury cell process in chlorine production. | (T) |
| Pesticides: | | |
| K031 ........... | By-product salts generated in the production of MSMA and cacodylic acid. | (T) |
| K032 ........... | Wastewater treatment sludge from the production of chlordane. | (T) |
| K033 ........... | Wastewater and scrub water from the chlorination of cyclopentadiene in the production of chlordane. | (T) |
| K034 ........... | Filter solids from the filtration of hexachlorocyclopentadiene in the production of chlordane. | (T) |
| K035 ........... | Wastewater treatment sludge generated in the production of creosote. | (T) |
| K036 ........... | Still bottoms from toluene reclamation distillation in the production of disulfoton. | (T) |
| K037 ........... | Wastewater treatment sludges from the production of disulfoton. | (T) |
| K038 ........... | Wastewater from the washing and stripping of phorate production. | (T) |
| K039 ........... | Filter cake from the filtration of diethylphosphorodithoic acid in the production of phorate. | (T) |
| K040 ........... | Wastewater treatment sludge from the production of phorate. | (T) |
| K041 ........... | Wastewater treatment sludge from the production of toxaphene. | (T) |
| K042 ........... | Heavy ends or distillation residues from the distillation of tetrachlorobenzene in the production of 2,4,5-T. | (T) |
| K043 ........... | 2,6-Dichlorophenol waste fromthe production of 2,4-D. | (T) |
| K097 ........... | Vacuum stripper discharge from the chlordane chlorinator in the production of chlordane. | (T) |
| K098 ........... | Untreated process wastewater from the production of toxaphene. | (T) |
| K099 ........... | Untreated wastewater from the production of 2,4-D. | (T) |
| K123 ........... | Process wastewater (including supernates, filtrates, and washwaters) from the production of ethylenebisdithiocarbamic acid and its salt. | (T) |
| K124 ........... | Reactor vent scrubber water from the production of ethylenebisdithiocarbamic acid and its salt. | (C,T) |
| K125 ........... | Filtration, evaporation, and centrifugation solids from the production of ethylenebisdithiocarbamic acid and its salts. | (T) |
| K126 ........... | Baghouse dust and floor sweepings in milling and packaging operation from the production of formulation of ethylenebisdithiocarbamic acid and its salts. | (T) |
| K131 ........... | Wastewater from the reactor and spent sulfuric acid from the acid dryer from the production of methyl bromide. | (C,T) |

## Appendix B2. Hazardous and toxic wastes *(continued)*

| Industry and EPA hazardous waste No. | Hazardous waste | Hazard code |
|---|---|---|
| K132 | Spent absorbent and wastewater separator solids from the production of methyl bromide. | (T) |
| Explosives: | | |
| K044 | Wastewater treatment sludges from the manufacturing and processing of explosives. | (R) |
| K045 | Spent carbon from the treatment of wastewater containing explosives. | (R) |
| K046 | Wastewater treatment sludges from the manufacturing, formulation and loading of lead-based initiating compounds. | (T) |
| K047 | Pink/red water from TNT operations. | (R) |
| Petroleum refining: | | |
| K048 | Dissolved air flotation (DAF) float from the petroleum refining industry. | (T) |
| K049 | Slop oil emulsion solids from the petroleum refining industry. | (T) |
| K050 | Heat exchanger bundle cleaning sludge from the petroleum refining industry. | (T) |
| K051 | API separator sludge from the petroleum refining industry. | (T) |
| K052 | Tank bottoms (leaded from the petroleum refining industry. | (T) |
| Iron and steel: | | |
| K061 | Emission control dust/sludge from the primary production of steel in electric furnaces. | (T) |
| K062 | Spent pickle liquor generated by steel finishing operations of facilities within the iron and steel industry (SIC Codes 331 and 332). | (C,T) |
| Primary Copper: | | |
| K064 | Acid plant blowdown slurry/sludge resulting from the thickening of blowdown slurry from primary copper production. | (T) |
| Primary Lead: | | |
| K065 | Surface impoundment solids contained in and dredged from surface impoundments at primary lead smelting facilities. | (T) |
| Primary zinc: | | |
| K066 | Sludge from treatment of process wastewater and/or acid plant blowdown from primary zinc production. | (T) |
| Primary aluminum: | | |
| K088 | Spent potliners from primary aluminum reduction. | (T) |
| Ferroalloys: | | |
| K090 | Emission control dust or sludge from ferrochromiumsilicon production. | (T) |
| K091 | Emission control dust or sludge from ferrochromium production. | (T) |

**Appendix B2. Hazardous and toxic wastes** *(continued)*

| Industry and EPA hazardous waste No. | Hazardous waste | Hazard code |
|---|---|---|
| Secondary lead: | | |
| K069 ........... | Emission control dust/sludge from secondary lead smelting. (Note: This listing is stayed administratively for sludge generated from secondary acid scrubber systems. The stay will remain in effect until further administrative action is taken. If EPA takes further action effecting this stay, EPA will publish a notice of the action in the Federal Register. | (T) |
| K100 ........... | Waste leaching solution from acid leaching of emission control dust/sludge from secondary lead smelting. | (T) |
| Veterinary pharmaceuticals: | | |
| K084 ........... | Wastewater treatment sludges generated during the production of veterinary pharmaceuticals from arsenic or organo-arsenic compounds. | (T) |
| K101 ........... | Distillation tar residues from the distillation of aniline-based compounds in the production of veterinary pharmaceuticals from arsenic or organo-arsenic compounds. | (T) |
| K102 ........... | Residue from the use of activitated carbon for decolorization in the production of veterinary pharmaceuticals from arsenic or organo-arsenic compounds. | (T) |
| Ink formulation: | | |
| K086 ........... | Solvent washes and sludges, caustic washes and sludges, or water washes and sludges from cleaning tubs and equipment used in the formulation of ink from pigments, driers, soaps, and stabilizers containing chromium and lead. | (T) |
| Coking: | | |
| K060 ........... | Ammonia still lime sludge from coking operations. | (T) |
| K087 ........... | Decanter tank tar sludge from coking operations. | (T) |

[46 FR 4618, Jan. 16, 1991]

Editorial Note: For Federal Register citations affecting § 261.32, see the list of CFR Sections Affected in the Finding Aids section of this volume.

### § 261.33 Discarded commerical chemical products, off-specification species, container residues, and spill residues thereof.

The following materials or items are hazardous wastes if and when they are discarded or intedned to be discarded as described in § 261.2(a)(2)(i), when they are mixed with waste oil or used oil or other material and applied to the land for dust suppression or road treatment, when they are otherwise applied to

**Appendix B2. Hazardous and toxic wastes** *(continued)*

the land in lieu of their original intended use or when they are contained in products that are applied to the land in lieu of their original intended use, or when, in lieu of their original intended use, they are produced for use as (or as a component of) a fuel, distributed for use as a fuel, or burned as a fuel.

(a) Any commercial chemical product, or manufacturing chemical intermediate having the generic name listed in paragraph (e) or (f) of this section.

(b) Any off-specification commercial chemical product or manufacturing chemical intermediate which, if it met specifications, would have the generic name listed in paragraph (e) or (f) of this section.

(c) Any residue remaining in a container or in an inner liner removed from a container that has held any commercial chemical product or manufacturing chemical intermediate having the generic name listed in paragraphs (e) or (f) of this section, unless the container is empty as defined in 261.7 (b) of this chapter.

[*Comment:* Unless the residue is being beneficially used or reused, or legitimately recycled or reclaimed; or being accumulated, stored, transported or treated prior to such use, re-use, recycling or reclamation, EPA considers the residue to be intended for discard, and thus, a hazardous waste. An example of a legitimate re-use of the residue would be where the residue remains in the container and the container is used to hold the same commercial chemical product or manufacturing chemical intermediate it previously held. An example of the discard of the residue would be where the drum is sent to a drum reconditioner who reconditions the drum but discards the residue.]

(d) Any residue or contaminated soil, water or other debris resulting from the cleanup of a spill into or on any land or water of any commercial chemical product or manufacturing chemical intermediate having the generic name listed in paragraph (e) or (f) of this section, or any residue or contaminated soil, water or other debris resulting from the cleanup of a spill, into or on any land or water, of any off-specification chemical product and manufacturing chemical intermediate which, if it met specifications, would have the generic name listed in paragraph (e) or (f) of this section.

[Comment: The phrase "commercial chemical product or manufacturing chemical, intermediate having the generic name listed in . . ." refers to a chemical substance which is manufactured or formulated for commercial or manufacturing use which consists of the commerically pure grade of the chemical, any technical grades of the chemical that are produced or marketed, and all formulations in which the chemical is the sole active ingredient. It does not refer to a material, such as a manufacturing process waste that contains any of the substanced listed in paragraph (e) or (f). Where a manufacturing process waste is deemed to be a hazardous waste because it contains a substance listed in paragraph (e)

**Appendix B2. Hazardous or toxic wastes** *(continued)*

or (f), such waste will be listed in either 261.31 or § 261.32 or will be identified as a hazardous waste by the characteristics set forth in subpart C of this part.]

(e) The commercial chemical products, manufacturing chemical intermediates or off-specification commercial chemical products or manufacturing chemical intermediates referred to in paragraphs (a) through (d) of this section, are identified as acute hazardous wastes (H) and are subject to be the small quantity exclusion defined in § 261.5(e).

[Comment: For the convenience of the regulated community the primary hazardous properties of these materials have been indicated by the letters T (Toxicity), and R (Reactivity). Absence of a letter indicates that the compound only is listed for acute toxicity.]

These wastes and their corresponding EPA Hazardous Waste Numbers are:

| Hazardous waste No. | Chemical abstracts No. | Substance |
|---|---|---|
| P023 | 107-20-0 | Acetaldehyde, chloro- |
| P002 | 591-08-2 | Acetamide, N-(aminothioxomethyl)- |
| P057 | 640-19-7 | Acetamide, 2-fluoro- |
| P058 | 62-74-8 | Acetic acid, fluoro-, sodium salt |
| P002 | 591-08-2 | 1-Acetyl-2-thiourea |
| P003 | 107-02-8 | Acrolein |
| P070 | 116-06-3 | Aldicarb |
| P004 | 309-00-2 | Aldrin |
| P005 | 107-18-6 | Allyl alcohol |
| P006 | 20859-73-8 | Aluminum Phosphide (R,T) |
| P007 | 2763-96-4 | 5-(Aminomethyl)-3-isoxazolol |
| P008 | 504-24-5 | 4-Aminopyridine |
| P009 | 131-74-8 | Ammonium picrate (R) |
| P119 | 7803-55-6 | Ammonium vanadate |
| P099 | 506-61-6 | Argentate(1)-, bis(cyano-C)-, potassium |
| P010 | 7778-39-4 | Arsenic acid, $H_3AsO_4$ |
| P012 | 1327-53-3 | Arsenic oxide $As_2O_3$ |
| P011 | 1303-28-2 | Arsenic oxide $As_2O_3$ |
| P011 | 1303-28-2 | Arsenic pentoxide |
| P012 | 1327-53-3 | Arsenic trioxide |
| P038 | 692-42-2 | Arsine, diethyl- |
| P036 | 696-28-6 | Arsonous dichloride, phenyl- |
| P054 | 151-56-4 | Aziridine |
| P067 | 75-55-8 | Aziridine, 2-methyl- |
| P013 | 542-62-1 | Barium cyanide |
| P024 | 106-47-8 | Benzenamine, 4-chloro- |
| P077 | 100-01-6 | Benzenamine, 4-nitro- |

**Appendix B2. Hazardous and toxic wastes** *(continued)*

| Hazardous waste No. | Chemical abstracts No. | Substance |
|---|---|---|
| P028 | 100-44-7 | Benzene, (chloromethyl)- |
| P042 | 51-43-4 | 1,2-Benzenediol, 4-[1-hydroxy-2-(methylamino)ethyl]-, (R)- |
| P046 | 122-09-8 | Benzeneethanamine, alpha, alpha-dimethyl- |
| P014 | 108-98-5 | Benzenethiol |
| P001 | [1]81-81-2 | 2H-1-Benzopyran-2-one, 4-hydroxy-3-(-3-oxo-1-phenylbutyl)-, & salts, when present at concentrations greater than 0.3% |
| P028 | 100-44-7 | Benzyl chloride |
| P015 | 7440-41-7 | Beryllium |
| P017 | 598-31-2 | Bromoacetone |
| P018 | 357-57-3 | Brucine |
| P045 | 39196-18-4 | 2-Butanone, 3,3-dimethyl-1-(-methylthio)-, O-[methylamino)carbony] oxime |
| P021 | 592-01-8 | Calcium cyanide |
| P021 | 592-01-8 | Calcium cyanide $Ca(CN)_2$. |
| P022 | 75-15-0 | Carbon disulfide |
| P095 | 75-44-5 | Carbonic dischloride |
| P023 | 107-20-0 | Chloroacetaldehyde |
| P024 | 106-47-8 | p-Chloroaniline |
| P026 | 5344-82-1 | 1-(o-Chlorophenyl)thiourea |
| P027 | 542-76-7 | 3-Chloropropionitrile |
| P029 | 544-92-3 | Copper cyanide |
| P029 | 544-92-3 | Copper cyanide Cu(CN) |
| P030 | .......... | Cyanides (soluble cyanide salts), not otherwise specified |
| P031 | 460-19-5 | Cyanogen |
| P033 | 506-77-4 | Cyanogen chloride |
| P033 | 506-77-4 | Cyanogen chloride (CN)CI |
| P034 | 131-89-5 | 2-Cyclohexyl-4,6-dinitrophenol |
| P016 | 542-88-1 | Dichloromethyl ether |
| P036 | 696-28-6 | Dichlorophenylarsine |
| P037 | 60-57-1 | Dieldrin |
| P038 | 692-42-2 | Diethylarsine |
| P041 | 311-45-5 | Diethyl-p-nitrophenyl phosphate |
| P040 | 297-97-2 | O,O-Diethyl O-pyrazinyl phosphorothioate |
| P043 | 55-91-4 | Diisopropylfluorophosphate (DFP) |

[1] See footnote on page 173.

**Appendix B2. Hazardous and toxic wastes** *(continued)*

| Hazardous waste No. | Chemical abstracts No. | Substance |
|---|---|---|
| P004 | 309-00-2 | 1,4,5,8-Dimethanonapthalene, 1,2,3,4,10,10-hexa-chloro-1,4,4a,5,8,8a,-hexahydro-,(1alpha, 4alpha,4abeta,5alpha,8alpha,8abeta)- |
| P060 | 465-73-6 | 1,4,5,8-Dimethanonaphthalene, 1,2,3,4,10,10-hexa-chloro-1,4,4a,5,8,8a-hexahydro-,(1alpha,4alpha, 4abeta,5beta,8beta,8abeta)- |
| P037 | 60-57-1 | 2,7:3,6-Dimethanonaphth[2,3-b]oxirene, 3,4,5,6,9,9-Hexachloro-1a,2,2a,3,6,6a,7,7a-octahydro-, (1aalpha,2beta,2aalpha,3beta,6beta,6aalpha, 7beta,7aalpha)- |
| P051 | [1]72-20-8 | 2,7:3,6-Dimethanonaphth [2,3-b]oxirene, 3,4,5,6, 9,9-hexachloro- 1a,2,2a,3,6,6a,7,7a-octahydro-, (1aalpha,2beta,2abeta,3alpha,6alpha,6abeta, 7beta,7aalpha)-, & metabolites |
| P044 | 60-51-5 | Dimethoate |
| P046 | 122-09-8 | alpha,aplha-Dimethylphenethylamine |
| P047 | [1]534-52-1 | 4,6-Dinitro-o-cresol, & salts |
| P048 | 51-28-5 | 2,4-Dinitrophenol |
| P020 | 88-85-7 | Dinoseb |
| P085 | 152-16-9 | Diphosphoramide, octamethyl- P111 107-49-3 Diphosphoric acid, tetraethyl ester |
| P111 | 107-49-3 | Diphosphoric acid, tetraethyl ester |
| P039 | 298-04-4 | Disulfoton |
| P049 | 541-53-7 | Dithiobiuret |
| P050 | 115-29-7 | Endosulfan |
| P088 | 145-73-3 | Endothall |
| P051 | 72-20-8 | Endrin |
| P051 | 72-20-8 | Endrin, & metabolites |
| P042 | 51-43-4 | Epinephrine |
| P031 | 460-19-5 | Ethanedinitrile |
| P066 | 16752-77-5 | Ethanimidothioic acid, N-[[(methylamino)carbonyl]oxy]-, methyl ester |
| P101 | 107-12-0 | Ethyl cyanide |
| P054 | 151-56-4 | Ethyleneimine |
| P097 | 52-85-7 | Famphur |
| P056 | 7782-41-4 | Fluorine |
| P057 | 640-19-7 | Fluoroacetamide |
| P058 | 62-74-8 | Fluoroacetic acid, sodium salt |
| P065 | 628-86-4 | Fulminic acid, mercury(2+) salt (R,T) |
| P059 | 76-44-8 | Heptachlor |
| P062 | 757-58-4 | Hexaethyl tetraphosphate |
| P116 | 79-19-6 | Hydrazinecarbothioamide |

[1] See footnote on page 173.

**Appendix B2. Hazardous and toxic wastes** *(continued)*

| Hazardous waste No. | Chemical abstracts No. | Substance |
|---|---|---|
| P068 | 60-34-4 | Hydrazine, methyl- |
| P063 | 74-90-8 | Hydrocyanic acid |
| P063 | 74-90-8 | Hydrogen cyanide |
| P096 | 7803-51-2 | Hydrogen phosphide |
| P060 | 465-73-6 | Isodrin |
| P007 | 2763-96-4 | 3(2H)-Isoxazolone, 5-(aminomethyl)- |
| P092 | 62-38-4 | Mercury, (acetato-O)phenyl- |
| P065 | 628-86-4 | Mercury fulminate (R,T) |
| P082 | 62-75-9 | Methanamine, N-methyl-N-nitroso- |
| P064 | 624-83-9 | Methane, isocyanato- |
| P016 | 542-88-1 | Methane, oxybis[chloro- |
| P112 | 509-14-8 | Methane, tetranitro- (R) |
| P118 | 75-70-7 | Methanethiol, trichloro- |
| P050 | 115-29-7 | 6,9-Methano-2,4,3-benzodioxathiepin, 6,7,8,9,10,10-hexachloro-1,5,5a,6,9,9a-hexahydro-, 3-oxide |
| P059 | 76-44-8 | 4,7-Methano-1H-indene, 1,4,5,6,7,8,8-heptachloro-3a,4,7,7a-tetrahydro- |
| P066 | 16752-77-5 | Methomyl |
| P068 | 60-34-4 | Methyl hydrazine |
| P064 | 624-83-9 | Methyl isocyanate |
| P069 | 75-86-5 | 2-Methyllactonitrile |
| P071 | 298-00-0 | Methyl parathion |
| P072 | 86-88-4 | alpha-Naphthylthiourea |
| P073 | 13463-39-3 | Nickel carbonyl |
| P073 | 13463-39-3 | Nickel carbonyl $Ni(CO)_4$, (T-4)- |
| P074 | 557-19-7 | Nickel cyanide |
| P074 | 557-19-7 | Nickel cynaide $Ni(CN)_2$ |
| P075 | [1]54-11-5 | Nicotine & salts |
| P076 | 10102-43-9 | Nitric oxide |
| P077 | 100-01-6 | p-Nitroaniline |
| P078 | 10102-44-0 | Nitrogen dioxide |
| P076 | 10102-43-9 | Nitrogen oxide NO |
| P078 | 10102-44-0 | Nitrogen oxide $NO_2$ |
| P081 | 55-63-0 | Nitroglycerine (R) |
| P082 | 62-75-9 | N-Nitrosodimethylamine |
| P084 | 4549-40-0 | N-Nitrosomethylvinylamine |
| P085 | 152-16-9 | Octamethylpyrophosphoramide |
| P087 | 20816-12-0 | Osmium oxide $OsO_4$, (T-4)- |
| P087 | 20816-12-0 | Osmium tetroxide |
| P088 | 145-73-3 | 7-Oxabicyclo[2.2.1]heptane-2,3-dicarboxylic acid |
| P089 | 56-38-2 | Parathion |

[1] See footnote on page 173.

**Appendix B2. Hazardous and toxic wastes** *(continued)*

| Hazardous waste No. | Chemical abstracts No. | Substance |
|---|---|---|
| P034 | 131-89-5 | Phenol, 2-cyclohexyl-4,6-dinitro- |
| P048 | 51-28-5 | Phenol,2,4-dinitro- |
| P047 | [1]534-52-1 | Phenol, 2-methyl-4,6-dinitro-, & salts |
| P020 | 88-85-7 | Phenol, 2-(1-methylpropyl)-4,6-dinitro- |
| P009 | 131-74-8 | Phenol, 2,4,6-trinitro-, ammonium salt (R) |
| P092 | 62-38-4 | Phenylmercury acetate |
| P093 | 103-85-5 | Phenylthiourea |
| P094 | 298-02-2 | Phorate |
| P095 | 75-44-5 | Phosgene |
| P096 | 7803-51-2 | Phosphine |
| P041 | 311-45-5 | Phosphoric acid, diethyl 4-nitrophenyl ester |
| P039 | 298-04-4 | Phosphorodithioic acid, O,O-diethyl S-[2-(ethylthio)ethyl] ester |
| P094 | 298-02-2 | Phosphorodithioic acid, O,O-diethyl S-[(ethylthio)methyl] ester |
| P044 | 60-51-5 | Phosphorodithioic acid, O,O-dimethyl S-[2-(methylamino)-2-oxoethyl] ester |
| P043 | 55-91-4 | Phosphorofluoridic acid, bis(1-methylethyl) ester |
| P089 | 56-38-2 | Phosphorothioic acid, O,O-diethyl O-(4-nitrophenyl) ester |
| P040 | 297-97-2 | Phosphorothioic acid, O,O-diethyl O-pyrazinyl ester |
| P097 | 52-85-7 | Phosphorothioic acid, O-[4-[(dimethylamino) sulfonyl]phenyl] O,O-dimethyl ester |
| P071 | 298-00-0 | Phosphorothioic acid, O,O,-dimethyl O-(4-nitrophenyl) ester |
| P110 | 78-00-2 | Plumbane, tetraethyl- |
| P098 | 151-50-8 | Potassium cyanide |
| P098 | 151-50-8 | Potassium cyanide K(CN) |
| P099 | 506-61-6 | Potassium silver cyanide |
| P070 | 116-06-3 | Propanal, 2-methyl-2-(methylthio)-, O-[(methylamino)carbonl]oxime |
| P101 | 107-12-0 | Propanenitrile |
| P027 | 542-76-7 | Propanenitrile, 3-chloro- |
| P069 | 75-86-5 | Propanenitrile,2-hydroxy-2-methyl- |
| P081 | 55-63-0 | 1,2,3-Propanetriol, trinitrate (R) |
| P017 | 598-31-2 | 2-Propanone, 1-bromo- |
| P102 | 107-19-7 | Propargyl alcohol |
| P003 | 107-02-8 | 2-Propenal |
| P005 | 107-18-6 | 2-Propen-1-ol |
| P067 | 75-55-8 | 1,2-Propylenimine |

[1] See footnote on page 173.

**Appendix B2. Hazardous and toxic wastes** *(continued)*

| Hazardous waste No. | Chemical abstracts No. | Substance |
|---|---|---|
| P102 | 107-19-7 | 2-Propyn-1-ol |
| P008 | 504-24-5 | 4-Pyridinamine |
| P075 | [1]54-11-5 | Pyridine,3-(1-methyl-2-pyrrolidinyl)-(S)-, & salts |
| P114 | 12039-52-0 | Selenious acid, dithallium(1 + ) salt |
| P103 | 630-10-4 | Selenourea |
| P104 | 506-64-9 | Silver cyanide |
| P104 | 506-64-9 | Silver cyanide Ag(CN) |
| P105 | 26628-22-8 | Sodium azide |
| P106 | 143-33-9 | Sodium cyanide |
| P106 | 143-33-9 | Sodium cyanide Na(CN) |
| P108 | [1]57-24-9 | Strychnidin-10-one, & salts |
| P018 | 357-57-3 | Strychnidin-10-one,2,3-dimethoxy- |
| P108 | [1]57-24-9 | Strychnine, & salts |
| P115 | 7446-18-6 | Sulfuric acid, dithallium(1 + ) salt |
| P109 | 3689-24-5 | Tetraethyldithiopyrophosphate |
| P110 | 78-00-2 | Tetraethyl lead |
| P111 | 107-49-3 | Tetraethyl pyrophosphate |
| P112 | 509-14-8 | Tetranitromethane (R) |
| P062 | 757-58-4 | Tetraphosphoric acid, hexaethyl ester |
| P113 | 1314-32-5 | Thallic oxide |
| P113 | 1314-32-5 | Thallium oxide $Tl_2O_3$ |
| P114 | 12039-52-0 | Thallium(I) selenite |
| P115 | 7446-18-6 | Thallium(I) sulfate |
| P109 | 3689-24-5 | Thiodiphosphoric acid, tetraethyl ester |
| P045 | 39196-18-4 | Thiofanox |
| P049 | 541-53-7 | Thioimidodicarbonic diamide $[(H_2N)C(S)]_2NH$ |
| P014 | 108-98-5 | Thiophenol |
| P116 | 79-19-6 | Thiosemicarbazide |
| P026 | 5344-82-1 | Thiourea, (2-chlorophenyl)- |
| P072 | 86-88-4 | Thiourea, 1-naphthalenyl- |
| P093 | 103-85-5 | Thiourea, phenyl- |
| P123 | 8001-35-2 | Toxaphene |
| P118 | 75-70-7 | Trichloromethanethiol |
| P119 | 7803-55-6 | Vanadic acid, ammonium salt |
| P120 | 1314-62-1 | Vanadium oxide $V_2O_5$ |
| P120 | 1314-62-1 | Vanadium pentoxide |
| P084 | 4549-40-0 | Vinylamine, N-methyl-N-nitroso- |
| P001 | [1]81-81-2 | Warfarin, & salts, when present at concentrations greater than 0.3% |

[1] See footnote on page 173

**Appendix B2. Hazardous and toxic wastes** *(continued)*

| Hazardous waste No. | Chemical abstracts No. | Substance |
|---|---|---|
| P121 | 557-21-1 | Zinc cyanide |
| P121 | 557-21-1 | Zinc cyanide $Zn(CN)_2$ |
| P122 | 1314-84-7 | Zinc phosphide $Zn_3P_2$, when present at concentrations greater than 10% (R,T) |

[1] CAS Number given for parent compound only.

(f) The commercial chemical products, manufacturing chemical intermediates, or off-specification commercial chemical products reffered to in paragraphs (a) through (d) of this section, are identified as toxic wastes (T), unless otherwise designated and are subject to the small quantity generator exclusion defined in 261.5 (a) and (g).

[Comment: For the convenience of the regulated community, the primary hazardous properties of these materials have been indicated by the letters T (Toxicity), R (Reactivity), I (Ignitability) and C (Corrosivity). Absence of a letter indicates that the compound is only listed for toxicity.]

These wastes and their corresponding EPA Hazardous Waste Numbers are:

| Hazardous waste No. | Chemical abstracts No. | Substance |
|---|---|---|
| U001 | 75-07-0 | Acetaldehyde (I) |
| U034 | 75-87-6 | Acetaldehyde, trichloro- |
| U187 | 62-44-2 | Acetamide, N-(4-ethoxyphenyl)- |
| U005 | 53-96-3 | Acetamide, N-9H-fluoren-2-yl- |
| U240 | [1]94-75-7 | Acetic acid, (2,4-dichlorophenoxy)-, salts & esters |
| U112 | 141-78-6 | Acetic acid ethyl ester (I) |
| U144 | 301-04-2 | Acetic acid, lead(2 + ) salt |
| U214 | 563-68-8 | Acetic acid, thallium(1 + ) salt |
| see F027 | 93-76-5 | Acetic acid, (2,4,5-trichlorophenoxy)- |
| U002 | 67-64-1 | Acetone (I)) |
| U003 | 75-05-8 | Acetonitrile (I,T) |
| U004 | 98-86-2 | Acetophenone |
| U005 | 53-96-3 | 2-Acetylaminofluorene |
| U006 | 75-36-5 | Acetyl chloride (C,R,T) |
| U007 | 79-06-1 | Acrylamide |
| U008 | 79-10-7 | Acrylic acid (I) |
| U009 | 107-13-1 | Acrylonitrile |

**Appendix B2. Hazardous and toxic wastes** *(continued)*

| Hazardous waste No. | Chemical abstracts No. | Substance |
|---|---|---|
| U011 | 61-82-5 | Amitrole |
| U012 | 62-53-3 | Aniline (I,T) |
| U136 | 75-60-5 | Arsinic acid, dimethyl- |
| U014 | 492-80-8 | Auramine |
| U015 | 115-02-6 | Azaserine |
| U010 | 50-07-7 | Azirino[2′,3′:3,4]pyrrolo[1,2-a]indole-4,7-dione, 6-amino-8-[[(aminocarbonyl)oxy]methyl]-1,1a, 2,8,8a,8b-hexahydro-8a-methoxy-5-methyl-, [1aS-(1aalpha, 8beta,8aalpha,8balpha)]- |
| U157 | 56-49-5 | Benz[j]aceanthrylene, 1,2-dihydro-3-methyl- |
| U016 | 255-51-4 | Benz[c]acridine |
| U017 | 98-87-3 | Benzal chloride |
| U192 | 23950-58-5 | Benzamide, 3,5-dichloro-N-(1,1-dimethyl-2-propynyl)- |
| U018 | 56-55-3 | Benz[a]anthracene |
| U094 | 57-97-6 | Benz[a]anthracene,7,12-dimethyl- |
| U012 | 62-53-3 | Benzenamine (I,T) |
| U014 | 492-80-8 | Benzenamine, 4,4′-carbonimidoylbis[N,N-dimethyl- |
| U049 | 3165-93-3 | Benzenamine, 4-chloro-2-methyl-, hydrochloride |
| U093 | 60-11-7 | Benzenamine, N,N-dimethyl-4-(phenylazo)- |
| U328 | 95-53-4 | Benzenamine, 2-methyl- |
| U353 | 106-49-0 | Benzenamine, 4-methyl- |
| U158 | 101-14-4 | Benzenamine, 4,4′-methylenebis[2-chloro- |
| U222 | 636-21-5 | Benzenamine, 2-methyl-, hydrochloride |
| U181 | 99-55-8 | Benzenamine, 2-methyl-5-nitro- |
| U019 | 71-43-2 | Benzene (I,T) |
| U038 | 510-15-6 | Benzeneacetic acid, 4-chloro-alpha-(4-chlorophenyl)-alpha-hydroxy-, ehtyl ester |
| U030 | 101-55-3 | Benzene, 1-bromo-4-phenoxy- |
| U035 | 305-03-3 | Benzenebutanoic acid, 4-[bis(2-chloroethy)amino]- |
| U037 | 108-90-7 | Benzene, chloro- |
| U221 | 25376-45-8 | Benzenediamine, ar-methyl- |
| U028 | 117-81-7 | 1,2-Benzenedicarboxylic acid, bis(2-ethylhexyl) ester |
| U069 | 84-74-2 | 1,2-Benzenedicarboxylic acid, dibutyl ester |
| U088 | 84-66-2 | 1,2-Benzenedicarboxylic acid, diethyl ester |
| U102 | 131-11-3 | 1,2-Benzenedicarboxylic acid, dimethyl ester |
| U107 | 117-84-0 | 1,2-Benzenedicarboxylic acid, dioctyl ester |
| U070 | 95-50-1 | Benzene, 1,2-dichloro- |
| U071 | 541-73-1 | Benzene, 1,3-dichloro- |
| U072 | 106-46-7 | Benzene, 1,4-dichloro- |

**Appendix B2. Hazardous and toxic wastes** *(continued)*

| Hazardous waste No. | Chemical abstracts No. | Substance |
|---|---|---|
| U060 | 72-54-8 | Benzene, 1,1′-(2,2-dichloroethylidene)bis[4-chloro- |
| U017 | 98-87-3 | Benzene, (dichloromethyl)- |
| U223 | 26471-62-5 | Benzene, 1,3-diisocyanatomethyl- (R,T) |
| U239 | 1330-20-7 | Benzene, dimethyl- (I,T) |
| U201 | 108-46-3 | 1,3-Benzenediol |
| U127 | 118-74-1 | Benzene, hexachloro- |
| U056 | 110-82-7 | Benzene, hexahydro- (I) |
| U220 | 108-88-3 | Benzene, methyl- |
| U105 | 121-14-2 | Benzene, 1-methyl-2,4-dinitro- |
| U106 | 606-20-2 | Benzene, 2-methyl-1,3-dinitro- |
| U055 | 98-82-8 | Benzene, (1-methylethyl)- (I) |
| U169 | 98-95-3 | Benzene, nitro- |
| U183 | 608-93-5 | Benzene, pentachloro- |
| U185 | 82-68-8 | Benzene, pentachloronitro- |
| U020 | 98-09-9 | Benzenesulfonic acid chloride (C,R) |
| U020 | 98-09-9 | Benzenesulfonyl chloride (C,R) |
| U207 | 95-94-3 | Benzene, 1,2,4,5-tetrachloro- |
| U061 | 50-29-3 | Benzene, 1,1′-(2,2,2-trichloroethylidene) bis[4-chloro- |
| U247 | 72-43-5 | Benzene, 1,1′-(2,2,2-trichloroethylidene) bis[4-methoxy- |
| U023 | 98-07-7 | Benzene, (trichloromethyl)- |
| U234 | 99-35-4 | Benzene, 1,3,5-trinitro- |
| U021 | 92-87-5 | Benzidine |
| U202 | [1]81-07-2 | 1,2-Benzisothiazol-3(2H)-one, 1,1-dioxide, & salts |
| U203 | 94-59-7 | 1,3-Benzodioxole, 5-(2-propenyl)- |
| U141 | 120-58-1 | 1,3-Benzodioxole, 5-(1-propenyl) |
| U090 | 94-58-6 | 1,3-Benzodioxole, 5-propyl- |
| U064 | 189-55-9 | Benzo[rst]pentaphene |
| U248 | [1]81-81-2 | 2H-1-Benzopyran-2-one, 4-hydroxy-3-(3-oxo-1-phenyl-butyl)-, & salts when present at concentrations of 0.3% or less |
| U022 | 50-32-8 | Benzo[a]pyrene |
| U197 | 106-51-4 | p-Benzoquinone |
| U023 | 98-07-7 | Benzotrichloride (C,R,T) |
| U085 | 1464-53-5 | 2,2′-Bioxirane |
| U021 | 92-87-5 | [1,1′-Biphenyl]-4,4′-diamine |
| U073 | 91-94-1 | [1,1′-Biphenyl]-4,4′-diamine, 3,3′-dichloro- |
| U091 | 119-90-4 | [1,1′-Biphenyl]-4,4′-diamine, 3,3′-dimethoxy- |
| U095 | 119-93-7 | [1,1′-Biphenyl]-4,4′-diamine, 3,3′-dimethyl- |
| U225 | 75-25-2 | Bromoform |

[1] See footnote on page 185.

**Appendix B2. Hazardous and toxic wastes** *(continued)*

| Hazardous waste No. | Chemical abstracts No. | Substance |
|---|---|---|
| U030 | 101-55-3 | 4-Bromophenyl phenyl ether |
| U128 | 87-68-3 | 1,3-Butadiene, 1,1,2,3,4,4-hexachloro- |
| U172 | 924-16-3 | 1-Butanamine, N-butyl-N-nitroso- |
| U031 | 71-36-3 | 1-Butanol (I) |
| U159 | 78-93-3 | 2-Butanone (I,T) |
| U160 | 1338-23-4 | 2-Butanone, peroxide (R,T) |
| U053 | 4170-30-3 | 2-Butenal |
| U074 | 764-41-0 | 2-Butene, 1,4-dichloro- (I,T) |
| U143 | 303-34-4 | 2-Butenoic acid, 2-methyl-, 7-[&2,3-dihydroxy-2-(1-methoxyethyl-3-methyl-1-oxobutoxy]methyl]-2,3,5,7a-tetrahydro-1H-pyrrolizin-1-yl ester, [1S-[1alpha(Z),7(2S*,3R*),7aalpha]]- |
| U031 | 71-36-3 | n-Butyl alcohol (I) |
| U136 | 75-60-5 | Cacodylic acid |
| U032 | 13765-19-0 | Calcium chromate |
| U238 | 51-79-6 | Carbamic acid, ethyl ester |
| U178 | 615-53-2 | Carbamic acid, methylnitroso-, ethyl ester |
| U097 | 79-44-7 | Carbamic chloride, dimethyl- |
| U114 | [1]111-54-6 | Carbamodithioic acid, 1,2-ethanediylbis-, salts & esters |
| U062 | 2303-16-4 | Carbamothioic acid, bis(1-methylethyl)-, S(2,3-dichloro-2-propenyl) ester |
| U215 | 6533-73-9 | Carbonic acid, dithallium(1 + ) salt) |
| U033 | 353-50-4 | Carbonic difluoride |
| U156 | 79-22-1 | Carbonochloridic acid, methyl ester (I,T) |
| U033 | 353-50-4 | Carbon oxyfluoride (R,T) |
| U211 | 56-23-5 | Carbon tetrachloride |
| U034 | 75-87-6 | Chloral |
| U035 | 305-03-3 | Chlorambucil |
| U036 | 57-74-9 | Chlordane, alpha & gamma isomers |
| U026 | 494-03-1 | Chlornaphazin |
| U037 | 108-90-7 | Chlorobenzene |
| U038 | 510-15-6 | Chlorobenzilate |
| U039 | 59-50-7 | p-chloro-m-cresol |
| U042 | 110-75-8 | 2-Chloroethyl vinyl ether |
| U044 | 67-66-3 | Chloroform |
| U046 | 107-30-2 | Chloromethyl methyl ether |
| U047 | 91-58-7 | beta-Chloronaphthalene |
| U048 | 95-57-8 | o-Chlorophenol |
| U049 | 3165-93-3 | 4-Chloro-o-toluidine, hydrochloride |
| U032 | 13765-19-0 | Chromic acid $H_2CrO_4$, calcium salt |
| U050 | 218-01-9 | Chrysene |

[1] See footnote on page 185.

**Appendix B2. Hazardous and toxic wastes** *(continued)*

| Hazardous waste No. | Chemical abstracts No. | Substance |
|---|---|---|
| U051 | ........... | Creosote |
| U052 | 1319-77-3 | Cresol (Cresylic acid) |
| U053 | 4170-30-3 | Crotonaldehyde |
| U055 | 98-82-8 | Cumene (I) |
| U246 | 506-68-3 | Cyanogen bromide (CN)Br |
| U197 | 106-51-4 | 2,5-Cylohexadiene-1,4-dione |
| U056 | 110-82-7 | Cyclohexane (I) |
| U129 | 58-89-9 | Cylohexane, 1,2,3,4,5,6-hexachloro-, (1alpha,2alpha,3beta,4alpha,5alpha,6beta)- |
| U057 | 108-94-1 | Cyclohexanone (I) |
| U130 | 77-47-4 | 1,3-Cyclopentadiene, 1,2,3,4,5,5-hexachloro- |
| U058 | 50-18-0 | Cyclophosphamide |
| U240 | [1]94-75-7 | 2,4-D, salts & esters |
| U059 | 20830-81-3 | Daunomycin |
| U060 | 72-54-8 | DDD |
| U061 | 50-29-3 | DDT |
| U062 | 2303-16-4 | Diallate |
| U063 | 53-70-3 | Dibenz[a,h]anthracene |
| U064 | 189-55-9 | Dibenzo[a,i]pyrene |
| U066 | 96-12-8 | 1,2-Dibromo-3-chloropropane |
| U069 | 84-74-2 | Dibutyl phthalate |
| U070 | 95-50-1 | o-Dichlorobenzene |
| U071 | 541-73-1 | m-Dichlorobenzene |
| U072 | 106-46-7 | p-Dichlorobenzene |
| U073 | 91-94-1 | 3,3′-Dichlorobenzidine |
| U074 | 764-41-0 | 1,4-Dichloro-2-butene (I,T) |
| U075 | 75-71-8 | Dichlorodifluoromethane |
| U078 | 75-35-4 | 1,1-Dichloroethylene |
| U079 | 156-60-5 | 1,2-Dichloroethylene |
| U025 | 111-44-4 | Dichloroethyl ether |
| U027 | 108-60-1 | Dichloroisopropyl ether |
| U024 | 111-91-1 | Dichloromethoxy ethane |
| U081 | 120-83-2 | 2,4-Dichlorophenol |
| U082 | 87-65-0 | 2,6-Dichlorophenol |
| U084 | 542-75-6 | 1,3-Dichloropropene |
| U085 | 1464-53-5 | 1,2:3,4-Diepoxybutane (I,T) |
| U108 | 123-91-1 | 1,4-Diethyleneoxide |
| U028 | 117-81-7 | Diethylhexyl phthalate |
| U086 | 1615-80-1 | N,N′-Diethylhydrazine |
| U087 | 3288-58-2 | O,O-Diethyl S-methyl dithiophosphate |
| U088 | 84-66-2 | Diethyl phthalate |

[1] See footnote on page 185.

**Appendix B2. Hazardous and toxic wastes** *(continued)*

| Hazardous waste No. | Chemical abstracts No. | Substance |
|---|---|---|
| U089 | 56-53-1 | Diethylstilbesterol |
| U090 | 94-58-6 | Dihydrosafrole |
| U091 | 119-90-4 | 3,3′-Dimethoxybenzidine |
| U092 | 124-40-3 | Dimethylamine (I) |
| U093 | 60-11-7 | p-Dimethylaminoazobenzene |
| U094 | 57-97-6 | 7,12-Dimethylbenz[a]anthracene |
| U095 | 119-93-7 | 3,3′Dimethylbenzidine |
| U096 | 80-15-9 | alpha,alpha-Dimethylbenzylhydroperoxide (R) |
| U097 | 79-44-7 | Dimethylcarbamoyl chloride |
| U098 | 57-14-7 | 1,1-Dimethylhydrazine |
| U099 | 540-73-8 | 1,2-Dimethylhydrazine |
| U101 | 105-67-9 | 2,4-Dimethylphenol |
| U102 | 131-11-3 | Dimethyl phthalate |
| U103 | 77-78-1 | Dimethyl sulfate |
| U105 | 121-14-2 | 2,4-Dinitrotoluene |
| U106 | 606-20-2 | 2,6-Dinitrotoluene |
| U107 | 117-84-0 | Di-n-octyl phthalate |
| U108 | 123-91-1 | 1,4-Dioxane |
| U109 | 122-66-7 | 1,2-Diphenylhydrazine |
| U110 | 142-84-7 | Dipropylamine (I) |
| U111 | 621-64-7 | Di-n-propylnitorsamine |
| U041 | 106-89-8 | Epichlorohydrin |
| U001 | 75-07-0 | Ethanal (I) |
| U174 | 55-18-5 | Ethanamine, N-ethyl-N-nitroso- |
| U155 | 91-80-5 | 1,2-Ethanediamine, N,N-dimethyl-N′-2-pyridinyl-N′-(2-thienylmethyl)- |
| U067 | 106-93-4 | Ethane, 1,2-dibromo- |
| U076 | 75-34-3 | Ethane, 1,1-dichloro- |
| U077 | 107-06-2 | Ethane, 1,2-dichloro- |
| U131 | 67-72-1 | Ethane, hexachloro- |
| U024 | 111-91-1 | Ethane, 1,1′-[methylenebis(oxy)]bis[2-chloro- |
| U117 | 60-29-7 | Ethane, 1,1′-oxybis-(I) |
| U025 | 111-44-4 | Ethane, 1,1′-oxybis[2-chloro- |
| U184 | 76-01-7 | Ethane, pentachloro- |
| U208 | 630-20-6 | Ethane, 1,1,1,2-tetrachloro- |
| U209 | 79-34-5 | Ethane, 1,1,2,2-tetrachloro- |
| U218 | 62-55-5 | Ethanethioamide |
| U226 | 71-55-6 | Ethane, 1,1,1-trichloro- |
| U227 | 79-00-5 | Ethane, 1,1,2-trichloro- |
| U359 | 110-80-5 | Ethanol, 2-ethoxy- |
| U173 | 1116-54-7 | Ethanol, 2,2′-(nitrosoimino)bis- |

**Appendix B2. Hazardous and toxic wastes** *(continued)*

| Hazardous waste No. | Chemical abstracts No. | Substance |
|---|---|---|
| U004 | 98-86-2 | Ethanone, 1-phenyl- |
| U043 | 75-01-4 | Ethane, chloro- |
| U042 | 110-75-8 | Ethene, (2-chloroethoxy)- |
| U078 | 75-35-4 | Ethene, 1,1-dichloro- |
| U079 | 156-60-5 | Ethene, 1,2-dichloro-, (E) |
| U210 | 127-18-4 | Ethene, tetrachloro- |
| U228 | 79-01-6 | Ethene, trichloro- |
| U112 | 141-78-6 | Ethyl acetate (I) |
| U113 | 140-88-5 | Ethyl acrylate (I) |
| U238 | 51-79-6 | Ethyl carbamate (urethane) |
| U117 | 60-29-7 | Ethyl ether (I) |
| U114 | [1]111-54-6 | Ethylenebisdithiocarbamic acid, salts & esters |
| U067 | 106-93-4 | Ethylene dibromide |
| U077 | 107-06-2 | Ethylene dichloride |
| U359 | 110-80-5 | Ethylene glycol monoethyl ether |
| U115 | 75-21-8 | Ethylene oxide (I,T) |
| U116 | 96-45-7 | Ethylenethiourea |
| U076 | 75-34-3 | Ethylidene dichloride |
| U118 | 97-63-2 | Ethyl methacrylate |
| U119 | 62-50-0 | Ethyl methanesulfonate |
| U120 | 206-44-0 | Fluoranthene |
| U122 | 50-00-0 | Formaldehyde |
| U123 | 64-18-6 | Formic acid (C,T) |
| U124 | 110-00-9 | Furan (I) |
| U125 | 98-01-1 | 2-Furancarboxaldehyde (I) |
| U147 | 108-31-6 | 2,5-Furandione |
| U213 | 109-99-9 | Furan, tetrahydro-(I) |
| U125 | 98-01-1 | Furfural (I) |
| U124 | 110-00-9 | Furfuran (I) |
| U206 | 18883-66-4 | Glucopyranose, 2-deoxy-2-(3-methyl-3-nitrosoureido)-, D- |
| U206 | 18883-66-4 | D-Glucose, 2-deoxy-2-[[(methylnitrosoamino)-carbonyl]amino]- |
| U126 | 765-34-4 | Glycidylaldehyde |
| U163 | 70-25-7 | Guanidine, N-methyl-N′-nitro-N-nitroso- |
| U127 | 118-74-1 | Hexachlorobenzene |
| U128 | 87-68-3 | Hexachlorobutadiene |
| U130 | 77-47-4 | Hexachlorocyclopentadiene |
| U131 | 67-72-1 | Hexachloroethane |
| U132 | 70-30-4 | Hexachlorophene |
| U243 | 1888-71-7 | Hexachloropropene |

[1] See footnote on page 185.

**Appendix B2. Hazardous and toxic wastes** *(continued)*

| Hazardous waste No. | Chemical abstracts No. | Substance |
|---|---|---|
| U133 | 302-01-2 | Hydrazine (R,T) |
| U086 | 1615-80-1 | Hydrazine, 1,2-diethyl- |
| U098 | 57-14-7 | Hydrazine, 1,1-dimethyl- |
| U099 | 540-73-8 | Hydrazine, 1,2-dimethyl- |
| U109 | 122-66-7 | Hydrazine, 1,2-diphenyl- |
| U134 | 7664-39-3 | Hydrofluoric acid (C,T) |
| U134 | 7664-39-3 | Hydrogen fluoride (C,T) |
| U135 | 7783-06-4 | Hydrogen sulfide |
| U135 | 7783-06-4 | Hydrogen sulfide $H_2S$ |
| U096 | 80-15-9 | Hydroperoxide, 1-methyl-1-phenylethyl- (R) |
| U116 | 96-45-7 | 2-Imidazolidinethione |
| U137 | 193-39-5 | Indeno [1,2,3-cd]pyrene |
| U190 | 85-44-9 | 1,3-Isobenzofurandione |
| U140 | 78-83-1 | Isobutyl alcohol (I,T) |
| U141 | 120-58-1 | Isosafrole |
| U142 | 143-50-0 | Kepone |
| U143 | 303-34-4 | Lasiocarpine |
| U144 | 301-04-2 | Lead acetate |
| U146 | 1335-32-6 | Lead, bis(acetato-O)tetrahydroxytri- |
| U145 | 7446-27-7 | Lead phosphate |
| U146 | 1335-32-6 | Lead subacetate |
| U129 | 58-89-9 | Lindane |
| U163 | 70-25-7 | MNNG |
| U147 | 108-31-6 | Maleic anhydride |
| U148 | 123-33-1 | Maleic hydrazide |
| U149 | 109-77-3 | Malononitrile |
| U150 | 148-82-3 | Melphalan |
| U151 | 7439-97-6 | Mercury |
| U152 | 126-98-7 | Methacrylonitrile (I,T) |
| U092 | 124-40-3 | Methanamine, N-methyl- (I) |
| U029 | 74-83-9 | Methane, bromo- |
| U045 | 74-87-3 | Methane, chloro- (I,T) |
| U046 | 107-30-2 | Methane, chloromethoxy- |
| U068 | 74-95-3 | Methane, dibromo- |
| U080 | 75-09-2 | Methane, dichloro- |
| U075 | 75-71-8 | Methane, dichlorodifluoro- |
| U138 | 74-88-4 | Methane, iodo- |
| U119 | 62-50-0 | Methanesulfonic acid, ethyl ester |
| U211 | 56-23-5 | Methane, tetrachloro- |
| U153 | 74-93-1 | Methanethiol (I,T) |
| U225 | 75-25-2 | Methane, tribromo- |

**Appendix B2. Hazardous and toxic wastes** *(continued)*

| Hazardous waste No. | Chemical abstracts No. | Substance |
|---|---|---|
| U044 | 67-66-3 | Methane, trichloro- |
| U121 | 75-69-4 | Methane, trichlorofluoro- |
| U036 | 57-74-9 | 4,7-Methano-1H-indene, 1,2,4,5,6,7,8,8-octachloro-2,3,3a,4,7,7a-hexahydro- |
| U154 | 67-56-1 | Methanol (I) |
| U155 | 91-80-5 | Methapyrilene |
| U142 | 143-50-0 | 1,3,4-Metheno-2H-cyclobuta[cd]pentalen-2-one, 1,1a,3,3a,4,5,5,5a,5b,6-decachlorooctahydro- |
| U247 | 72-43-5 | Methoxychlor |
| U154 | 67-56-1 | Methyl alcohol (I) |
| U029 | 74-83-9 | Methyl bromide |
| U186 | 504-60-9 | 1-Methylbutadiene (I) |
| U045 | 74-87-3 | Methyl chloride (I,T) |
| U156 | 79-22-1 | Methyl chlorocarbonate (I,T) |
| U226 | 71-55-6 | Methyl chloroform |
| U157 | 56-49-5 | 3-Methylcholanthrene |
| U158 | 101-14-4 | 4,4′-Methylenebis (2-chloroanliline) |
| U068 | 74-95-3 | Methylene bromide |
| U080 | 75-09-2 | Methylene chloride |
| U159 | 78-93-3 | Methyl ethyl ketone (MEK) (I,T) |
| U160 | 1338-23-4 | Methyl ethyl ketone peroxide (R,T) |
| U138 | 74-88-4 | Methyl iodide |
| U161 | 108-10-1 | Methyl isobutyl ketone (I) |
| U162 | 80-62-6 | Methyl methacrylate (I,T) |
| U161 | 108-10-1 | 4-Methyl-2-pentanone (I) |
| U164 | 56-04-2 | Methylthiouracil |
| U010 | 50-07-7 | Mitomycin C |
| U059 | 20830-81-3 | 5,12-Naphthacenedione, 8-acetyl-10-[(3-amino-2,3,6-trideoxy)-alpha-L-Iyxo-Hexopyranosy)oxy]-7,8,9,10-tetrahydro-6,8,11-trihydroxy-1-methoxy-,(8S-cis)- |
| U167 | 134-32-7 | 1-Naphthalenamine |
| U168 | 91-59-8 | 2-Naphthalenamine |
| U026 | 494-03-1 | Naphthalenamine, N,N′-bis(2-chloroethyl)- |
| U165 | 91-20-3 | Naphthalene |
| U047 | 91-58-7 | Naphthalene, 2-chloro- |
| U166 | 130-15-4 | 1,4-Naphthalenedione |
| U236 | 72-57-1 | 2,7-Naphthalenedisulfonic acid, 3,3′-[(3,3′-dimethyl [1,1′-biphenyl]-4,4′-diyl)bis(azo)bis [5-amino-4-hydroxy]-, tetrasodium salt |
| U166 | 130-15-4 | 1,4-Naphthoquinone |

**Appendix B2. Hazardous and toxic wastes** *(continued)*

| Hazardous waste No. | Chemical abstracts No. | Substance |
|---|---|---|
| U167 | 134-32-7 | alpha-Naphthylamine |
| U168 | 91-59-8 | beta-Naphthylamine |
| U217 | 10102-45-1 | Nitric acid, thallium(1 + ) salt |
| U169 | 98-95-3 | Nitrobenzene (I,T) |
| U170 | 100-02-7 | p-Nitrophenol |
| U171 | 79-46-9 | 2-Nitropropane (I,T) |
| U172 | 924-16-3 | N-Nitrosodi-n-butylamine |
| U173 | 1116-54-7 | N-Nitrosodiethanolamine |
| U174 | 55-18-5 | N-Nitrosodiethylamine |
| U176 | 759-73-9 | N-Nitroso-N-ethylurea |
| U177 | 684-93-5 | N-Nitroso-N-methylurea |
| U178 | 615-53-2 | N-Nitroso-N-methylurethane |
| U179 | 100-75-4 | N-Nitrosopiperidine |
| U180 | 930-55-2 | N-Nitrosopyrrolidine |
| U181 | 99-55-8 | 5-Nitro-o-toluidine |
| U193 | 1120-71-4 | 1,2-Oxathiolane, 2,2-dioxide |
| U058 | 50-18-0 | 2H-1,3,2-Oxazaphosphorin-2-amine, N,N-bis(2-chloroethyl)tetrahydro-, 2-oxide |
| U115 | 75-21-8 | Oxirane (I,T) |
| U126 | 765-34-4 | Oxiranecarboxyaldehyde |
| U041 | 106-89-8 | Oxirane, (chloromethyl)- |
| U182 | 123-63-7 | Paraldehyde |
| U183 | 608-93-5 | Pentachlorobenzene |
| U184 | 76-01-7 | Pentachloroethane |
| U185 | 82-68-8 | Pentachloronitrobenzene (PCNB) |
| See F027 | 87-86-5 | Pentachlorophenol |
| U161 | 108-10-1 | Pentanol, 4-methyl- |
| U186 | 504-60-9 | 1,3-Pentadiene (I) |
| U187 | 62-44-2 | Phenacetin |
| U188 | 108-95-2 | Phenol |
| U048 | 95-57-8 | Phenol, 2-chloro- |
| U039 | 59-50-7 | Phenol, 4-chloro-3-methyl- |
| U081 | 120-83-2 | Phenol, 2,3-dichloro- |
| U082 | 87-65-0 | Phenol, 2,6,-dichloro- |
| U089 | 56-53-1 | Phenol, 4,4′-(1,2-diethyl-1,2-ethenediyl)bis-, (E)- |
| U101 | 105-67-9 | Phenol, 2,4-dimethyl- |
| U052 | 1319-77-3 | Phenol, methyl- |
| U132 | 70-30-4 | Phenol, 2,2′-methylenebis[3,4,6-trichloro- |
| U170 | 100-02-7 | Phenol, 4-nitro- |
| See F027 | 87-86-5 | Phenol, pentachloro- |
| See F027 | 58-90-2 | Phenol, 2,3,4,6-tetrachloro- |

**Appendix B2. Hazardous and toxic wastes** *(continued)*

| Hazardous waste No. | Chemical abstracts No. | Substance |
|---|---|---|
| See F027 | 95-95-4 | Phenol, 2,4,5-trichloro- |
| See F027 | 88-06-2 | Phenol, 2,4,6-trichloro- |
| U150 | 148-82-3 | L-Phenylalanine, 4-[bis(2-chloroethyl)amino]- |
| U145 | 7446-27-7 | Phosphoric acid, lead(2 + ) salt (2:3) |
| U087 | 3288-58-2 | Phosphorodithioic acid, O,O-diethyl S-methyl ester |
| U189 | 1314-80-3 | Phosphorus sulfide (R) |
| U190 | 85-44-9 | Phthalic anhydride |
| U191 | 109-06-8 | 2-Picoline |
| U179 | 100-75-4 | Piperidine, 1-nitroso- |
| U192 | 23950-58-5 | Pronamide |
| U194 | 107-10-8 | 1-Propanamine (I,T) |
| U111 | 621-64-7 | 1-Propanamine, N-nitroso-N-propyl |
| U110 | 142-84-7 | 1-Propanamine, N-propyl- (I) |
| U066 | 96-12-8 | Propane, 1,2-dibromo-3-chloro- |
| U083 | 78-87-5 | Propane, 1,2-dichloro- |
| U149 | 109-77-3 | Propanedinitrile |
| U171 | 79-46-9 | Propane, 2-nitro- (I,T) |
| U027 | 108-60-1 | Propane, 2,2′-oxybis[2-chloro- |
| U193 | 1120-71-4 | 1,3-Propane sultone |
| See F027 | 93-72-1 | Propanoic acid, 2-(2,4,5-trichlorophenoxy)- |
| U235 | 126-72-7 | 1-Propanol, 2,3-dibromo-,phosphate (3:1) |
| U140 | 78-83-1 | 1-Propanol, 2-methyl- (I,T) |
| U002 | 67-64-1 | 2-Propanone (I) |
| U007 | 79-06-1 | 2-Propenamide |
| U084 | 542-75-6 | 1-Propene, 1,3-dichloro- |
| U243 | 1888-71-7 | 1-Propene, 1,1,2,3,3,3-hexachloro- |
| U009 | 107-13-1 | 2-Propenenitrile |
| U152 | 126-98-7 | 2-Propenenitrile, 2-methyl- (I,T) |
| U008 | 79-10-7 | 2-Propenoic acid (I) |
| U113 | 140-88-5 | 2-Propenoic acid, ethyl ester (I) |
| U118 | 97-63-2 | 2-Propenoic acid, 2-methyl-, ethyl ester |
| U162 | 80-62-6 | 2-Propenoic acid, 2-methyl-, methyl ester (I,T) |
| U194 | 107-10-8 | n-Propylamine (I,T) |
| U083 | 78-87-5 | Propylene dichloride |
| U148 | 123-33-1 | 3,6-Pyridazinedione, 1,2-dihydro- |
| U196 | 110-86-1 | Pyridine |
| U191 | 109-06-8 | Pyridine, 2-methyl- |
| U237 | 66-75-1 | 2,4-(1H,3H)-Pyrimidinedione, 5-[bis(2-chloroethyl)amino]- |
| U164 | 56-04-2 | 4(1H)-Pyrimidinone, 2,3-dihydro-6-methyl-2-thioxo- |
| U180 | 930-55-2 | Pyrrolidine, 1-nitroso- |

**Appendix B2. Hazardous and toxic wastes** *(continued)*

| Hazardous waste No. | Chemical abstracts No. | Substance |
|---|---|---|
| U200 | 50-55-5 | Reserpine |
| U201 | 108-46-3 | Resorcinol |
| U202 | [1]81-07-2 | Saccharin, & salts |
| U203 | 94-59-7 | Safrole |
| U204 | 7783-00-8 | Selenious acid |
| U204 | 7783-00-8 | Selenium dioxide |
| U205 | 7488-56-4 | Selenium sulfide |
| U205 | 7488-56-4 | Selenium sulfide $SeS_2$ (R,T) |
| U015 | 115-02-6 | L-Serine, diazoacetate (ester) |
| See F027 | 93-72-1 | Silvex (2,4,5-TP) |
| U206 | 18883-66-4 | Streptozotocin |
| U103 | 77-78-1 | Sulfuric acid, dimethyl ester |
| U189 | 1314-80-3 | Sulfur phosphide (R) |
| See F027 | 93-76-5 | 2,4,5-T |
| U207 | 95-94-3 | 1,2,4,5-Tetrachlorobenzene |
| U208 | 630-20-6 | 1,1,1,2-Tetrachloroethane |
| U209 | 79-34-5 | 1,1,2,2-Tetrachloroethane |
| U210 | 127-18-4 | Tetrachloroethylene |
| See F027 | 58-90-2 | 2,3,4,6-Tetrachlorophenol |
| U213 | 109-99-9 | Tetrahydrofuran (I) |
| U214 | 563-68-8 | Thallium(I) acetate |
| U215 | 65333-73-9 | Thallium(I) carbonate |
| U216 | 7791-12-0 | Thallium(I) chloride |
| U216 | 7791-12-0 | Thallium chloride Tlcl |
| U217 | 10102-45-1 | Thallium(I) nitrate |
| U218 | 62-55-5 | Thioacetamide |
| U153 | 74-93-1 | Thiomethanol (I,T) |
| U244 | 137-26-8 | Thioperoxydicarbonic diamide $[(H_2N)C(S)]_2S_2$ tetramethyl- |
| U219 | 62-56-6 | Thiourea |
| U244 | 137-26-8 | Thiram |
| U220 | 108-88-3 | Toluene |
| U221 | 25376-45-8 | Toluenediamine |
| U223 | 26471-62-5 | Toluene diisocyanate (R,T) |
| U328 | 95-53-4 | o-Toluidine |
| U353 | 106-49-0 | p-Toluidine |
| U222 | 636-21-5 | o-Toluidine hydrochloride |
| U011 | 61-82-5 | 1H-1,2,4-Triazol-3-amine |
| U227 | 79-00-5 | 1,1,2-Trichloroethane |
| U228 | 79-01-6 | Trichloroethylene |
| U121 | 75-69-4 | Trichloromonofluoromethane |

[1] See footnote on page 185.

**Appendix B2. Hazardous and toxic wastes** *(concluded)*

| Hazardous waste No. | Chemical abstracts No. | Substance |
|---|---|---|
| See F027 | 95-95-4 | 2,4,5-Trichlorophenol |
| See F027 | 88-06-2 | 2,4,6-Trichlorophenol |
| U234 | 99-35-4 | 1,3,5-Trinitrobenzene (R,T) |
| U182 | 123-63-7 | 1,3,5-Trioxane, 2,4,6-trimethyl- |
| U235 | 126-72-7 | Tris(2,3-dibromopropyl) phosphate |
| U236 | 72-57-1 | Trypan blue |
| U237 | 66-75-1 | Uracil mustard |
| U176 | 759-73-9 | Urea, N-ethyl-N-nitroso- |
| U177 | 684-93-5 | Urea, N-methyl-N-nitroso- |
| U043 | 75-01-4 | Vinyl chloride |
| U248 | [1]81-81-2 | Warfarin, & salts, when present at concentrations of 0.3% or less. |
| U239 | 1330-20-7 | Xylene (I) |
| U200 | 50-55-5 | Yohimban-16-carboxylic acid, 11,17-dimethoxy-18-[(3,4,5-trimethoxybenzoyl)oxy]-, methyl ester, (3beta, 16beta,17alpha,18beta,20alpha)- |
| U249 | 1314-84-7 | Zinc phosphide $Zn_3P_2$, when present at concentrations of 10% or less |

[1] CAS Number given for parent compound only.

(Approved by the Office of Management and Budget under control number 2050-0047)[45 FR 78529, 78541, Nov. 25,1980]

Editorial Note: For Federal Register citations affecting § 261.33, see the List of CFR Sections Affected in the Finding Aids section.

**Appendix C1. Periodic Table of the Elements**

| Element | Symbol | Atomic Number | Column, Row | Atomic Weight |
|---|---|---|---|---|
| Actinium | Ac | 89 | c,7 | (227) |
| Aluminum | Al | 13 | m,3 | 26.982 |
| Americium | Am | 95 | j,9A | (243) |
| Antimony | Sb | 51 | o,5 | 121.750 |
| Argon | Ar | 18 | r,3 | 39.948 |
| Arsenic | As | 33 | o,4 | 74.922 |
| Astatine | At | 85 | q,6 | (210) |
| Barium | Ba | 56 | b,6 | 137.340 |
| Berkelium | Bk | 97 | l,9A | (247) |
| Berylium | Be | 4 | b,2 | 9.012 |
| Bismuth | Bi | 83 | o,6 | 208.980 |
| Boron | B | 5 | m,2 | 10.811 |
| Bromine | Br | 35 | q,4 | 79.909 |
| Cadmium | Cd | 48 | l,5 | 112.400 |
| Calcium | Ca | 20 | b,4 | 40.080 |
| Californium | Cf | 98 | m,9A | (251) |
| Carbon | C | 6 | n,2 | 12.011 |
| Cerium | Ce | 58 | e,8L | 140.120 |
| Cesium | Cs | 55 | a,6 | 132.905 |
| Chlorine | Cl | 17 | q,3 | 35.453 |
| Chromium | Cr | 24 | f,4 | 51.996 |
| Cobalt | Co | 27 | i,4 | 58.933 |
| Copper | Cu | 29 | k,4 | 63.540 |
| Curium | Cm | 96 | k,9A | (247) |
| Dysprosium | Dy | 66 | m,8L | 162.500 |
| Einsteinium | Es | 99 | n,9A | (254) |
| Erbium | Er | 68 | o,8L | 167.260 |
| Europium | Eu | 63 | j,8L | 151.960 |
| Fermium | Fm | 100 | o,9A | (253) |
| Fluorine | F | 9 | q,2 | 18.998 |
| Francium | Fr | 87 | a,7 | (223) |
| Gadolinium | Gd | 64 | k,8L | 157.250 |
| Gallium | Ga | 31 | m,4 | 69.720 |
| Germanium | Ge | 32 | n,4 | 72.590 |
| Gold | Au | 79 | k,6 | 196.967 |

*(continued)*

**Appendix C1.** *(continued)*

| Element | Symbol | Atomic Number | Column, Row | Atomic Weight |
|---|---|---|---|---|
| Hafnium | Hf | 72 | d,6 | 178.490 |
| (Hahnium) | (Ha) | 105 | e,7 | (260) |
| Helium | He | 2 | r,1 | 4.003 |
| Holmium | Ho | 67 | n,8L | 164.930 |
| Hydrogen | H | 1 | a,1 | 1.008 |
| Indium | In | 49 | m,5 | 114.820 |
| Iodine | I | 53 | q,5 | 126.904 |
| Iridium | Ir | 77 | i,6 | 192.200 |
| Iron | Fe | 26 | h,4 | 55.847 |
| Krypton | Kr | 36 | r,4 | 83.800 |
| Lanthanum | La | 57 | c,6 | 138.910 |
| (Lawrencium) | (Lw) | 103 | r,9A | (257) |
| Lead | Pb | 82 | n,6 | 207.190 |
| Lithium | Li | 3 | a,2 | 6.939 |
| Lutetium | Lu | 71 | r,8L | 174.970 |
| Magnesium | Mg | 12 | b,3 | 24.312 |
| Manganese | Mn | 25 | g,4 | 54.938 |
| Mendelevium | Md | 101 | p,9A | (256) |
| Mercury | Hg | 80 | l,6 | 200.590 |
| Molybdenum | Mo | 42 | f,5 | 95.940 |
| Neodymium | Nd | 60 | g,8L | 144.240 |
| Neon | Ne | 10 | r,2 | 20.183 |
| Neptunium | Np | 93 | h,9A | (237) |
| Nickel | Ni | 28 | j,4 | 58.710 |
| Niobium | Nb | 41 | e,5 | 92.906 |
| Nitrogen | N | 7 | o,2 | 14.007 |
| Nobelium | No | 102 | q,9A | (254) |
| Osmium | Os | 76 | h,6 | 190.200 |
| Oxygen | O | 8 | p,2 | 15.999 |
| Palladium | Pd | 46 | j,5 | 106.400 |
| Phosphorus | P | 15 | o,3 | 30.974 |
| Platinum | Pt | 78 | j,6 | 195.090 |
| Plutonium | Pu | 94 | i,9A | (242) |
| Polonium | Po | 84 | p,6 | (210) |

*(continued)*

**Appendix C1.** *(concluded)*

| Element | Symbol | Atomic Number | Column, Row | Atomic Weight |
|---|---|---|---|---|
| Potassium | K | 19 | a,4 | 39.102 |
| Praseodymium | Pr | 59 | f,8L | 140.907 |
| Promethium | Pm | 61 | h,8L | (147) |
| Protactinium | Pa | 91 | f,9A | 231.036 |
| Radium | Ra | 88 | b,7 | 226.025 |
| Radon | Rn | 86 | r,6 | (222) |
| Rhenium | Re | 75 | g,6 | 186.200 |
| Rhodium | Rh | 45 | i,5 | 102.905 |
| Rubidium | Rb | 37 | a,5 | 85.470 |
| Ruthenium | Ru | 44 | h,5 | 101.070 |
| (Rutherfordium) | (Rf) | 104 | d,7 | (261) |
| Samarium | Sm | 62 | i,8L | 150.350 |
| Scandium | Sc | 21 | c,4 | 44.956 |
| Selenium | Se | 34 | p,4 | 78.960 |
| Silicon | Si | 14 | n,3 | 28.086 |
| Silver | Ag | 47 | k,5 | 107.870 |
| Sodium | Na | 11 | a,3 | 22.990 |
| Strontium | Sr | 38 | b,5 | 87.620 |
| Sulfur | S | 16 | p,3 | 32.064 |
| Tantalum | Ta | 73 | e,6 | 180.948 |
| Technetium | Tc | 43 | g,5 | (99) |
| Tellurium | Te | 52 | p,5 | 127.600 |
| Terbium | Tb | 65 | l,8L | 158.925 |
| Thallium | Tl | 81 | m,6 | 204.370 |
| Thorium | Th | 90 | e,9A | 232.038 |
| Thulium | Tm | 69 | p,8L | 168.934 |
| Tin | Sn | 50 | n,5 | 118.690 |
| Titanium | Ti | 22 | d,4 | 47.900 |
| Tungsten (see Wolfram) | | | | |
| Uranium | U | 92 | g,9A | 238.030 |
| Vanadium | V | 23 | e,4 | 50.94 |
| Wolfram (Tungsten) | W | 74 | f,6 | 183.850 |
| Xenon | Xe | 54 | r,5 | 131.300 |
| Ytterbium | Yb | 70 | q,8L | 173.040 |
| Yttrium | Y | 39 | c,5 | 88.905 |
| Zinc | Zi | 30 | l,4 | 65.370 |
| Zirconium | Zr | 40 | d,5 | 91.220 |

**Appendix C2a. Heating values of organic compounds.** All but the last three columns of this table are from Enthalpy of Combustion of Selected Organic Compounds in the CRC Handbook of Chemistry and Physics, 73rd Edition, Lide, D.R., Editor, CRC Press, Boca Raton, Florida, 1992. With Permission. *(Reference 11)*

This table gives the standard molar enthalpy (heat) of combustion at 25 C (77 F) of selected compounds containing carbon, hydrogen, nitrogen, and oxygen. The products of combustion are taken to be $CO_2$ (gas), $N_2$ (gas), and $H_2$2O (liquid) in their standard state. The molar mass is included in the table to facilitate conversion from a molar to mass basis.

Note that the quantity tabulated here is the negative of the enthalpy of combustion; thus the tabulated number gives the heat released upon combustion of one mole of the compound.

Substances are arranged by molecular formula in the Hill order. See Appendices B1 and C2c of this book, or Appendix B of Reference 11, for an index of molecular formula from alphabetically-arranged compound names. Source = "Standard Thermodynamic Properties of Chemical Substances" in Reference 11.

| Molecular formula | Name | *M* g/mol | Enthalpy of combustion $-\Delta_C H°$, kJ/mol crystal | liquid | gas | Higher or gross heat of combustion Btu/lb crystal | liquid | Btu/cu ft gas @ 60 F |
|---|---|---|---|---|---|---|---|---|
| C | Carbon | 12.011 | 393.5 | | 1110.2 | 14080 | | |
| $CN_4O_8$ | Tetranitromethane | 196.033 | | 431.9 | 475.5 | | 950 | |
| CO | Carbon Monoxide | 28.010 | | | 283.0 | | | 320 |
| CHN | Hydrogen cyanide | 27.026 | | 645.3 | 671.5 | | 10260 | 760 |
| $CH_2N_2$ | Cyanamide | 42.040 | 738.1 | | | 7550 | | |
| $CH_2O$ | Formaldehyde | 30.026 | | | 570.7 | | | 650 |
| $CH_2O_2$ | Formic acid | 46.026 | | 254.6 | 300.7 | | 2380 | 340 |
| $CH_3NO$ | Formamide | 45.041 | | 568.3 | | | 5420 | |
| $CH_3NO_2$ | Nitromethane | 61.040 | | 709.2 | 747.6 | | 4990 | 850 |
| $CH_3NO_3$ | Methyl nitrate | 77.040 | | 663.3 | 697.6 | | 3700 | 790 |
| $CH_4$ | Methane | 16.043 | | | 890.8 | | | 1010 |
| $CH_4N_2$ | Ammonium cyanide | 44.056 | 965.6 | | | 9420 | | |
| $CH_4N_2O$ | Urea | 60.056 | 631.6 | | 719.4 | 4520 | | 820 |
| $CH_4O$ | Methanol | 32.042 | | 726.1 | 763.7 | | 9740 | 870 |
| $CH_5N$ | Methylamine | 31.057 | | 1060.8 | 1085.6 | | 14680 | 1230 |
| $CH_5NO_3$ | Ammonium hydrogen carbonate | 79.056 | 258.7 | | | 1410 | | |
| $CH_6N_2$ | Methylhydrazine | 46.072 | | 1305.0 | 1345.3 | | 12180 | 1530 |
| $C_2N_2$ | Cyanogen | 52.036 | | 1072.9 | 1093.7 | | 8860 | 1240 |
| $C_2H_2$2 | Acetylene | 26.038 | | | 1301.1 | | | 1480 |

*(continued)*

**Appendix C2a.** *(continued)*

| Molecular formula | Name | $M$ g/mol | Enthalpy of combustion $-\Delta_C H°$, kJ/mol crystal | liquid | gas | Higher or gross heat of combustion Btu/lb crystal | liquid | Btu/cu ft gas @ 60 F |
|---|---|---|---|---|---|---|---|---|
| $C_2H_2O$ | Ketene | 42.037 | | 1005.0 | 1025.4 | | 10280 | 1160 |
| $C_2H_2O_2$ | Glyoxal | 58.037 | | | 860.8 | | | 980 |
| $C_2H_2O_4$ | Oxalic acid | 90.036 | 251.1 | | 349.1 | 1200 | | 400 |
| $C_2H_3N$ | Acetonitrile | 41.053 | | 1247.2 | 1280.1 | | 13060 | 1450 |
| $C_2H_3N$ | Isocyanomethane | 41.053 | | 1333.0 | 1364.8 | | 13960 | 1550 |
| $C_2H_3NO$ | Methylisocyanate | 57.052 | | 1123.8 | | | 8470 | |
| $C_2H_4$ | Ethylene | 28.054 | | | 1411.2 | | | 1600 |
| $C_2H_4O$ | Acetaldehyde | 44.053 | | 1166.9 | 1192.5 | | 11390 | 1350 |
| $C_2H_4O$ | Ethylene oxide | 44.053 | | 1280.9 | 1306.1 | | 12500 | 1480 |
| $C_2H_4O_2$ | Acetic acid | 60.053 | | 874.2 | 925.9 | | 6260 | 1050 |
| $C_2H_4O_2$ | Methyl formate | 60.053 | | 972.6 | 1003.2 | | 6960 | 1140 |
| $C_2H_5N$ | Ethyleneimine | 43.068 | | 1593.5 | 1628.1 | | 15910 | 1850 |
| $C_2H_5NO$ | Acetamide | 59.068 | 1184.6 | | 1263.3 | 8620 | | 1430 |
| $C_2H_5NO_2$ | Nitroethane | 75.067 | | 1357.7 | 1399.3 | | 7770 | 1590 |
| $C_2H_5NO_2$ | Glycine | 75.067 | 973.1 | | 1109.5 | 5570 | | 1260 |
| $C_2H_6$ | Ethane | 30.070 | | | 1560.7 | | | 1770 |
| $C_2H_6O$ | Dimethyl ether | 46.069 | | | 1460.4 | | | 1660 |
| $C_2H_6O$ | Ethanol | 46.069 | | 1366.8 | 1409.4 | | 12750 | 1600 |
| $C_2H_6O_2$ | Ethylene glycol | 62.068 | | 1189.2 | 1257.0 | | 8240 | 1430 |
| $C_2H_7N$ | Dimethylamine | 45.084 | | 1743.5 | 1768.9 | | 16620 | 2010 |
| $C_2H_7N$ | Ethylamine | 45.084 | | 1713.3 | 1739.9 | | 16340 | 1970 |
| $C_2H_8N_2$ | 1,1-Dimethylhydrazine | 60.099 | | 1979.2 | 2014.2 | | 14160 | 2280 |
| $C_2H_8N_2$ | 1,2-Ethanediamine | 60.099 | | 1867.3 | 1912.7 | | 13360 | 2170 |
| $C_2H_8N_2O_4$ | Ammonium oxalate | 124.097 | 807.3 | | | 2800 | | |
| $C_3H_3N$ | Propenenitrile | 53.064 | | 1756.4 | 1789.9 | | 14230 | 2030 |
| $C_3H_3NO$ | Oxazole | 69.063 | | 1561.3 | 1593.8 | | 9720 | 1810 |
| $C_3H_3NO$ | Isoxazole | 69.063 | | 1651.4 | 1687.9 | | 10280 | 1910 |
| $C_3H_4$ | Propyne | 40.065 | | | 1937.1 | | | 2200 |
| $C_3H_4$ | Allene | 40.065 | | | 1942.7 | | | 2200 |
| $C_3H_4$ | Cyclopropene | 40.065 | | | 2029.3 | | | 2300 |
| $C_3H_4N_2$ | Imidazole | 68.078 | 1810.7 | | | 11430 | | |
| $C_3H_4O_2$ | Propenoic acid | 72.064 | | 1368.4 | | | 8160 | |
| $C_3H_4O_2$ | 2-Oxetanone | 72.064 | | 1422.3 | 1469.3 | | 8480 | 1670 |

*(continued)*

**Appendix C2a.** *(continued)*

| Molecular formula | Name | $M$ g/mol | Enthalphy of combustion $-\Delta_C H°$, kJ/mol | | | Higher or gross heat of combustion Btu/lb | | Btu/cu ft |
|---|---|---|---|---|---|---|---|---|
| | | | crystal | liquid | gas | crystal | liquid | gas @ 60 F |
| $C_3H_5N$ | Propanenitrile | 55.079 | | 1910.6 | 1946.6 | | 14910 | 2210 |
| $C_3H_5N_3O_9$ | Trinitroglycerol | 227.088 | | 1524.2 | 1624.2 | | 2890 | 1840 |
| $C_3H_6$ | Propene | 42.081 | | 2039.7 | 2058.0 | | 20840 | 2330 |
| $C_3H_6$ | Cyclopropane | 42.081 | | | 2091.3 | | | 2370 |
| $C_3H_6O$ | Acetone | 58.080 | | 1789.9 | 1820.7 | | 13250 | 2070 |
| $C_3H_6O$ | Allyl alcohol | 58.080 | | 1866.2 | 1913.5 | | 13810 | 2170 |
| $C_3H_6O$ | Propanal | 58.080 | | 1822.7 | 1852.4 | | 13490 | 2100 |
| $C_3H_6O$ | Methyloxirane | 58.080 | | 1915.4 | 1943.3 | | 14180 | 2200 |
| $C_3H_6O_2$ | Methyl acetate | 74.079 | | 1592.2 | 1626.1 | | 9240 | 1840 |
| $C_3H_6O_2$ | Propanoic acid | 74.079 | | 1527.3 | 1584.5 | | 8860 | 1800 |
| $C_3H_6O_2$ | 1,3-Dioxolane | 74.079 | | 1704.5 | 1740.0 | | 9890 | 1970 |
| $C_3H_6O_3$ | Trioxane | 90.079 | 1515.5 | | 1572.1 | 7230 | | 1780 |
| $C_3H_7N$ | Cyclopropylamine | 57.095 | | 2226.7 | 2257.9 | | 16760 | 2560 |
| $C_3H_7NO$ | *N,N*-Dimethylformamide | 73.095 | | 1941.6 | 1989.2 | | 11420 | 2260 |
| $C_3H_7NO_2$ | 1-Nitropropane | 89.094 | | 2013.7 | 2057.1 | | 9720 | 2330 |
| $C_3H_7NO_2$ | 2-Nitropropane | 89.094 | | 2000.6 | 2041.9 | | 9650 | 2320 |
| $C_3H_7NO_2$ | 1-Alanine | 89.094 | 1576.9 | | 1715.0 | 7610 | | 1950 |
| $C_3H_7NO_2$ | d-Alanine | 89.094 | 1619.7 | | | 7810 | | |
| $C_3H_7NO_3$ | 1-Serine | 105.094 | 1448.2 | | | 5920 | | |
| $C_3H_8$ | Propane | 44.097 | | | 2219.2 | | | 2520 |
| $C_3H_8O$ | 1-Propanol | 60.096 | | 2021.3 | 2068.8 | | 14460 | 2350 |
| $C_3H_8O$ | 2-Propanol | 60.096 | | 2005.8 | 2051.1 | | 14350 | 2330 |
| $C_3H_8O$ | Ethyl methyl ether | 60.096 | | | 2107.5 | | | 2390 |
| $C_3H_8O_2$ | 1,2-Propylene glycol | 76.095 | | 1838.2 | 1902.6 | | 10380 | 2160 |
| $C_3H_8O_2$ | 1,3-Propylene glycol | 76.095 | | 1859.0 | 1931.8 | | 10500 | 2190 |
| $C_3H_8O_2$ | Dimethoxymethane | 76.095 | | 1946.2 | 1975.5 | | 10990 | 2240 |
| $C_3H_8O_3$ | Glycerol | 92.095 | | 1655.4 | 1741.2 | | 7730 | 1970 |
| $C_3H_9N$ | Propylamine | 59.111 | | 2365.3 | 2396.7 | | 17200 | 2720 |
| $C_3H_9N$ | Isopropylamine | 59.111 | | 2354.5 | 2383.1 | | 17120 | 2700 |
| $C_3H_9N$ | Trimethylamine | 59.111 | | 2421.1 | 2443.1 | | 17610 | 2770 |
| $C_4H_2O_3$ | Maleic anhydride | 98.058 | 1390.1 | | 1461.6 | 6090 | | 1660 |
| $C_4H_4N_2$ | Pyridazine | 80.089 | | 2370.5 | 2424.0 | | 12720 | 2750 |
| $C_4H_4N_2$ | Pyrimidine | 80.089 | | 2291.6 | 2341.6 | | 12300 | 2660 |

*(continued)*

**Appendix C2a.** *(continued)*

| Molecular formula | Name | *M* g/mol | Enthalpy of combustion $-\Delta_C H°$, kJ/mol | | | Higher or gross heat of combustion Btu/lb | | Btu/cu ft |
|---|---|---|---|---|---|---|---|---|
| | | | crystal | liquid | gas | crystal | liquid | gas @ 60 F |
| $C_4H_4N_2$ | Succinonitrile | 80.089 | | 2285.4 | 2355.4 | | 12270 | 2670 |
| $C_4H_4N_2O_2$ | Uracil | 112.088 | 1716.3 | | 1842.8 | 6580 | | 2090 |
| $C_4H_4O$ | Furan | 68.075 | | 2083.4 | 2110.8 | | 13160 | 2390 |
| $C_4H_4O_2$ | Diketene | 84.075 | | 1912.6 | 1955.5 | | 9780 | 2220 |
| $C_4H_4O_3$ | Succinic anhydride | 100.074 | 1537.9 | | | 6610 | | |
| $C_4H_4O_4$ | Fumaric acid | 116.073 | 1334.0 | | 1469.9 | 4940 | | 1670 |
| $C_4H_4O_4$ | Maleic acid | 116.073 | 1356.3 | | 1466.3 | 5020 | | 1660 |
| $C_4H_5N$ | Cyclopropanecarbonitrile | 67.090 | | 2429.4 | 2470.4 | | 15570 | 2800 |
| $C_4H_5N$ | Pyrrole | 67.090 | | 2351.7 | 2396.9 | | 15070 | 2720 |
| $C_4H_5N_3O$ | Cytosine | 111.103 | 2067.3 | | | 8000 | | |
| $C_4H_6$ | 1,2-Butadiene | 54.092 | | 2570.5 | 2593.8 | | 20430 | 2940 |
| $C_4H_6$ | 1,3-Butadiene | 54.092 | | 2519.4 | 2541.5 | | 20020 | 2880 |
| $C_4H_6$ | 1-Butyne | 54.092 | | 2573.4 | 2596.7 | | 20450 | 2950 |
| $C_4H_6$ | 2-Butyne | 54.092 | | 2550.6 | 2577.2 | | 20270 | 2920 |
| $C_4H_6$ | Cyclobutene | 54.092 | | | 2588.2 | | | 2940 |
| $C_4H_6O$ | *trans*-2-Butenal | 70.091 | | 2292.8 | 2330.9 | | 14060 | 2640 |
| $C_4H_6O$ | Divinyl ether | 70.091 | | 2391.7 | 2417.9 | | 14670 | 2740 |
| $C_4H_6O_2$ | Methyl acrylate | 86.090 | | 2069.3 | 2098.5 | | 10330 | 2380 |
| $C_4H_6O_2$ | Vinyl acetate | 86.090 | | 2151.4 | 2116.6 | | 10740 | 2400 |
| $C_4H_6O_3$ | Acetic anhydride | 102.090 | | 1807.1 | 1859.0 | | 7610 | 2110 |
| $C_4H_6O_4$ | Syccinic acid | 118.089 | 1491.0 | | 1608.5 | 5430 | | 1820 |
| $C_4H_6O_4$ | Dimethyl oxalate | 118.089 | 1675.2 | | 1722.6 | 6100 | | 1950 |
| $C_4H_7N$ | Butanenitrile | 69.106 | | 2568.6 | 2608.0 | | 15980 | 2960 |
| $C_4H_7N$ | 2-Methylpropanenitrile | 69.106 | | 2560.6 | 2597.7 | | 15930 | 2950 |
| $C_4H_7NO$ | 2-Pyrrolidone | 85.106 | | 2288.2 | | | 11560 | |
| $C_4H_7NO_4$ | 1-Aspartic acid | 133.104 | 1601.1 | | | 5170 | | |
| $C_4H_8$ | 1-Butene | 56.107 | | 2696.9 | 2717.5 | | 20660 | 3080 |
| $C_4H_8$ | *cis*-2-Butene | 56.107 | | 2687.7 | 2710.3 | | 20590 | 3070 |
| $C_4H_8$ | *trans*-2-Butene | 56.107 | | 2684.4 | 2706.0 | | 20570 | 3070 |
| $C_4H_8$ | Isobutene | 56.107 | | 2679.9 | 2700.5 | | 20530 | 3060 |
| $C_4H_8$ | Cyclobutane | 56.107 | | 2721.1 | 2745.8 | | 20850 | 3110 |
| $C_4H_8$ | Methylcyclopropane | 56.107 | | 2719.1 | | | 20830 | |
| $C_4H_8N_2O_3$ | 1-Asparagine | 132.119 | 1928.0 | | | 6270 | | |

*(continued)*

**Appendix C2a.** *(continued)*

| Molecular formula | Name | $M$ g/mol | Enthalpy of combustion $-\Delta_C H°$, kJ/mol crystal | liquid | gas | Higher or gross heat of combustion Btu/lb crystal | liquid | Btu/cu ft gas @ 60 F |
|---|---|---|---|---|---|---|---|---|
| $C_4H_8O$ | 1,2-Epoxybutane | 72.107 | | 2548.5 | | | 15190 | |
| $C_4H_8O$ | Butanal | 72.107 | | 2478.2 | 2512.6 | | 14770 | 2850 |
| $C_4H_8O$ | Isobutanal | 72.107 | | 2470.0 | 2501.6 | | 14730 | 2840 |
| $C_4H_8O$ | 2-Butanone | 72.107 | | 2444.1 | 2478.7 | | 14570 | 2810 |
| $C_4H_8O$ | Tetrahydrofuran | 72.107 | | 2501.2 | 2533.2 | | 14910 | 2870 |
| $C_4H_8O$ | Ethyl vinyl ether | 72.107 | | 2550.0 | 2576.6 | | 15200 | 2920 |
| $C_4H_8O_2$ | 1,3-Dioxane | 88.106 | | 2337.7 | 2375.1 | | 11410 | 2690 |
| $C_4H_8O_2$ | 1,4-Dioxane | 88.106 | | 2363.5 | 2401.6 | | 11530 | 2720 |
| $C_4H_8O_2$ | Ethyl acetate | 88.106 | | 2238.1 | 2273.3 | | 10920 | 2580 |
| $C_4H_8O_2$ | Propyl formate | 88.106 | | 2217.1 | | | 10820 | |
| $C_4H_8O_2$ | Butanoic acid | 88.106 | | 2183.6 | 2241.6 | | 10650 | 2540 |
| $C_4H_9N$ | Pyrrolidine | 71.122 | | 2819.3 | 2856.9 | | 17040 | 3240 |
| $C_4H_9NO$ | *N,N*-Dimethylacetamide | 87.122 | | 2582.0 | | | 12740 | |
| $C_4H_9NO_3$ | 1-Threonine | 119.120 | 2053.1 | | | 7410 | | |
| $C_4H_{10}$ | Butane | 58.123 | | 2856.6 | 2877.6 | | 21130 | 3260 |
| $C_4H_{10}$ | Isobutane | 58.123 | | 2849.7 | 2869.0 | | 21080 | 3250 |
| $C_4H_{10}O$ | 1-Butanol | 74.123 | | 2675.9 | 2728.2 | | 15520 | 3090 |
| $C_4H_{10}O$ | 2-Butanol | 74.123 | | 2660.6 | 2710.3 | | 15430 | 3070 |
| $C_4H_{10}O$ | 2-Methyl-2-propanol | 74.123 | | 2644.0 | 2690.7 | | 15330 | 3050 |
| $C_4H_{10}O$ | 2-Methyl-1-propanol | 74.123 | | 2668.5 | 2719.3 | | 15480 | 3080 |
| $C_4H_{10}O$ | Diethyl ether | 74.123 | | 2723.9 | 2751.1 | | 15800 | 3120 |
| $C_4H_{10}O$ | Methyl propyl ether | 74.123 | | 2737.2 | 2765.0 | | 15870 | 3140 |
| $C_4H_{10}O$ | Isopropyl methyl ether | 74.123 | | 2724.5 | 2751.2 | | 15800 | 3120 |
| $C_4H_{10}O_2$ | 1,3-Butanediol | 90.122 | | 2502.2 | 2570.0 | | 11940 | 2910 |
| $C_4H_{10}O_2$ | 1,4-Butanediol | 90.122 | | 2499.9 | 2576.5 | | 11920 | 2920 |
| $C_4H_{10}O_3$ | Diethylene glycol | 106.122 | | 2374.7 | 2432.0 | | 9620 | 2760 |
| $C_4H_{11}N$ | Butylamine | 73.138 | | 3018.4 | 3054.1 | | 17740 | 3460 |
| $C_4H_{11}N$ | Isobutylamine | 73.138 | | 3013.5 | 3047.4 | | 17710 | 3460 |
| $C_4H_{11}N$ | *sec*-Butylamine | 73.138 | | 3008.6 | 3041.2 | | 17680 | 3450 |
| $C_4H_{11}N$ | *tert*-Butylamine | 73.138 | | 2995.5 | 3025.2 | | 17610 | 3430 |
| $C_4H_{11}N$ | Diethylamine | 73.138 | | 3042.4 | 3073.6 | | 7880 | 3490 |
| $C_5H_4N_4O$ | Hypoxanthine | 136.113 | 2428.4 | | | 7670 | | |
| $C_5H_4N_4O_2$ | Xanthine | 152.113 | 2159.6 | | | 6100 | | |

*(continued)*

**Appendix C2a.** *(continued)*

| Molecular formula | Name | $M$ g/mol | Enthalpy of combustion $-\Delta_cH^\circ$, kJ/mol crystal | liquid | gas | Higher or gross heat of combustion Btu/lb crystal | liquid | Btu/cu ft gas @ 60 F |
|---|---|---|---|---|---|---|---|---|
| $C_5H_4N_4O_3$ | Uric acid | 168.112 | 1920.4 | | | 4910 | | |
| $C_5H_4O_2$ | Furfural | 96.086 | | 2337.6 | 2388.2 | | 10460 | 2710 |
| $C_5H_5N_5$ | Adenine | 135.128 | 2778.1 | | 2886.9 | 8840 | | 3270 |
| $C_5H_5N$ | Pyridine | 79.101 | | 2782.3 | 2822.5 | | 15120 | 3200 |
| $C_5H_5N_5O$ | Guanine | 151.128 | 2498.2 | | | 7110 | | |
| $C_5H_6$ | *cis*-3-Penten-1-yne | 66.103 | | | 2906.4 | | | 3300 |
| $C_5H_6$ | *trans*-3-Penten-1-yne | 66.103 | | 3053.2 | | | 19860 | |
| $C_5H_6$ | 1,3-Cyclopentadiene | 66.103 | | 2930.9 | 2959.3 | | 19060 | 3360 |
| $C_5H_6N_2O_2$ | Thymine | 126.115 | 2362.2 | | 2496.3 | 8050 | | 2830 |
| $C_5H_6O_2$ | Furfuryl alcohol | 98.101 | | 2548.8 | 2613.2 | | 11170 | 2960 |
| $C_5H_7N$ | Cyclobutanecarbonitrile | 81.117 | | 3071.0 | 3111.1 | | 16270 | 3530 |
| $C_5H_8$ | 2-Methyl-1,3-butadiene | 68.119 | | 3159.1 | 3186.4 | | 19940 | 3610 |
| $C_5H_8$ | *cis*-1,3-Pentadiene | 68.119 | | | 3192.3 | | | 3620 |
| $C_5H_8$ | *trans*-1,3-Pentadiene | 68.119 | | | 3187.0 | | | 3610 |
| $C_5H_8$ | 1,4-Pentadiene | 68.119 | | | 3216.5 | | | 3650 |
| $C_5H_8$ | Cyclopentene | 68.119 | | 3115.3 | 3144.8 | | 19660 | 3570 |
| $C_5H_8$ | Spiropentane | 68.119 | | 3268.6 | 3296.1 | | 20630 | 3740 |
| $C_5H_8O$ | Cyclopentanone | 84.118 | | 2875.2 | 2918.8 | | 14690 | 3310 |
| $C_5H_8O_2$ | 2,4-Pentanedione | 100.117 | | 2687.1 | 2730.3 | | 11540 | 3100 |
| $C_5H_8O_4$ | Glutaric acid | 132.116 | 2150.9 | | | 7000 | | |
| $C_5H_9N$ | Pentanenitrile | 83.133 | | 3220.7 | 3264.3 | | 16650 | 3700 |
| $C_5H_9N$ | 2,2-Dimethylpropanenitrile | 83.133 | | 3214.0 | 3251.3 | | 16620 | 3690 |
| $C_5H_9NO$ | *N*-Methyl-2-pyrrolidone | 99.133 | | 2991.6 | | | 12970 | |
| $C_5H_9NO_2$ | 1-Proline | 115.132 | 2741.6 | | 2887.6 | 10240 | | 3280 |
| $C_5H_9NO_4$ | 1-Glutamic acid | 147.131 | 2244.1 | | | 6560 | | |
| $C_5H_{10}$ | 1-Pentene | 70.134 | | 3349.8 | 3375.4 | 20530 | | 3830 |
| $C_5H_{10}$ | *cis*-2-Pentene | 70.134 | | 3343.0 | 3369.1 | | 20490 | 3820 |
| $C_5H_{10}$ | *trans*-2-Pentene | 70.134 | | 3338.5 | 3364.8 | | 20460 | 3820 |
| $C_5H_{10}$ | 2-Methyl-1-butene | 70.134 | | 3335.7 | 3361.4 | | 20450 | 3810 |
| $C_5H_{10}$ | 2-Methyl-2-butene | 70.134 | | 3328.1 | 3354.9 | | 20400 | 3810 |
| $C_5H_{10}$ | 3-Methyl-1-butene | 70.134 | | 3345.2 | 3369.1 | | 20500 | 3820 |
| $C_5H_{10}$ | Cyclopentane | 70.134 | | 3291.6 | 3320.3 | | 20180 | 3770 |
| $C_5H_{19}N_2O_3$ | 1-Glutamine | 146.146 | 2570.3 | | | 7560 | | |

*(continued)*

**Appendix C2a.** *(continued)*

| Molecular formula | Name | $M$ g/mol | Enthalpy of combustion $-\Delta_C H°$, kJ/mol crystal | liquid | gas | Higher or gross heat of combustion Btu/lb crystal | liquid | Btu/cu ft gas @ 60 F |
|---|---|---|---|---|---|---|---|---|
| $C_5H_{10}O$ | Cyclopentanol | 86.134 | | 3096.6 | 3154.1 | | 15450 | 3580 |
| $C_5H_{10}O$ | 2-Pentanone | 86.134 | | 3099.4 | 3137.7 | | 15470 | 3560 |
| $C_5H_{10}O$ | 3-Pentanone | 86.134 | | 3100.2 | 3138.8 | | 15470 | 3560 |
| $C_5H_{10}O$ | 3-Methyl-2-butanone | 86.134 | | 3097.3 | 3134.2 | | 15460 | 3550 |
| $C_5H_{10}O$ | 3,3-Dimethyloxetane | 86.134 | | 3214.5 | 3248.5 | | 16040 | 3680 |
| $C_5H_{10}O$ | Tetrahydropyran | 86.134 | | 3138.4 | 3173.3 | | 15660 | 3600 |
| $C_5H_{10}O$ | Pentanal | 86.134 | | 3129.4 | 3108.2 | | 15620 | 3530 |
| $C_5H_{10}O_2$ | Isopropyl acetate | 102.133 | | 2877.8 | 2915.0 | | 12110 | 3310 |
| $C_5H_{10}O_2$ | Ethyl propanoate | 102.133 | | 2894.0 | 2933.1 | | 12180 | 3330 |
| $C_5H_{10}O_2$ | Pentanoic acid | 102.133 | | 2837.3 | 2904.8 | | 11940 | 3290 |
| $C_5H_{10}O_2$ | 3-Methylbutanoic acid | 102.133 | | 2835.1 | 2886.7 | | 11930 | 3270 |
| $C_5H_{10}O_2$ | 2-Methylbutanoic acid | 102.133 | | 2842.2 | | | 11960 | |
| $C_5H_{10}O_2$ | Tetrahydrofurfuryl alcohol | 102.133 | | 2961.0 | 3027.5 | | 12460 | 3430 |
| $C_5H_{10}O_3$ | Diethyl carbonate | 118.133 | | 2715.2 | 2758.8 | | 9880 | 3130 |
| $C_5H_{11}N$ | Piperidine | 85.149 | | 3453.2 | 3492.4 | | 17430 | 3960 |
| $C_5H_{11}N$ | Cyclopentylamine | 85.149 | | 3444.5 | 3484.7 | | 17390 | 3950 |
| $C_5H_{11}NO_2$ | 1-Valine | 117.148 | 2921.7 | | 3084.5 | 10720 | | 3500 |
| $C_5H_{12}$ | Pentane | 72.150 | | 3509.0 | 3535.6 | | 20910 | 4010 |
| $C_5H_{12}$ | Isopentane | 72.150 | | 3504.0 | 3528.8 | | 20880 | 4000 |
| $C_5H_{12}$ | Neopentane | 72.150 | | 3492.3 | 3514.4 | | 20810 | 3990 |
| $C_5H_{12}O$ | Butyl methyl ether | 88.150 | | 3391.9 | 3424.4 | | 16540 | 3880 |
| $C_5H_{12}O$ | *tert*-Butyl methyl ether | 88.150 | | 3368.9 | 3399.0 | | 16430 | 3860 |
| $C_5H_{12}O$ | Ethyl propyl ether | 88.150 | | 3378.9 | 3410.3 | | 16480 | 3870 |
| $C_5H_{12}O$ | 1-Pentanol | 88.150 | | 3330.9 | 3387.8 | | 16240 | 3840 |
| $C_5H_{12}O$ | 2-Pentanol | 88.150 | | 3317.3 | 3369.8 | | 16180 | 3820 |
| $C_5H_{12}O$ | 3-Pentanol | 88.150 | | 3313.6 | 3365.3 | | 16160 | 3820 |
| $C_5H_{12}O$ | 2-Methyl-1-butanol | 88.150 | | 3325.9 | 3380.5 | | 16220 | 3830 |
| $C_5H_{12}O$ | 2-Methyl-2-butanol | 88.150 | | 3303.0 | 3351.7 | | 16110 | 3800 |
| $C_5H_{12}O$ | 3-Methyl-1-butanol | 88.150 | | 3326.1 | 3381.2 | | 16220 | 3830 |
| $C_5H_{12}O$ | 3-Methyl-2-butanol | 88.150 | | 3315.9 | 3367.3 | | 16170 | 3820 |
| $C_5H_{12}O$ | 2,2-Dimethyl-1-propanol | 88.150 | | 3283.1 | | | 16010 | |
| $C_5H_{12}O_2$ | 1,5-Pentanediol | 104.149 | | 3151.0 | 3233.4 | | 13010 | 3670 |
| $C_5H_{12}O_2$ | Diethoxymethane | 104.149 | | 3232.1 | 3267.7 | | 13340 | 3710 |

*(continued)*

**Appendix C2a.** *(continued)*

| Molecular formula | Name | *M* g/mol | Enthalpy of combustion $-\Delta_c H°$, kJ/mol crystal | liquid | gas | Higher or gross heat of combustion Btu/lb crystal | Btu/lb liquid | Btu/cu ft gas @ 60 F |
|---|---|---|---|---|---|---|---|---|
| $C_5H_{12}O_4$ | Pentaerythritol | 136.148 | 2761.9 | | 2905.8 | 8720 | | 3300 |
| $C_6H_4O_2$ | *p*-Benzoquinone | 108.097 | 2747.0 | | 2809.8 | 10920 | | 3190 |
| $C_6H_5NO_2$ | Nitrobenzene | 123.111 | | 3088.1 | 3143.1 | | 10780 | 3560 |
| $C_6H_6$ | Benzene | 78.114 | | 3267.6 | 3301.2 | | 17980 | 3740 |
| $C_6H_6N_2O_2$ | *o*-Nitroaniline | 138.126 | 3192.5 | 3209.2 | 3282.4 | 9940 | 9990 | 3720 |
| $C_6H_6N_2O_2$ | *m*-Nitroaniline | 138.126 | 3180.3 | 3204.2 | 3277.0 | 9900 | 9970 | 3720 |
| $C_6H_6N_2O_2$ | *p*-Nitroaniline | 138.126 | 3176.6 | 3197.9 | 3277.4 | 9890 | 9950 | 3720 |
| $C_6H_6O$ | Phenol | 94.113 | 3053.5 | | 3122.2 | 13950 | | 3540 |
| $C_6H_6O_2$ | *p*-Hydroquinone | 110.112 | 2854.1 | | 2953.3 | 11140 | | 3350 |
| $C_6H_7N$ | Aniline | 93.128 | | 3392.8 | 3449.0 | | 15660 | 3910 |
| $C_6H_7N$ | 2-Methylpyridine | 93.128 | | 3418.2 | 3460.7 | | 15780 | 3930 |
| $C_6H_7N$ | 3-Methylpyridine | 93.128 | | 3423.4 | 3467.9 | | 15800 | 3930 |
| $C_6H_7N$ | 4-Methylpyridine | 93.128 | | 3420.7 | 3465.6 | | 15790 | 3930 |
| $C_6H_7N$ | 1-Cyclopentenecarbonitrile | 93.128 | | 3473.0 | 3518.0 | | 16030 | 3990 |
| $C_6H_8N_2$ | Adiponitrile | 108.143 | | 3589.5 | 3653.9 | | 14270 | 4140 |
| $C_6H_8N_2$ | Phenylhydrazine | 108.143 | | 3645.4 | 3707.3 | | 14490 | 4200 |
| $C_6H_8N_2$ | *o*-Phenylenediamine | 108.143 | 3504.1 | | | 13930 | | |
| $C_6H_8N_2$ | *m*-Phenylenediamine | 108.143 | 3496.6 | | | 13900 | | |
| $C_6H_8N_2$ | *p*-Phenylenediamine | 108.143 | 3507.5 | | | 13940 | | |
| $C_6H_9N$ | Cyclopentanecarbonitrile | 95.144 | | 3648.0 | 3690.3 | | 16480 | 4190 |
| $C_6H_9NO_3$ | Triacetamide | 143.142 | | 3036.8 | 3097.2 | | 9120 | 3510 |
| $C_6H_9N_3O_2$ | 1-Histidine | 155.157 | 3180.6 | | | 8810 | | |
| $C_6H_{10}$ | Cyclohexene | 82.145 | | 3751.7 | 3785.2 | | 19630 | 4290 |
| $C_6H_{10}$ | 1,5-Hexadiene | 82.145 | | 3844.3 | 3874.3 | | 20120 | 4390 |
| $C_6H_{10}$ | 3,3-Dimethyl-1-butyne | 82.145 | | 3868.6 | | | 20240 | |
| $C_6H_{10}O$ | Cyclohexanone | 98.145 | | 3519.0 | 3564.1 | | 15410 | 4040 |
| $C_6H_{10}O_2$ | Methyl cyclobutanecarboxylate | 114.144 | | 3395.2 | 3434.9 | | 12790 | 3900 |
| $C_6H_{10}O_3$ | Propanoic anhydride | 130.144 | | 3111.1 | 3163.7 | | 10280 | 3590 |
| $C_6H_{10}O_4$ | Adipic acid | 146.143 | 2795.9 | | 2925.2 | 8220 | | 3320 |
| $C_6H_{10}O_4$ | Diethyl oxalate | 146.143 | | 2984.7 | 3048.2 | | 8780 | 3460 |
| $C_6H_{11}NO$ | Caprolactam | 113.160 | 3603.7 | | 3686.9 | 13690 | | 4180 |
| $C_6H_{12}$ | Cyclohexane | 84.161 | | 3919.6 | 3952.6 | | 20020 | 4480 |
| $C_6H_{12}$ | Methylcyclopentane | 84.161 | | 3938.1 | 3969.8 | | 20110 | 4500 |

*(continued)*

**Appendix C2a.** *(continued)*

| Molecular formula | Name | M g/mol | Enthalpy of combustion $-\Delta_C H°$, kJ/mol crystal | liquid | gas | Higher or gross heat of combustion Btu/lb crystal | Btu/lb liquid | Btu/cu ft gas @ 60 F |
|---|---|---|---|---|---|---|---|---|
| $C_6H_{12}$ | Ethylcyclobutane | 84.161 | | 4017.0 | 4049.7 | | 20520 | 4590 |
| $C_6H_{12}$ | 1-Hexene | 84.161 | | 4001.8 | 4032.5 | | 20440 | 4570 |
| $C_6H_{12}$ | *cis*-2-Hexene | 84.161 | | 3992.1 | 4023.7 | | 20390 | 4560 |
| $C_6H_{12}$ | *trans*-2-Hexene | 84.161 | | 3990.5 | 4022.1 | | 20380 | 4560 |
| $C_6H_{12}$ | 2-Methyl-1-pentene | 84.161 | | 3986.0 | 4016.6 | | 20360 | 4560 |
| $C_6H_{12}$ | 2-Methyl-2-pentene | 84.161 | | 3977.5 | 4009.1 | | 20320 | 4550 |
| $C_6H_{12}$ | 4-Methyl-1-pentene | 84.161 | | 3996.0 | 4024.7 | | 20410 | 4560 |
| $C_6H_{12}$ | 4-Methyl-*cis*-2-pentene | 84.161 | | 3989.0 | 4018.5 | | 20370 | 4560 |
| $C_6H_{12}$ | 4-Methyl-*trans*-2-pentene | 84.161 | | 3984.5 | 4014.5 | | 20350 | 4550 |
| $C_6H_{12}$ | 2,3-Dimethyl-1-butene | 84.161 | | 3982.7 | 4013.4 | | 20340 | 4550 |
| $C_6H_{12}$ | 2,3-Dimethyl-2-butene | 84.161 | | 3974.5 | 4007.8 | | 20300 | 4550 |
| $C_6H_{12}$ | 2-Ethyl-1-butene | 84.161 | | 3988.9 | 4020.0 | | 20370 | 4560 |
| $C_6H_{12}O$ | Cyclohexanol | 100.161 | | 3727.8 | 3789.8 | | 16000 | 4300 |
| $C_6H_{12}O$ | 2-Hexanone | 100.161 | | 3754.0 | 3796.2 | | 16110 | 4310 |
| $C_6H_{12}O$ | 3-Hexanone | 100.161 | | 3755.8 | 3797.7 | | 16120 | 4310 |
| $C_6H_{12}O$ | 2-Methyl-3-pentanone | 100.161 | | 3750.1 | 3789.9 | | 16090 | 4300 |
| $C_6H_{12}O$ | 3,3-Dimethyl-2-butanone | 100.161 | | 3747.4 | 3785.3 | | 16080 | 4290 |
| $C_6H_{12}O_2$ | Butyl acetate | 116.160 | | 3546.8 | 3590.4 | | 13130 | 4070 |
| $C_6H_{12}O_2$ | Methyl pentanonate | 116.160 | | 3561.8 | 3604.8 | | 13180 | 4090 |
| $C_6H_{12}O_2$ | Methyl 2,2-dimethylpropanoate | 116.160 | | 3546.0 | 3581.7 | | 13120 | 4060 |
| $C_6H_{12}O_2$ | Hexanoic acid | 116.160 | | 3492.2 | 3564.1 | | 12920 | 4040 |
| $C_6H_{13}$ | N Cyclohexylamine | 99.176 | | 4071.3 | 4114.1 | | 17650 | 4670 |
| $C_6H_{13}NO_2$ | 1-Leucine | 131.175 | 3581.6 | | 3732.2 | 11740 | | 4230 |
| $C_6H_{13}NO_2$ | *d*-Leucine | 131.175 | 3581.7 | | | 11740 | | |
| $C_6H_{13}NO_2$ | 1-Isoleucine | 131.175 | 3581.1 | | | 11740 | | |
| $C_6H_{14}$ | Hexane | 86.177 | | 4163.2 | 4194.8 | | 20770 | 4760 |
| $C_6H_{14}$ | 2-Methylpentane | 86.177 | | 4157.3 | 4187.1 | | 20740 | 4750 |
| $C_6H_{14}$ | 3-Methylpentane | 86.177 | | 4159.5 | 4189.8 | | 20750 | 4750 |
| $C_6H_{14}$ | 2,2-Dimethylbutane | 86.177 | | 4148.1 | 4175.8 | | 20690 | 4740 |
| $C_6H_{14}$ | 2,3-Dimethylbutane | 86.177 | | 4154.5 | 4183.6 | | 20720 | 4750 |
| $C_6H_{14}N_2$ | Azopropane | 114.191 | | 4373.4 | 4413.4 | | 16460 | 5010 |
| $C_6H_{14}N_2O_2$ | Lysine | 146.189 | 3683.2 | | | 10830 | | |
| $C_6H_{14}N_4O_2$ | d-Arginine | 174.203 | 3738.4 | | | 9230 | | |

(continued)

**Appendix C2a.** *(continued)*

| Molecular formula | Name | $M$ g/mol | Enthalpy of combustion $-\Delta_c H°$, kJ/mol | | | Higher or gross heat of combustion Btu/lb | | Btu/cu ft |
|---|---|---|---|---|---|---|---|---|
| | | | crystal | liquid | gas | crystal | liquid | gas @ 60 F |
| $C_6H_{14}O$ | Dipropyl ether | 102.177 | | 4033.1 | 4069.0 | | 16970 | 4620 |
| $C_6H_{14}O$ | Diisopropyl ether | 102.177 | | 4010.4 | 4042.7 | | 16870 | 4590 |
| $C_6H_{14}O$ | 1-Hexanol | 102.177 | | 3984.4 | 4046.1 | | 16760 | 4590 |
| $C_6H_{14}O$ | 2-Hexanol | 102.177 | | 3969.9 | | | 16700 | |
| $C_6H_{14}O$ | 4-Methyl-2-pentanol | 102.177 | | 3967.2 | | | 16690 | |
| $C_6H_{14}O_2$ | 1,1-Diethoxyethane | 118.176 | | 3870.5 | 3908.4 | | 14080 | 4430 |
| $C_6H_{14}O_2$ | 1,2-Diethoxyethane | 118.176 | | 3910.5 | 3953.7 | | 14220 | 4480 |
| $C_6H_{14}O_2$ | 1,6-Hexanediol | 118.176 | 3792.0 | 3817.5 | 3900.7 | 13790 | 13890 | 4420 |
| $C_6H_{14}O_3$ | Trimethylolpropane | 134.175 | 3611.0 | | | 11570 | | |
| $C_6H_{14}O_4$ | Triethylene glycol | 150.175 | | 3557.7 | 3636.9 | | 10180 | 4120 |
| $C_6H_{15}N$ | Triethylamine | 101.192 | | 4377.1 | 4412.0 | | 18590 | 5000 |
| $C_6H_{15}N$ | Dipropylamine | 101.192 | | 4348.7 | 4388.7 | | 18470 | 4980 |
| $C_6H_{15}N$ | Diisopropylamine | 101.192 | | 4326.3 | 4360.8 | | 18380 | 4950 |
| $C_7H_5N$ | Benzonitrile | 103.123 | | 3632.3 | 3684.8 | | 15140 | 4180 |
| $C_7H_6O$ | Benzaldehyde | 106.124 | | 3525.1 | 3575.4 | | 14280 | 4060 |
| $C_7H_6O_2$ | Benzoic acid | 122.123 | 3226.9 | | 3318.0 | 11360 | | 3760 |
| $C_7H_6O_3$ | Salicylic acid | 138.123 | 3022.2 | | 3117.3 | 9410 | | 3540 |
| $C_7H_7NO$ | Benzamide | 121.139 | 3552.4 | | | 12610 | | |
| $C_7H_7NO_2$ | *o*-Nitrotoluene | 137.138 | | 3745.3 | | | 11740 | |
| $C_7H_7NO_2$ | *m*-Nitrotoluene | 137.138 | | 3723.5 | | | 11670 | |
| $C_7H_7NO_2$ | *p*-Nitrotoluene | 137.138 | 3706.9 | | 3786.0 | 11620 | | 4290 |
| $C_7H_8$ | Toluene | 92.141 | | 3910.3 | 3948.3 | | 18240 | 4480 |
| $C_7H_8O$ | *o*-Cresol | 108.140 | 3693.3 | | 3769.3 | 14680 | | 4280 |
| $C_7H_8O$ | *m*-Cresol | 108.140 | | 3703.9 | 3765.6 | | 14720 | 4270 |
| $C_7H_8O$ | *p*-Cresol | 108.140 | 3698.6 | | 3772.5 | 14700 | | 4280 |
| $C_7H_8O$ | Benzyl alcohol | 108.140 | | 3737.2 | 3797.5 | | 14860 | 4310 |
| $C_7H_8O$ | Anisole | 108.140 | | 3783.1 | 3830.0 | | 15040 | 4340 |
| $C_7H_9N$ | *o*-Methyl aniline | 107.155 | | 4034.5 | 4097.2 | | 16190 | 4650 |
| $C_7H_9N$ | *m*-Methyl aniline | 107.155 | | 4032.7 | 4095.4 | | 16180 | 4650 |
| $C_7H_9N$ | *p*-Methyl aniline | 107.155 | 4017.3 | | 4096.1 | 16120 | | 4650 |
| $C_7H_9N$ | 1-Cyclohexenecarbonitrile | 107.155 | | 4088.9 | 4142.4 | | 16400 | 4700 |
| $C_7H_9N$ | 2,3-Dimethylpyridine | 107.155 | | 4060.2 | 4109.1 | | 16290 | 4660 |
| $C_7H_9N$ | 2,4-Dimethylpyridine | 107.155 | | 4057.0 | 4104.7 | | 16280 | 4660 |

*(continued)*

**Appendix C2a.** *(continued)*

| Molecular formula | Name | $M$ g/mol | Enthalpy of combustion $-\Delta_C H°$, kJ/mol crystal | liquid | gas | Higher or gross heat of combustion Btu/lb crystal | liquid | Btu/cu ft gas @ 60 F |
|---|---|---|---|---|---|---|---|---|
| $C_7H_9N$ | 2,5-Dimethylpyridine | 107.155 | | 4059.5 | 4107.3 | | 16290 | 4660 |
| $C_7H_9N$ | 2,6-Dimethylpyridine | 107.155 | | 4053.5 | 4099.5 | | 16260 | 4650 |
| $C_7H_9N$ | 3,4-Dimethylpyridine | 107.155 | | 4059.1 | 4110.9 | | 16280 | 4660 |
| $C_7H_9N$ | 3,5-Dimethylpyridine | 107.155 | | 4063.3 | 4113.6 | | 16300 | 4670 |
| $C_7H_{11}N$ | Cyclohexanecarbonitrile | 109.171 | | 4279.4 | 4331.4 | | 16850 | 4910 |
| $C_7H_{12}$ | 1-Methylbicyclo(3,1,0)hexane | 96.172 | | 4436.4 | 4471.1 | | 19830 | 5070 |
| $C_7H_{14}$ | Cycloheptane | 98.188 | | 4598.8 | 4637.3 | | 20130 | 5260 |
| $C_7H_{14}$ | Methylcyclohexane | 98.188 | | 4565.3 | 4600.7 | | 19990 | 5220 |
| $C_7H_{14}$ | Ethylcyclopentane | 98.188 | | 4592.0 | 4628.5 | | 20100 | 5250 |
| $C_7H_{14}$ | *cis*-1,3-Dimethylcyclopentane | 98.188 | | 4585.3 | 4619.5 | | 20070 | 5240 |
| $C_7H_{14}$ | *cis*-1,2-Dimethylcyclopentane | 98.188 | | 4590.1 | 4625.9 | | 20100 | 5250 |
| $C_7H_{14}$ | *trans*-1,2-Dimethylcyclopentane | 98.188 | | 4584.2 | 4618.8 | | 20070 | 5240 |
| $C_7H_{14}$ | 1-Heptene | 98.188 | | 4657.5 | 4693.1 | | 20390 | 5320 |
| $C_7H_{14}$ | *cis*-2-Heptene | 98.188 | | 4650.3 | | | 20360 | |
| $C_7H_{14}$ | *cis*-3-Heptene | 98.188 | | 4651.1 | | | 20360 | |
| $C_7H_{14}O$ | 2,4-Dimethyl-3-pentanone | 114.188 | | 4402.5 | 4444.1 | | 16570 | 5040 |
| $C_7H_{14}O$ | 2,2-Dimethyl-3-pentanone | 114.188 | | 4399.3 | 4441.7 | | 16560 | 5040 |
| $C_7H_{14}O$ | 1-Heptanal | 114.188 | | 4443.9 | 4491.6 | | 16730 | 5090 |
| $C_7H_{14}O$ | *cis*-2-Methylcyclohexanol | 114.188 | | 4365.2 | 4428.4 | | 16430 | 5020 |
| $C_7H_{14}O$ | *trans*-2-Methylcyclohexanol | 114.188 | | 4339.7 | 4402.9 | | 16340 | 4990 |
| $C_7H_{14}O$ | *cis*-3-Methylcyclohexanol | 114.188 | | 4339.3 | 4404.5 | | 16340 | 5000 |
| $C_7H_{14}O$ | *trans*-3-Methylcyclohexanol | 114.188 | | 4361.0 | 4426.3 | | 16420 | 5020 |
| $C_7H_{14}O$ | *cis*-4-Methylcyclohexanol | 114.188 | | 4342.2 | 4407.9 | | 16350 | 5000 |
| $C_7H_{14}O$ | *trans*-4-Methylcyclohexanol | 114.188 | | 4322.1 | 4388.2 | | 16270 | 4980 |
| $C_7H_{14}O_2$ | Ethyl 2,2-dimethylpropanoate | 130.187 | | 4178.2 | 4219.4 | | 13800 | 4790 |
| $C_7H_{14}O_2$ | Methyl hexanoate | 130.187 | | 4215.2 | 4262.8 | | 13920 | 4830 |
| $C_7H_{14}O_2$ | Ethyl pentanoate | 130.187 | | 4202.4 | 4248.5 | | 13880 | 4820 |
| $C_7H_{14}O_2$ | Ethyl 3-methylbutanoate | 130.187 | | 4184.5 | 4228.4 | | 13820 | 4800 |
| $C_7H_{14}O_2$ | Heptanoic acid | 130.187 | | 4145.2 | 4219.2 | | 13690 | 4790 |
| $C_7H_{16}$ | Heptane | 100.204 | | 4817.0 | 4853.5 | | 20660 | 5500 |
| $C_7H_{16}$ | 2-Methylhexane | 100.204 | | 4811.7 | 4846.6 | | 20640 | 5500 |
| $C_7H_{16}$ | 3-Methylhexane | 100.204 | | 4814.8 | 4849.9 | | 20660 | 5500 |
| $C_7H_{16}$ | 2,2-Dimethylpentane | 100.204 | | 4802.9 | 4835.3 | | 20600 | 5480 |

*(continued)*

**Appendix C2a.** *(continued)*

| Molecular formula | Name | $M$ g/mol | Enthalpy of combustion $-\Delta_C H°$, kJ/mol crystal | liquid | gas | Higher or gross heat of combustion Btu/lb crystal | liquid | Btu/cu ft gas @ 60 F |
|---|---|---|---|---|---|---|---|---|
| $C_7H_{16}$ | 2,3-Dimethylpentane | 100.204 | | 4808.1 | 4842.3 | | 20630 | 5490 |
| $C_7H_{16}$ | 2,4-Dimethylpentane | 100.204 | | 4806.6 | 4839.5 | | 20620 | 5490 |
| $C_7H_{16}$ | 3,3-Dimethylpentane | 100.204 | | 4807.0 | 4840.0 | | 20620 | 5490 |
| $C_7H_{16}$ | 3-Ethylpentane | 100.204 | | 4816.4 | 4851.6 | | 20660 | 5500 |
| $C_7H_{16}$ | 2,2,3-Trimethylbutane | 100.204 | | 4804.7 | 4836.7 | | 20610 | 5490 |
| $C_7H_{16}O$ | 1-Heptanol | 116.203 | | 4637.9 | 4704.8 | | 17160 | 5340 |
| $C_8H_4O_3$ | Phthalic anhydride | 148.118 | 3259.6 | | 3348.3 | 9460 | | 3800 |
| $C_8H_6O_4$ | Phthalic acid | 166.133 | 3223.6 | | | 8340 | | |
| $C_8H_6O_4$ | Isophthalic acid | 166.133 | 3202.6 | | 3309.3 | 8290 | | 3750 |
| $C_8H_6O_4$ | Terephthalic acid | 166.133 | 3189.5 | | 3287.7 | 8250 | | 3730 |
| $C_8H_8$ | Styrene | 104.152 | | 4395.2 | 4439.3 | | 18140 | 5040 |
| $C_8H_8O$ | Acetophenone | 120.151 | | 4148.9 | 4204.7 | | 14840 | 4770 |
| $C_8H_8O_2$ | *o*-Toluic acid | 136.150 | 3874.9 | | | 12230 | | |
| $C_8H_8O_2$ | *m*-Toluic acid | 136.150 | 3865.3 | | | 12200 | | |
| $C_8H_8O_2$ | *p*-Toluic acid | 136.150 | 3862.2 | | | 12190 | | |
| $C_8H_8O_2$ | Methyl benzoate | 136.150 | | 3947.9 | 4003.5 | | 12460 | 4540 |
| $C_8H_{10}$ | Ethylbenzene | 106.167 | | 4564.9 | 4607.1 | | 18480 | 5230 |
| $C_8H_{10}$ | *o*-Xylene | 106.167 | | 4552.8 | 4596.3 | | 18430 | 5210 |
| $C_8H_{10}$ | *m*-Xylene | 106.167 | | 4551.8 | 4594.5 | | 18430 | 5210 |
| $C_8H_{10}$ | *p*-Xylene | 106.167 | | 4552.8 | 4595.2 | | 18430 | 5210 |
| $C_8H_{10}O$ | *o*-Ethylphenol | 122.167 | | 4368.4 | 4432.0 | | 15370 | 5030 |
| $C_8H_{10}O$ | *m*-Ethylphenol | 122.167 | | 4362.9 | 4431.1 | | 15350 | 5030 |
| $C_8H_{10}O$ | *p*-Ethylphenol | 122.167 | 4352.8 | | 4433.1 | 15320 | | 5030 |
| $C_8H_{10}O$ | 2,3-Xylenol | 122.167 | 4336.1 | | 4420.0 | 15260 | | 5010 |
| $C_8H_{10}O$ | 2,4-Xylenol | 122.167 | | 4348.5 | 4414.3 | | 15300 | 5010 |
| $C_8H_{10}O$ | 2,5-Xylenol | 122.167 | 4330.6 | | 4415.6 | 15240 | | 5010 |
| $C_8H_{10}O$ | 2,6-Xylenol | 122.167 | 4339.8 | | 4415.4 | 15270 | | 5010 |
| $C_8H_{10}O$ | 3,4-Xylenol | 122.167 | 4334.9 | | 4420.6 | 15250 | | 5010 |
| $C_8H_{10}O$ | 3,5-Xylenol | 122.167 | 4332.8 | | 4415.7 | 15250 | | 5010 |
| $C_8H_{10}O$ | Phenetole | 122.167 | | 4424.6 | 4475.6 | | 15570 | 5080 |
| $C_8H_{11}N$ | *N,N*-Dimethylaniline | 121.182 | | 4767.8 | 4820.6 | | 16910 | 5470 |
| $C_8H_{11}N$ | *N*-Ethylaniline | 121.182 | | 4724.1 | 4776.4 | | 16760 | 5420 |
| $C_8H_{15}N$ | Octanenitrile | 125.214 | | 5184.5 | 5241.3 | | 17800 | 5940 |

*(continued)*

**Appendix C2a.** *(continued)*

| Molecular formula | Name | $M$ g/mol | Enthalphy of combustion $-\Delta_C H°$, kJ/mol crystal | liquid | gas | Higher or gross heat of combustion Btu/lb crystal | liquid | Btu/cu ft gas @ 60 F |
|---|---|---|---|---|---|---|---|---|
| $C_8H_{16}$ | Cyclooctane | 112.215 | | 5267.0 | 5310.3 | | 20180 | 6020 |
| $C_8H_{16}$ | Propylcyclopentane | 112.215 | | 5245.9 | 5287.0 | | 20100 | 6000 |
| $C_8H_{16}$ | 1-Ethyl-1-methylcyclopentane | 112.215 | | 5240.9 | | | 20080 | |
| $C_8H_{16}$ | Ethylcyclohexane | 112.215 | | 5222.8 | 5263.0 | | 20010 | 5970 |
| $C_8H_{16}$ | 1,1-Dimethylcyclohexane | 112.215 | | 5216.0 | 5253.8 | | 19980 | 5960 |
| $C_8H_{16}$ | *cis*-1,2-Dimethylcyclohexane | 112.215 | | 5222.9 | 5262.6 | | 20010 | 5970 |
| $C_8H_{16}$ | *trans*-1,2-Dimethylcyclohexane | 112.215 | | 5216.5 | 5254.8 | | 19980 | 5960 |
| $C_8H_{16}$ | *cis*-1,3-Dimethylcyclohexane | 112.215 | | 5211.8 | 5250.1 | | 19970 | 5950 |
| $C_8H_{16}$ | *trans*-1,3-Dimethylcyclohexane | 112.215 | | 5219.0 | 5258.2 | | 19990 | 5960 |
| $C_8H_{16}$ | *cis*-1,4-Dimethylcyclohexane | 112.215 | | 5219.1 | 5258.1 | | 19990 | 5960 |
| $C_8H_{16}$ | *trans*-1,4-Dimethylcyclohexane | 112.215 | | 5212.3 | 5250.2 | | 19970 | 5950 |
| $C_8H_{16}O$ | 2,2,4-Trimethyl-3-pentanone | 128.214 | | 5053.1 | 5096.4 | | 16940 | 5780 |
| $C_8H_{16}O$ | 2-Ethylhexanal | 128.214 | | 5086.2 | 5135.1 | | 17050 | 5820 |
| $C_8H_{16}O_2$ | Octanoic acid | 144.214 | | 4798.7 | 4880.4 | | 14300 | 5540 |
| $C_8H_{16}O_2$ | 2-Ethylhexanoic acid | 144.214 | | 4799.6 | 4875.2 | | 14310 | 5530 |
| $C_8H_{16}O_2$ | Methyl heptanoate | 144.214 | | 4867.6 | 4918.8 | | 14510 | 5580 |
| $C_8H_{18}$ | Octane | 114.231 | | 5470.5 | 5512.0 | | 20590 | 6250 |
| $C_8H_{18}$ | 2-Methylheptane | 114.231 | | 5465.6 | 5505.2 | | 20570 | 6240 |
| $C_8H_{18}$ | 3-Methylheptane | 114.231 | | 5468.3 | 5508.1 | | 20580 | 6250 |
| $C_8H_{18}$ | 4-Methylheptane | 114.231 | | 5469.0 | 5508.6 | | 20580 | 6250 |
| $C_8H_{18}$ | 2,2-Dimethylhexane | 114.231 | | 5458.7 | 5496.0 | | 20540 | 6230 |
| $C_8H_{18}$ | 2,3-Dimethylhexane | 114.231 | | 5468.0 | 5506.8 | | 20580 | 6250 |
| $C_8H_{18}$ | 2,4-Dimethylhexane | 114.231 | | 5463.6 | 5501.4 | | 20560 | 6240 |
| $C_8H_{18}$ | 2,5-Dimethylhexane | 114.231 | | 5460.2 | 5498.1 | | 20550 | 6240 |
| $C_8H_{18}$ | 3,3-Dimethylhexane | 114.231 | | 5463.1 | 5500.6 | | 20560 | 6240 |
| $C_8H_{18}$ | 3,4-Dimethylhexane | 114.231 | | 5468.8 | 5507.8 | | 20580 | 6250 |
| $C_8H_{18}$ | 3-Ethylhexane | 114.231 | | 5470.2 | 5509.9 | | 20590 | 6250 |
| $C_8H_{18}$ | 3-Ethyl-2-methylpentane | 114.231 | | 5471.0 | 5509.6 | | 20590 | 6250 |
| $C_8H_{18}$ | 3-Ethyl-3-methylpentane | 114.231 | | 5467.8 | 5505.8 | | 20580 | 6240 |
| $C_8H_{18}$ | 2,2,3-Trimethylpentane | 114.231 | | 5463.7 | 5500.6 | | 20560 | 6240 |
| $C_8H_{18}$ | 2,2,4-Trimethylpentane | 114.231 | | 5461.4 | 5496.6 | | 20550 | 6230 |
| $C_8H_{18}$ | 2,3,3-Trimethylpentane | 114.231 | | 5467.1 | 5504.3 | | 20570 | 6240 |
| $C_8H_{18}$ | 2,3,4-Trimethylpentane | 114.231 | | 5465.6 | 5503.3 | | 20570 | 6240 |

*(continued)*

**Appendix C2a.** *(continued)*

| Molecular formula | Name | $M$ g/mol | Enthalpy of combustion $-\Delta_c H°$, kJ/mol crystal | liquid | gas | Higher or gross heat of combustion Btu/lb crystal | liquid | Btu/cu ft gas @ 60 F |
|---|---|---|---|---|---|---|---|---|
| $C_8H_{18}$ | 2,2,3,3-Tetramethylbutane | 114.231 | 5451.7 | | 5495.0 | 20520 | | 6230 |
| $C_8H_{18}N_2$ | Azobutane | 142.244 | | 5680.5 | 5729.8 | | 17170 | 6500 |
| $C_8H_{18}O$ | 1-Octanol | 130.230 | | 5294.1 | 5365.1 | | 17480 | 6090 |
| $C_8H_{18}O$ | 2-Ethyl-1-hexanol | 130.230 | | 5287.8 | 5355.3 | | 17450 | 6070 |
| $C_8H_{18}O$ | Dibutyl ether | 130.230 | | 5342.7 | 5387.2 | | 17640 | 6110 |
| $C_8H_{18}O$ | Di-*sec*-butyl ether | 130.230 | | 5319.1 | 5359.7 | | 17560 | 6080 |
| $C_8H_{18}O$ | Di-*tert*-butyl ether | 130.230 | | 5321.0 | 5358.6 | | 17560 | 6080 |
| $C_8H_{18}O_5$ | Tetraethylene glycol | 194.228 | | 4738.9 | 4837.6 | | 10490 | 5490 |
| $C_8H_{19}N$ | Dibutylamine | 129.246 | | 5657.5 | 5706.9 | | 18820 | 6470 |
| $C_8H_{19}N$ | Diisobutylamine | 129.246 | | 5645.0 | 5684.3 | | 18780 | 6450 |
| $C_9H_7N$ | Isoquinoline | 129.161 | | 4686.5 | | | 15600 | |
| $C_9H_8$ | Indene | 116.163 | | 4795.5 | 4848.3 | | 17750 | 5500 |
| $C_9H_{10}$ | Cyclopropylbenzene | 118.178 | | 5071.0 | 5121.2 | | 18450 | 5810 |
| $C_9H_{10}$ | Indan | 118.178 | | 4982.2 | 5031.4 | | 18120 | 5710 |
| $C_9H_{11}NO_2$ | 1-Phenylalanine | 165.192 | 4646.8 | | 4800.8 | 12090 | | 5450 |
| $C_9H_{11}NO_3$ | 1-Tyrosine | 181.191 | 4428.6 | | | 10510 | | |
| $C_9H_{12}$ | Cumene | 120.194 | | 5215.5 | 5260.6 | | 18650 | 5970 |
| $C_9H_{12}$ | *o*-Ethyltoluene | 120.194 | | 5210.2 | 5257.9 | | 18630 | 5960 |
| $C_9H_{12}$ | *m*-Ethyltoluene | 120.194 | | 5207.9 | 5254.8 | | 18630 | 5960 |
| $C_9H_{12}$ | *p*-Ethyltoluene | 120.194 | | 5206.8 | 5253.4 | | 18620 | 5960 |
| $C_9H_{12}$ | Propylbenzene | 120.194 | | 5218.3 | 5264.5 | | 18660 | 5970 |
| $C_9H_{12}$ | 1,2,3-Trimethylbenzene | 120.194 | | 5198.1 | 5247.1 | | 18590 | 5950 |
| $C_9H_{12}$ | 1,2,4-Trimethylbenzene | 120.194 | | 5194.8 | 5242.8 | | 18580 | 5950 |
| $C_9H_{12}$ | Mesitylene | 120.194 | | 5193.2 | 5240.7 | | 18570 | 5940 |
| $C_9H_{14}O_6$ | Triacetin | 218.207 | | 4211.6 | 4293.6 | | 8300 | 4870 |
| $C_9H_{18}$ | Propylcyclohexane | 126.242 | | 5876.7 | 5921.6 | | 20010 | 6720 |
| $C_9H_{18}O$ | 2-Nonanone | 142.241 | | 5716.9 | 5773.4 | | 17280 | 6550 |
| $C_9H_{18}O$ | 5-Nonanone | 142.241 | | 5715.9 | 5769.2 | | 17270 | 6540 |
| $C_9H_{18}O$ | 2,6-Dimethyl-4-heptanone | 142.241 | | 5705.6 | 5756.5 | | 17240 | 6530 |
| $C_9H_{18}O_2$ | Nonanoic acid | 158.241 | | 5454.4 | 5536.8 | | 14820 | 6280 |
| $C_9H_{18}O_2$ | Methyl octanoate | 158.241 | | 5523.8 | 5580.3 | | 15010 | 6330 |
| $C_9H_{20}$ | Nonane | 128.258 | | 6125.2 | 6171.7 | | 20530 | 7000 |
| $C_9H_{20}$ | 2,2-Dimethylheptane | 128.258 | | 6111.7 | | | 20480 | |

*(continued)*

**Appendix C2a.** *(continued)*

| Molecular formula | Name | $M$ g/mol | Enthalpy of combustion $-\Delta_cH^\circ$, kJ/mol crystal | liquid | gas | Higher or gross heat of combustion Btu/lb crystal | liquid | Btu/cu ft gas @ 60 F |
|---|---|---|---|---|---|---|---|---|
| $C_9H_{20}$ | 2,2,5-Trimethylhexane | 128.258 | | 6106.6 | | | 20470 | |
| $C_9H_{20}$ | 2,3,5-Trimethylhexane | 128.258 | | 6115.9 | | | 20500 | |
| $C_9H_{20}$ | 2,2,3,3-Tetramethylpentane | 128.258 | | 6121.6 | 6162.8 | | 20520 | 6990 |
| $C_9H_{20}$ | 2,2,3,4-Tetramethylpentane | 128.258 | | 6122.2 | 6163.0 | | 20520 | 6990 |
| $C_9H_{20}$ | 3,3-Diethylpentane | 128.258 | | 6124.5 | 6167.6 | | 20530 | 7000 |
| $C_9H_{20}$ | 2,2,4,4-Tetramethylpentane | 128.258 | | 6119.9 | 6158.3 | | 20510 | 6980 |
| $C_9H_{20}$ | 2,3,3,4-Tetramethylpentane | 128.258 | | 6122.0 | 6163.8 | | 20520 | 6990 |
| $C_9H_{20}O$ | 1-Nonanol | 144.257 | | 5943.4 | | | 17710 | |
| $C_{10}H_8$ | Azulene | 128.174 | 5290.7 | | 5367.5 | 17740 | | 6090 |
| $C_{10}H_8$ | Naphthalene | 128.174 | 5156.3 | | 5228.7 | 17290 | | 5930 |
| $C_{10}H_8O$ | 1-Naphthol | 144.173 | 4957.4 | | 5048.5 | 14780 | | 5730 |
| $C_{10}H_8O$ | 2-Naphthol | 144.173 | | 4954.2 | 5048.4 | | 14770 | 5730 |
| $C_{10}H_{10}O_4$ | Dimethyl isophthalate | 194.187 | 4633.4 | | | 10260 | | |
| $C_{10}H_{10}O_4$ | Dimethyl terephthalate | 194.187 | 4631.7 | | | 10250 | | |
| $C_{10}H_{12}$ | 1,2,3,4-Tetrahydronaphthalene | 132.205 | | 5620.9 | 5676.1 | | 18280 | 6440 |
| $C_{10}H_{14}$ | Butylbenzene | 134.221 | | 5872.7 | 5922.8 | | 18810 | 6720 |
| $C_{10}H_{14}$ | Isobutylbenzene | 134.221 | | 5866.1 | 5914.4 | | 18790 | 6710 |
| $C_{10}H_{14}$ | *sec*-Butylbenzene | 134.221 | | 5869.5 | 5918.5 | | 18800 | 6710 |
| $C_{10}H_{14}$ | *tert*-Butylbenzene | 134.221 | | 5865.2 | 5913.3 | | 18780 | 6710 |
| $C_{10}H_{14}$ | *o*-Cymene | 134.221 | | 5862.6 | | | 18780 | |
| $C_{10}H_{14}$ | *m*-Cymene | 134.221 | | 5857.3 | | | 18760 | |
| $C_{10}H_{14}$ | *p*-Cymene | 134.221 | | 5857.9 | | | 18760 | |
| $C_{10}H_{14}$ | *o*-Diethylbenzene | 134.221 | | 5867.4 | | | 18790 | |
| $C_{10}H_{14}$ | *m*-Diethylbenzene | 134.221 | | 5862.4 | | | 18780 | |
| $C_{10}H_{14}$ | *p*-Diethylbenzene | 134.221 | | 5863.1 | | | 18780 | |
| $C_{10}H_{14}$ | 1,2,4,5-Tetramethylbenzene | 134.221 | 5816.0 | | | 18630 | | |
| $C_{10}H_{14}$ | 3-Ethyl-*o*-xylene | 134.221 | | 5855.4 | | | 18750 | |
| $C_{10}H_{14}O$ | Thymol | 150.221 | 5626.2 | | 5717.4 | 16100 | | 6480 |
| $C_{10}H_{16}$ | α-Pinene | 136.237 | | 6205.3 | 6250.0 | | 19580 | 7090 |
| $C_{10}H_{16}$ | β-Pinene | 136.237 | | 6214.0 | 6260.4 | | 19610 | 7100 |
| $C_{10}H_{16}O$ | Camphor | 152.236 | 5902.3 | | 5954.2 | 16670 | | 6750 |
| $C_{10}H_{18}$ | *cis*-Decahydronaphthalene | 138.253 | | 6288.2 | 6338.4 | | 19550 | 7190 |
| $C_{10}H_{18}$ | *trans*-Decahydronaphthalene | 138.253 | | 6277.0 | 6325.5 | | 19520 | 7170 |

*(continued)*

**Appendix C2a.** *(continued)*

| Molecular formula | Name | $M$ g/mol | Enthalpy of combustion $-\Delta_c H°$, kJ/mol crystal | liquid | gas | Higher or gross heat of combustion Btu/lb crystal | liquid | Btu/cu ft gas @ 60 F |
|---|---|---|---|---|---|---|---|---|
| $C_{10}H_{18}O_4$ | Sebacic acid | 202.251 | 5425.0 | | 5585.7 | 11530 | | 6340 |
| $C_{10}H_{19}N$ | Decanenitrile | 153.268 | | 6492.1 | 6559.0 | | 18210 | 7440 |
| $C_{10}H_{20}$ | Butylcyclohexane | 140.269 | | 6530.3 | 6580.1 | | 20010 | 7460 |
| $C_{10}H_{20}$ | 1-Decene | 140.269 | | 6619.6 | 6670.0 | | 20290 | 7570 |
| $C_{10}H_{20}$ | Decanoic acid | 172.268 | 6079.7 | 6109.1 | 6198.5 | 15170 | 15240 | 7030 |
| $C_{10}H_{22}$ | Decane | 142.285 | | 6778.3 | 6829.7 | | 20480 | 7750 |
| $C_{10}H_{22}$ | 2-Methylnonane | 142.285 | | 6769.4 | 6819.3 | | 20450 | 7730 |
| $C_{10}H_{22}$ | 5-Methylnonane | 142.285 | | 6771.3 | 6820.6 | | 20460 | 7740 |
| $C_{10}H_{22}O$ | 1-Decanol | 158.284 | | 6601.1 | 6682.8 | | 17930 | 7580 |
| $C_{11}H_{10}$ | 1-Methylnaphthalene | 142.200 | | 5814.1 | | | 17580 | |
| $C_{11}H_{10}$ | 2-Methylnaphthalene | 142.200 | 5802.7 | | 5864.5 | 17540 | | 6650 |
| $C_{11}H_{12}N_2O_2$ | 1-Tryptophan | 204.229 | 5628.3 | | | 11850 | | |
| $C_{11}H_{24}$ | Undecane | 156.312 | | 7431.4 | 7487.7 | | 20440 | 8490 |
| $C_{12}H_8$ | Acenaphthylene | 152.196 | 6052.1 | | 6125.1 | 17090 | | 6950 |
| $C_{12}H_9N$ | Carbazole | 167.210 | 6133.5 | | 6218.0 | 15770 | | 7050 |
| $C_{12}H_{10}$ | Acenaphthene | 154.211 | 6221.6 | | 6307.3 | 17340 | | 7150 |
| $C_{12}H_{10}$ | Biphenyl | 154.211 | 6250.7 | | 6332.7 | 17420 | | 7180 |
| $C_{12}H_{10}O$ | Diphenyl ether | 170.211 | 6119.2 | | 6203.3 | 15450 | | 7040 |
| $C_{12}H_{11}N$ | Diphenylamine | 169.226 | 6424.4 | | 6513.5 | 16320 | | 7390 |
| $C_{12}H_{12}N_2$ | *p*-Benzidine | 184.241 | 6507.8 | | | 15180 | | |
| $C_{12}H_{14}O_4$ | Diethyl phthalate | 222.241 | | 5946.3 | 6034.5 | | 11500 | 6840 |
| $C_{12}H_{16}$ | Cyclohexylbenzene | 160.259 | | 6932.2 | 6992.1 | | 18590 | 7930 |
| $C_{12}H_{18}$ | 5,7-Dodecadiyne | 162.275 | | 7476.1 | | | 19800 | |
| $C_{12}H_{18}$ | 3,9-Dodecadiyne | 162.275 | | 7492.4 | | | 19850 | |
| $C_{12}H_{18}$ | Hexamethylbenzene | 162.275 | 7133.1 | | 7207.8 | 18900 | | 8180 |
| $C_{12}H_{22}$ | Cyclohexylcyclohexane | 166.307 | | 7592.6 | 7650.6 | | 19630 | 8680 |
| $C_{12}H_{24}$ | 1-Dodecene | 168.323 | | 7925.9 | 7986.7 | | 20240 | 9060 |
| $C_{12}H_{24}O_2$ | Dodecanoic acid | 200.321 | 7377.5 | 7414.2 | 7510.1 | 15830 | 15910 | 8520 |
| $C_{12}H_{26}$ | Dodecane | 170.338 | | 8087.0 | 8148.2 | | 20410 | 9240 |
| $C_{12}H_{26}O$ | 1-Dodecanol | 186.338 | | 7909.4 | 8001.3 | | 18250 | 9080 |
| $C_{12}H_{27}N$ | Tributylamine | 185.353 | | 8299.2 | | | 19250 | |
| $C_{13}H_{10}O$ | Benzophenone | 182.222 | 6510.3 | | 6599.7 | 15360 | | 7490 |
| $C_{13}H_{12}$ | Diphenylmethane | 168.238 | 6902.1 | 6920.3 | 6969.6 | 17640 | 17680 | 7900 |

*(continued)*

**Appendix C2a.** *(concluded)*

| Molecular formula | Name | $M$ g/mol | Enthalphy of combustion $-\Delta_C H°$, kJ/mol crystal | liquid | gas | Higher or gross heat of combustion Btu/lb crystal | Btu/lb liquid | Btu/cu ft gas @ 60 F |
|---|---|---|---|---|---|---|---|---|
| $C_{13}H_{26}O_2$ | Methyl dodecanoate | 214.348 | | 8138.4 | 8216.6 | | 16320 | 9320 |
| $C_{13}H_{28}O$ | 1-Tridecanol | 200.365 | 8517.9 | | | 18270 | | |
| $C_{14}H_{10}$ | Anthracene | 178.233 | 7067.5 | | 7169.2 | 17050 | | 8130 |
| $C_{14}H_{10}$ | Phenanthrene | 178.233 | 7054.5 | | 7145.8 | 17010 | | 8100 |
| $C_{14}H_{10}$ | Diphenylacetylene | 178.233 | 7250.7 | | | 17490 | | |
| $C_{14}H_{10}O_2$ | Benzil | 210.232 | 6784.4 | | 6882.8 | 13870 | | 7810 |
| $C_{14}H_{12}$ | *cis*-Stilbene | 180.249 | | 7407.4 | 7476.4 | | 17670 | 8480 |
| $C_{14}H_{12}$ | *trans*-Stilbene | 180.249 | 7361.0 | | 7460.2 | 17560 | | 8460 |
| $C_{14}H_{14}$ | 1,1-Diphenylethane | 182.265 | | 7558.7 | | | 17830 | |
| $C_{14}H_{14}$ | 1,2-Diphenylethane | 182.265 | 7561.5 | | 7652.9 | 17830 | | 8680 |
| $C_{14}H_{27}N$ | Tetradecanenitrile | 209.375 | | 9107.6 | 9192.9 | | 18700 | 10430 |
| $C_{14}H_{28}O_2$ | Tetradecanoic acid | 228.375 | 8677.3 | 8722.0 | 8817.1 | 16330 | 16420 | 10000 |
| $C_{14}H_{30}O$ | 1-Tetradecanol | 214.392 | 9167.0 | 9216.0 | | 18380 | 18480 | |
| $C_{15}H_{30}O_2$ | Pentadecylic acid | 242.402 | 9328.4 | 9378.4 | 9491.1 | 16540 | 16630 | 10760 |
| $C_{16}H_{10}$ | Fluoranthene | 202.255 | 7915.2 | | 8014.3 | 16820 | | 9090 |
| $C_{16}H_{10}$ | Pyrene | 202.255 | 7850.8 | | 7951.0 | 16690 | | 9020 |
| $C_{16}H_{22}O_4$ | Dibutyl phthalate | 278.348 | | 8597.7 | 8689.4 | | 13280 | 9860 |
| $C_{16}H_{26}$ | Decylbenzene | 218.382 | | 9793.7 | 9873.4 | | 19280 | 11200 |
| $C_{16}H_{32}$ | 1-Hexadecene | 224.430 | | 10540.7 | 10620.9 | | 20190 | 12050 |
| $C_{16}H_{32}O_2$ | Hexadecanoic acid | 256.429 | 9977.9 | 10031.3 | 10132.3 | 16730 | 16820 | 11490 |
| $C_{16}H_{34}$ | Hexadecane | 226.446 | | 10699.2 | 10780.5 | | 20310 | 12230 |
| $C_{16}H_{34}O$ | 1-Hexadecanol | 242.445 | 10468.8 | | 10638.3 | 18560 | | 12070 |
| $C_{17}H_{34}O_2$ | Margaric acid | 270.456 | 10624.4 | 10683.2 | | 16890 | 16980 | |
| $C_{18}H_{12}$ | Chrysene | 228.293 | 8943.5 | | 9068.0 | 16840 | | 10280 |
| $C_{18}H_{36}O_2$ | Stearic acid | 284.483 | 11280.4 | 11343.4 | 11446.9 | 17050 | 17140 | 12980 |
| $C_{18}H_{38}$ | Octadecane | 254.500 | 11946.6 | | 12099.4 | 20180 | | 13720 |
| $C_{19}H_{36}O_2$ | Methyl oleate | 296.494 | | 11887.1 | 11971.7 | | 17230 | 13580 |
| $C_{20}H_{12}$ | Perylene | 252.315 | 9768.0 | | | 16640 | | |
| $C_{20}H_{40}O_2$ | Arachidic acid | 312.536 | 12574.9 | 12646.8 | 12774.4 | 17300 | 17390 | 14490 |
| $C_{22}H_{42}O_2$ | Brassidic acid | 338.574 | 13699.0 | | | 17390 | | |

**Appendix C2b: Heat of combustion of organic compounds.** *(Reproduced from the International Critical Tables [1929] by permission from the National Academy of Sciences.)*

The compounds are arranged in groups as follows:

1. (C, H), Hydrocarbons.
2. (C, H, O), Alcohols (carbinols).
3. (C, H, O), Carboxylic acids.
4. (C, H, O), Acid anhydrides and lactones.
5. (C, H, O), Carbohydrates (sugars, starch, cellulose, etc.).
6. (C, H, O), Other C, H, O-compounds.
7. Nitrogen compounds.
8. Halogen compounds.
9. Sulfur compounds.

| Formula | Name | kg cal/mol | Btu/scf |
|---|---|---|---|
| | **1. Hydrocarbons** | | |
| $CH_4$ | Methane *(gas)* | 210.8 | 1003.4 |
| $C_2H_2$ | Acetylene (ethine) *(gas)* | 312.0 | 1485.1 |
| $C_2H_4$ | Ethylene *(gas)* | 332 | 1580.3 |
| $C_2H_6$ | Ethane *(gas)* | 368.4 | 1753.6 |
| $C_3H_4$ | Allylene (propine) *(gas)* | 456.1 | 2171.0 |
| | | 473 | 2251.5 |
| $C_3H_6$ | Propylene | 490.2 | 2333.4 |
| $C_2H_6$ | Trimethylene *(gas)* | 496.8 | 2364.8 |
| $C_3H_8$ | Propane *(gas)* | 526.3 | 2505.2 |
| $C_4H_8$ | Isobutylene *(gas)* | 647.2 | 3080.7 |
| $C_4H_{10}$ | Isobutane (trimethylmethane) *(gas)* | 683.4 | 3253.0 |
| $C_5H_{10}$ | Amylene | 803.4 | 3824.2 |
| $C_5H_{10}$ | Cyclopentane | 783.6 | 3729.9 |
| $C_5H_{10}$ | Methylcyclobutane | 784.2 | 3732.8 |
| $C_5H_{10}$ | Trimethylethylene *(gas)* | 803.6 | 3825.1 |
| | *(liquid)* | 796 | 3789.0 |
| $C_5H_{12}$ | Isopentane *(gas)* | 843.5 | 4015.1 |
| $C_5H_{12}$ | *(liquid)* | 838.3 | 3990.3 |
| $C_5H_{12}$ | *n*-Pentane *(gas)* | 838.3 | 3990.3 |
| $C_5H_{12}$ | *(liquid)* | 833.4 | 3967.0 |
| $C_5H_{12}$ | Tetramethylmethane | 842.6 | 4010.8 |
| $C_6H_6$ | Benzene *(gas)* | 787.2 | 3747.1 |
| $C_6H_6$ | | 783.4 | 3729.0 |
| $C_6H_6$ | Dimethyldiacetylene (2, 4-hexadine) | 847.8 | 4035.5 |
| $C_6H_6$ | Dipropargyl (1, 5-hexadine) | 853.5 | 4062.7 |
| | | 882.9 | 4202.6 |

**Appendix C2b. Heat of combustion of organic compounds** *(continued)*

| Formula | Name | kg cal/mol | Btu/scf |
|---|---|---|---|
| | **1. Hydrocarbons** *(continued)* | | |
| $C_6H_8$ | Dihydrobenzene | 833.2 | 3966.0 |
| | | 847.8 | 4035.5 |
| $C_6H_{10}$ | Bicyclohexane (0, 1, 3) | 912.5 | 4343.5 |
| $C_6H_{10}$ | Diallyl *(gas)* | 928.1 | 4417.8 |
| | | 903.4 | 4300.2 |
| $C_6H_{10}$ | Dimethylmethylenecyclopropane | 898 | 4274.5 |
| $C_6H_{10}$ | Tetrahydrobenzene (cyclohexene) | 892 | 4245.9 |
| $C_6H_{12}$ | Cyclohexane | 939 | 4469.6 |
| $C_6H_{12}$ | Hexylene | 952.6 | 4534.4 |
| $C_6H_{12}$ | Methylcyclopentane | 937.9 | 4464.4 |
| $C_6H_{14}$ | Diisopropyl *(gas)* | 993.9 | 4731.0 |
| $C_6H_{14}$ | *n*-Hexane | 990.6 | 4715.3 |
| $C_7H_8$ | Toluene | 936 | 4455.4 |
| $C_7H_{12}$ | Bicycloheptane | 1030 | 4902.8 |
| $C_7H_{12}$ | Cycloheptene | 1049 | 4993.2 |
| $C_7H_{12}$ | Heptine-1 | 1091 | 5193.2 |
| $C_7H_{12}$ | Methylenecyclohexane | 1050 | 4998 |
| $C_7H_{12}$ | 1-Methyl-3-cyclohexene | 1043 | 4964.7 |
| $C_7H_{12}$ | 1-Methyl-1-cyclohexene | 1040 | 4950.4 |
| | | 1048 | 4988.5 |
| $C_7H_{14}$ | Cycloheptane | 1087 | 5174.1 |
| $C_7H_{14}$ | 1, 3-Dimethylcyclopentane | 1090 | 5188.4 |
| $C_7H_{14}$ | Methylcyclohexane | 1091 | 5193.2 |
| $C_7H_{16}$ | 2, 2-Dimethylpentane | 1148 | 5464.5 |
| $C_7H_{16}$ | 2, 3-Dimethylpentane | 1148 | 5464.5 |
| $C_7H_{16}$ | 2, 4-Dimethylpentane | 1148 | 5464.5 |
| $C_7H_{16}$ | 3, 3-Dimethylpentane | 1147 | 5459.7 |
| $C_7H_{16}$ | 3-Ethylpentane | 1149 | 5469.2 |
| $C_7H_{16}$ | *n*-Heptane | 1149 | 5469.2 |
| $C_7H_{16}$ | 2-Methylhexane | 1148 | 5464.5 |
| $C_7H_{16}$ | 3-Methylhexane | 1148 | 5464.5 |
| $C_7H_{16}$ | 2, 2, 3-Trimethylbutane | 1147 | 5459.7 |
| $C_8H_6$ | Phenylacetylene (phenylethine) | 1024 | 4874.2 |
| $C_8H_8$ | Styrene (phenylethylene) | 1046 | 4979.0 |
| $C_8H_{10}$ | Ethylbenzene | 1091 | 5193.2 |
| $C_8H_{10}$ | *o*-Xylene | 1093 | 5202.7 |
| $C_8H_{10}$ | *m*-Xylene | 1089 | 5183.6 |
| $C_8H_{10}$ | *p*-Xylene | 1089 | 5183.6 |

**Appendix C2b. Heat of combustion of organic compounds** *(continued)*

| Formula | Name | kg cal/mol | Btu/scf |
|---|---|---|---|
| | **1. Hydrocarbons** *(continued)* | | |
| $C_8H_{12}$ | 1, 4-Dimethylcyclohexa-1, 3-diene | 1152 | 5483.5 |
| $C_8H_{12}$ | 1, 3-Dimethyldihydrobenzene | 1148 | 5464.5 |
| $C_8H_{12}$ | 1, Methyl-3-methylene-1-cyclohexene | 1149 | 5469.2 |
| $C_8H_{14}$ | 1, 3-Dimethyl-3-cyclohexene | 1194 | 5683.4 |
| $C_8H_{14}$ | Ethylenecyclohexane | 1207 | 5745.3 |
| $C_8H_{14}$ | 1-Ethyl-1-cyclohexene | 1203 | 5726.3 |
| $C_8H_{14}$ | Laurolene | 1192 | 5673.9 |
| $C_8H_{14}$ | *i*-Laurolene (1, 1, 2-trimethylcyclopent-2-ene) | 1193 | 5678.7 |
| $C_8H_{16}$ | 1, 1-Dimethylcyclohexane | 1242 | 5911.9 |
| $C_8H_{16}$ | 1, 3-Dimethylcyclohexane | 1238 | 5892.9 |
| $C_8H_{16}$ | 1, 4-Dimethylcyclohexane | 1229 | 5850.0 |
| $C_8H_{16}$ | Diisobutylene | 1252 | 5959.5 |
| $C_8H_{16}$ | Methylcycloheptane | 1244 | 5921.4 |
| $C_8H_{16}$ | 1, 2, 4-Trimethylcyclopentane | 1245 | 5926.2 |
| $C_8H_{18}$ | 2, 5-Dimethylhexane | 1303 | 6202.3 |
| $C_8H_{18}$ | 3, 4-Dimethylhexane | 1303 | 6202.3 |
| $C_8H_{18}$ | 2-Ethylhexane | 1302 | 6197.5 |
| $C_8H_{18}$ | Hexamethylethane *(solid)* | 1301 | 6192.8 |
| $C_8H_{18}$ | 2-Methylheptane | 1306 | 6216.6 |
| $C_8H_{18}$ | *n*-Octane | 1305 | 6211.8 |
| $C_8H_{18}$ | 2, 2, 4-Trimethylpentane | 1303 | 6202.3 |
| $C_9H_{10}$ | $\alpha$-Methylstyrene | 1202 | 5721.5 |
| $C_9H_{10}$ | $\beta$-Methylstyrene (*p*-tolylethylene) | 1202 | 5721.5 |
| $C_9H_{12}$ | Isopropylbenzene | 1247 | 5935.7 |
| $C_9H_{12}$ | Mesitylene | 1243 | 5916.7 |
| $C_9H_{12}$ | *n*-Propylbenzene | 1246 | 5931.0 |
| $C_9H_{12}$ | Pseudocumene (1, 2, 4-trimethylbenzene) | 1241 | 5907.2 |
| $C_9H_{14}$ | 1-Methyl-4-ethylcyclohexa-1, 3-diene | 1310 | 6235.6 |
| $C_9H_{16}$ | 1-Isopropylcyclohex-1-ene | 1340 | 6378.4 |
| $C_9H_{18}$ | Ethylcycloheptane | 1406 | 6692.6 |
| $C_9H_{18}$ | Methyl-*l*-*n*-propylcyclopentane | 1401 | 6668.8 |
| $C_9H_{18}$ | 1, 3, 3-Trimethylcyclohexane | 1395 | 6640.2 |
| $C_9H_{18}$ | 1, 2, 3-Trimethylcyclohexane | 1396 | 6645.0 |
| $C_9H_{18}$ | 1, 3, 4-Trimethylcyclohexane | 1383 | 6583.1 |
| $C_{10}H_8$ | Naphthalene *(solid)* | 1231 | 5859.6 |
| $C_{10}H_{10}$ | $\Delta^1$-Dihydronaphthalene | 1300 | 6188 |
| $C_{10}H_{10}$ | $\Delta^2$-Dihydronaphthalene *(solid)* | 1298 | 6178.5 |
| $C_{10}H_{10}$ | 1-Phenyl-3-butine | 1340 | 6378.4 |

**Appendix C2b. Heat of combustion of organic compounds** *(continued)*

| Formula | Name | kg cal/mol | Btu/scf |
|---|---|---|---|
| | **1. Hydrocarbons** *(continued)* | | |
| $C_{10}H_{12}$ | α-β-Dimethylstyrene *(solid)* | 1357 | 6459.3 |
| $C_{10}H_{12}$ | β-Ethylstyrene | 1346 | 6407.0 |
| $C_{10}H_{12}$ | 1-Phenyl-3-butene | 1356 | 6454.6 |
| $C_{10}H_{12}$ | 1-Phenyl-2-butene | 1361 | 6478.4 |
| $C_{10}H_{12}$ | Tetrahydronaphthalene | 1340 | 6378.4 |
| $C_{10}H_{14}$ | *tert.*-Butylbenzene | 1400 | 6664.0 |
| $C_{10}H_{14}$ | Hexahydronaphthalene | 1419 | 6754.4 |
| $C_{10}H_{14}$ | 4-Isopropyltoluene (cymene) | 1405 | 6687.8 |
| $C_{10}H_{14}$ | 3-*n*-Propyltoluene | 1405 | 6687.8 |
| $C_{10}H_{14}$ | 3-Isopropyltoluene | 1409 | 6706.8 |
| $C_{10}H_{14}$ | 1, 2, 4, 5-Tetramethylbenzene (durol) *(solid)* | 1393 | 6630.7 |
| $C_{10}H_{16}$ | Borneocamphene | 1470 | 6997.2 |
| $C_{10}H_{16}$ | Camphene *(solid)* | 1468 | 6987.7 |
| $C_{10}H_{16}$ | *d*-"Citrene" | 1473 | 7011.5 |
| $C_{10}H_{16}$ | Cyclene (tricyclene) *(solid)* | 1467 | 6982.9 |
| $C_{10}H_{16}$ | 1, 5-Dimethyl-3-vinyl-1-cyclohexene | 1455 | 6925.8 |
| $C_{10}H_{16}$ | 1-Isobutenyl-1-cyclohexene | 1461 | 6954.4 |
| $C_{10}H_{16}$ | *d*-Limonene | 1471 | 7002.0 |
| $C_{10}H_{16}$ | *l*-Limonene | 1457 | 6935.3 |
| $C_{10}H_{16}$ | 1-Methyl-4-isopropylcyclohexa-1,3-diene | 1470 | 6997.2 |
| $C_{10}H_{16}$ | Octohydronaphthalene | 1461 | 6954.4 |
| $C_{10}H_{16}$ | *d*-α-Pinene (australene) | 1471 | 7002.0 |
| $C_{10}H_{16}$ | *l*-α-Pinene (terebenthene) | 1477 | 7030.5 |
| $C_{10}H_{16}$ | β-Pinolene | 1469 | 6992.4 |
| $C_{10}H_{16}$ | Sylvestrene | 1464 | 6968.6 |
| $C_{10}H_{16}$ | Terecamphene (inact.) | 1466 | 6978.2 |
| $C_{10}H_{16}$ | α-Terpinene | 1470 | 6997.2 |
| $C_{10}H_{18}$ | *cis*-Decahydronaphthalene | 1502 | 7149.5 |
| $C_{10}H_{18}$ | *trans*-Decahydronaphthalene | 1498 | 7130.5 |
| $C_{10}H_{18}$ | 1-Ethyl-5-dimethylcyclohexene | 1504 | 7159.0 |
| $C_{10}H_{18}$ | Fenchane | 1502 | 7149.5 |
| $C_{10}H_{18}$ | Menthene ($\Delta^3$-terpene) ($\Delta^3$-menthene) | 1523 | 7249.5 |
| $C_{10}H_{18}$ | Thujane | 1506 | 7168.6 |
| $C_{10}H_{20}$ | Diamylene | 1582 | 7530.3 |
| $C_{10}H_{20}$ | *p*-Menthane (1-isopropyl-4-methylcyclohexane) | 1514 | 7206.6 |
| $C_{10}H_{20}$ | Methyl-*l*-*n*-propyl-3-cyclohexane | 1502 | 7149.5 |
| $C_{10}H_{22}$ | Decane | 1610 | 7663.6 |
| $C_{10}H_{22}$ | Diisoamyl | 1615 | 7687.4 |

**Appendix C2b. Heat of combustion of organic compounds** *(continued)*

| Formula | Name | kg cal/mol | Btu/scf |
|---|---|---|---|
| | **1. Hydrocarbons** *(continued)* | | |
| $C_{11}H_{14}$ | 1-Phenyl-2-pentene | 1510 | 7187.6 |
| $C_{11}H_{16}$ | Pentamethylbenzene *(solid)* | 1554 | 7397.0 |
| $C_{11}H_{18}$ | 1, 5-Dimethyl-3-isopropene-1-cyclohexene | 1615 | 7687.4 |
| $C_{12}H_{10}$ | Acenaphthene *(solid)* (periethylenenaphthylene) | 1491 | 7097.2 |
| $C_{12}H_{10}$ | Diphenyl *(solid)* | 1493 | 7106.7 |
| $C_{12}H_{16}$ | β, β-Diethylstyrene | 1664 | 7920.6 |
| $C_{12}H_{18}$ | Hexamethylbenzene *(solid)* | 1711 | 8144.4 |
| $C_{12}H_{24}$ | Triisobutylene [$(CH_3)_2CHC(CH_2CH(CH_3)_2)$: $CHCH(CH_3)_2$] | 1858 | 8844.0 |
| $C_{13}H_{10}$ | Fluorene *(solid)* | 1584 | 7539.8 |
| $C_{13}H_{12}$ | Diphenylmethane *(solid)* | 1655 | 7877.8 |
| $C_{14}H_{10}$ | Anthracene *(solid)* | 1695 | 8068.2 |
| $C_{14}H_{10}$ | Phenanthrene *(solid)* | 1693 | 8058.7 |
| $C_{14}H_{12}$ | Stilbene (*sym.*-diphenylethylene) *(solid)* | 1765 | 8401.4 |
| $C_{14}H_{12}$ | Isostilbene | 1770 | 8425.2 |
| $C_{14}H_{14}$ | Dibenzyl *(solid)* | 1810 | 8615.6 |
| $C_{14}H_{26}$ | 3, 3-Dimethyldicyclohexy (*m*-hexahydroditolyl) | 2106 | 10024.6 |
| $C_{15}H_{14}$ | α-β-Methylphenylstyrene *(solid)* | 1929 | 9182.0 |
| $C_{16}H_{10}$ | Diphenyldiacetylene *(solid)* | 1975 | 9401.0 |
| $C_{16}H_{14}$ | Diphenylbutadiene *(solid)* | 2057 | 9791.3 |
| $C_{16}H_{14}$ | *cis-cis*-Diphenylbutadiene *(solid)* | 2035 | 9686.6 |
| $C_{16}H_{14}$ | *trans-trans*-Diphenylbutadiene *(solid)* | 2030 | 9662.8 |
| $C_{16}H_{34}$ | Hexadecane *(solid)* | 2559 | 12180.8 |
| $C_{18}H_{12}$ | Chrysene *(solid)* | 2139 | 10181.6 |
| $C_{18}H_{16}$ | Diphenylhexatriene *(solid)* | 2288 | 10890.9 |
| $C_{18}H_{18}$ | Dibenzylbutadiene *(solid)* | 2341 | 11143.2 |
| $C_{18}H_{18}$ | 1, 6-Diphenylhexa-1, 5-diene *(solid)* | 2342 | 11147.9 |
| $C_{18}H_{18}$ | Retene *(solid)* (methylisopropylphenanthrene) | 2307 | 10981.3 |
| $C_{18}H_{20}$ | 1, 4-Diphenyl-1-ethyl-3-butene | 2372 | 11290.7 |
| $C_{19}H_{15}$ | Triphenylmethyl *(solid)* | 2378 | 11319.3 |
| $C_{19}H_{16}$ | Triphenylmethane *(solid)* | 2384 | 11347.8 |
| $C_{20}H_{16}$ | Diphenylstyrene *(solid)* | 2496 | 11881.0 |
| $C_{20}H_{42}$ | Eicosane *(solid)* | 3183 | 15151.1 |
| $C_{24}H_{18}$ | 1, 3, 5-Triphenylbenzene | 2937 | 13980.1 |
| $C_{25}H_{20}$ | Tetraphenylmethane *(solid)* | 3102 | 14765.5 |
| $C_{28}H_{20}$ | Dianthracene *(solid)* | 3383 | 16103.1 |

**Appendix C2b. Heat of combustion of organic compounds** *(continued)*

| Formula | Name | kg cal/mol | Btu/scf |
|---|---|---|---|
| | **2. Alcohols** | | |
| $CH_4O$ | Methyl alcohol | 170.9 | 813.5 |
| $C_2H_6O$ | Ethyl alcohol | 328 | 1561.3 |
| $C_2H_6O_2$ | Glycol | 281.9 | 1341.8 |
| $C_3H_4O$ | Propargyi alcohol *(gas)* | 428.9 | 2041.6 |
| $C_3H_6O$ | Allyl alcohol | 442.4 | 2105.8 |
| $C_3H_8O$ | *n*-Propyl alcohol | 482 | 2294.3 |
| $C_3H_8O$ | Isopropyl alcohol | 474.8 | 2260.0 |
| $C_3H_8O_2$ | Propylene glycol | 431.0 | 2051.6 |
| $C_3H_8O_2$ | Isopropylene glycol | 436.1 | 2075.8 |
| $C_3H_8O_3$ | Glycerol | 397.0 | 1889.7 |
| $C_4H_{10}O$ | *n*-Butyl alcohol | 639 | 3041.6 |
| $C_4H_{10}O$ | Isobutyl alcohol (primary) | 638.2 | 3037.8 |
| $C_4H_{10}O$ | Trimethyl carbinol (*tert.*-butyl) | 629.3 | 2995.5 |
| $C_4H_{10}O_3$ | Diethylene glycol | 567 | 2698.9 |
| $C_4H_{10}O_4$ | Erythritol *(solid)* | 504.1 | 2399.5 |
| $C_5H_{10}O$ | Cyclobutyl carbinol | 748 | 3560.5 |
| $C_5H_{10}O$ | Ethyl vinyl carbinol | 753 | 3584.3 |
| $C_5H_{10}O_2$ | Dihydroxymethylcyclopropane | 708 | 3370.1 |
| $C_5H_{10}O_2$ | Cyclopentan-1, 2-diol *(cis)* | 696 | 3313.0 |
| $C_5H_{10}O_2$ | Cyclopentan-1, 2-diol *(trans)* | 694 | 3303.4 |
| $C_5H_{12}O$ | Amyl alcohol (?) | 787 | 3746.1 |
| $C_5H_{12}O$ | Amyl alcohol (ferm.) [mostly isobutyl carbinol] | 792 | 3769.9 |
| $C_5H_{12}O$ | Dimethyl ethyl carbinol | 785 | 3736.6 |
| $C_5H_{12}O_4$ | Pentaerythritol *(solid)* | 661 | 3146.4 |
| $C_5H_{12}O_5$ | Arabitol *(solid)* | 612 | 2913.1 |
| $C_6H_{12}O$ | Allyldimethyl carbinol | 886 | 4217.4 |
| | | 913 | 4345.9 |
| $C_6H_{12}O$ | Cyclohexanol | 891 | 4241.2 |
| $C_6H_{12}O$ | $\beta$-Methylcyclopentanol | 888 | 4226.9 |
| $C_6H_{12}O_2$ | Cyclohexan-1, 2-diol *(cis)* | 841 | 4003.2 |
| $C_6H_{12}O_2$ | Cyclohexan-1, 2-diol *(trans)* | 842 | 4007.9 |
| $C_6H_{12}O_5$ | Quercitol *(solid)* | 704 | 3351.0 |
| $C_6H_{12}O_6$ | Inositol (inosite) *(solid)* | 662 | 3151.1 |
| $C_6H_{14}O$ | Pinacolyl alcohol | 939 | 4469.6 |
| $C_6H_{14}O$ | Methyldiethyl carbinol | 927 | 4412.5 |
| $C_6H_{14}O_2$ | Pinacol *(solid)* | 897 | 4269.7 |
| $C_6H_{14}O_6$ | Dulcitol *(solid)* | 729 | 3470.0 |
| | | 723 | 3441.5 |

**Appendix C2b. Heat of combustion of organic compounds** *(continued)*

| Formula | Name | kg cal/mol | Btu/scf |
|---|---|---|---|
| | **2. Alcohols** *(continued)* | | |
| $C_6H_{14}O_6$ | *d*-Mannitol *(solid)* | 727 | 3460.5 |
| $C_7H_8O$ | Benzyl alcohol | 893 | 4250.7 |
| $C_7H_{12}O$ | Diallyl carbinol | 1028 | 4893.3 |
| $C_7H_{14}O$ | Allylmethylethyl carbinol | 1050 | 4998.0 |
| $C_7H_{14}O$ | Cycloheptanol | 1050 | 4998.0 |
| $C_7H_{14}O$ | Cyclohexyl carbinol | 1047 | 4983.7 |
| $C_7H_{14}O$ | 1, 3-Dimethylcyclopentan-2-ol | 1030 | 4902.8 |
| $C_7H_{14}O$ | 1, 3-Dimethylcyclopentan-3-ol | 1034 | 4921.8 |
| $C_7H_{14}O$ | 1-Eethylcyclopentan-2-ol | 1039 | 4945.6 |
| $C_7H_{14}O$ | *β*-Methylcyclohexanol | 1038 | 4940.9 |
| $C_7H_{14}O_2$ | *cis*-1-Methylcyclohexan-1, 2-diol | 992 | 4721.9 |
| | *trans*-1-Methylcyclohexan-1, 2-diol | 995 | 4736.2 |
| $C_7H_{16}O$ | *n*-Heptyl alcohol | 1104 | 5255.0 |
| $C_7H_{16}O$ | Triethyl carbinol | 1080 | 5140.8 |
| $C_7H_{16}O_7$ | Perseitol (*d*-mannoheptol) *(solid)* | 836 | 3979.4 |
| $C_8H_{14}O$ | Amylpropargyl alcohol ($C_5H_{11}$.C:C.$CH_2$OH.) | 1192 | 5673.9 |
| $C_8H_{14}O$ | Diallylmethyl carbinol | 1181 | 5621.6 |
| $C_8H_{14}O_2$ | Tetramethylbutindiol | 1142 | 5435.9 |
| $C_8H_{16}O$ | Allyldiethyl carbinol | 1207 | 5745.3 |
| $C_8H_{16}O$ | Allylmethylpropyl carbinol | 1202 | 5721.5 |
| $C_8H_{16}O$ | 1, 2-Dimethylcyclohexan-2-ol | 1196 | 5693.0 |
| $C_8H_{16}O$ | 1, 3-Dimethylcyclohexan-2-ol | 1196 | 5693.0 |
| $C_8H_{16}O$ | 1, 3-Dimethylcyclohexan-3-ol | 1192 | 5673.9 |
| $C_8H_{16}O$ | 1, 3-Dimethylcyclohexan-5-ol | 1183 | 5631.1 |
| $C_8H_{16}O_2$ | Tetramethylbutenediol (fumaroid) | 1176 | 5597.8 |
| $C_8H_{16}O_2$ | Tetramethylbutenediol (maleinoid) *(solid)* | 1172 | 5578.7 |
| $C_8H_{18}O$ | Octyl alcohol | 1262 | 6007.1 |
| $C_8H_{18}O$ | Methyldipropyl carbinol | 1233 | 5869.1 |
| $C_9H_8O$ | Phenylpropargyl alcohol ($C_6H_5$.C:C.$CH_2$OH.) | 1137 | 5412.1 |
| $C_9H_{10}O_2$ | Hydrinden-1, 2-diol *(cis)* | 1098 | 5226.5 |
| $C_9H_{10}O_2$ | Hydrinden-1, 2-diol *(trans)* | 1096 | 5217.0 |
| $C_9H_{16}O$ | Hexylpropargyl alcohol | 1340 | 6378.4 |
| $C_9H_{16}O$ | 1, 3, 5-Trimethylcyclohex-6-en-5-ol | 1295 | 6164.2 |
| $C_9H_{18}O$ | Allylmethyl-nbutyl carbinol | 1365 | 6497.4 |
| $C_9H_{18}O$ | Allylmethyl-tert.-butyl carbinol | 1363 | 6487.9 |
| $C_9H_{18}O$ | Cycloheptylmethyl carbinol | 1342 | 6387.9 |
| $C_9H_{18}O$ | 1-Methyl-3-ethylcyclohexan-3-ol | 1322 | 6292.7 |
| $C_9H_{20}O$ | Ethyldipropyl carbinol | 1386 | 6597.4 |

**Appendix C2b. Heat of combustion of organic compounds** *(continued)*

| Formula | Name | kg cal/mol | Btu/scf |
|---|---|---|---|
| | **2. Alcohols** *(continued)* | | |
| $C_{10}H_{12}O_2$ | 1,2,3,4-Tetrahydronaphthalenediols, all forms | 1250 | 5950.0 |
| $C_{10}H_{18}O$ | Borneol (synthetic) | 1466 | 6978.2 |
| $C_{10}H_{18}O$ | Borneol (Borneo camphor) | 1470 | 6997.2 |
| $C_{10}H_{18}O$ | α-Borneol *(solid)* | 1466 | 6978.2 |
| $C_{10}H_{18}O$ | *l*-Borneol | 1467 | 6982.9 |
| $C_{10}H_{18}O$ | Diallylpropyl carbinol | 1472 | 7006.7 |
| $C_{10}H_{18}O$ | Terpineol *(solid)* | 1469 to 1480 | 6992.4 to 7044.8 |
| $C_{10}H_{18}O$ | Thujyl alcohol | 1477 | 7030.5 |
| $C_{10}H_{20}O$ | Allyldipropyl carbinol | 1518 | 7225.7 |
| | | 1549 | 7373.2 |
| $C_{10}H_{20}O$ | Menthol *(solid)* | 1509 | 7182.8 |
| $C_{10}H_{22}O_3$ | Terpinehydrate *(solid)* | 1451 | 6906.8 |
| $C_{11}H_{22}O$ | Allylmethylhexyl carbinol | 1667 | 7934.9 |
| $C_{12}H_{16}O_2$ | 1-Phenylcyclohexan-1, 2-diol *(cis* and *trans)* | 1564 | 7444.6 |
| $C_{13}H_{12}O$ | Diphenyl carbinol *(solid)* | 1615 | 7687.4 |
| $C_{14}H_{14}O_2$ | Hydrobenzoin *(solid)* | 1723 | 8201.5 |
| $C_{14}H_{14}O_2$ | Isohydrobenzoin *(solid)* | 1728 | 8225.3 |
| $C_{14}H_{18}O$ | Amylphenylpropargyl alcohol | 1901 | 9048.8 |
| $C_{16}H_{34}O$ | Cetyl alcohol *(solid)* | 2504 | 11919.0 |
| $C_{19}H_{16}O$ | Triphenyl carbinol *(solid)* | 2341 | 11143.2 |
| $C_{21}H_{16}O$ | Diphenylphenylethinyl carbinol | 2572 | 12242.7 |
| | **3. C, H, O, Acids** | | |
| $CH_2O_2$ | Formic | 62.8 | 298.9 |
| $C_2H_2O_4$ | Oxalic *(solid)* | 60.1 | 286.1 |
| $C_2H_4O_2$ | Acetic | 209.4 | 996.7 |
| | | 207.1 | 985.8 |
| $C_2H_4O_3$ | Glycolic *(solid)* | 166.6 | 793.0 |
| $C_2H_4O_4$ | Dihydroxyacetic (glyozylic) *(solid)* | 125.5 | 597.4 |
| $C_3H_4O_2$ | Acrylic | 327 | 1556.5 |
| $C_3H_4O_3$ | Pyroracemic, pyruvic | 279.1 | 1328.5 |
| $C_3H_4O_4$ | Malonic *(solid)* | 206.8 | 984.4 |
| $C_3H_4O_5$ | Tartronic *(solid)* | 165.4 | 787.3 |
| $C_3H_4O_6$ | Mesoxalic (dihydroxymalonic) *(solid)* | 128.2 | 610.2 |
| $C_3H_6O_2$ | Propionic | 365 | 1737.4 |
| $C_3H_6O_3$ | Lactic *(solid)* | 326 | 1551.8 |

**Appendix C2b. Heat of combustion of organic compounds** *(continued)*

| Formula | Name | kg cal/mol | Btu/scf |
|---|---|---|---|
| | **3. C, H, O, Acids** *(continued)* | | |
| $C_4H_2O_4$ | Acetylenedicarboxylic *(solid)* | 305.9 | 1456.1 |
| $C_4H_4O_2$ | Tetrolic *(solid)* | 452.4 | 2153.4 |
| $C_4H_4O_4$ | Fumaric *(trans)* *(solid)* | 320 | 1523.2 |
| $C_4H_4O_4$ | Maleic *(cis)* *(solid)* | 326.1 | 1552.2 |
| $C_4H_6O_2$ | Crotonic *(solid)* | 477.7 | 2273.9 |
| $C_4H_6O_2$ | Trimethylenecarboxylic | 481 | 2289.6 |
| $C_4H_6O_4$ | Methylmalonic *(solid)* | 363 | 1727.9 |
| $C_4H_6O_4$ | Succinic *(solid)* | 356.8 | 1698.4 |
| $C_4H_6O_5$ | *l*-Malic | 320.1 | 1523.7 |
| $C_4H_6O_6$ | *d*-Tartaric *(solid)* | 275.1 | 1309.5 |
| $C_4H_6O_6$ | *dl*-Tartaric (anhydr.) *(solid)* | 273 | 1299.5 |
| $C_4H_6O_6$ | Mesotartaric *(solid)* | 276 | 1313.8 |
| $C_4H_8O_2$ | *n*-Butyric | 520 | 2475.2 |
| $C_4H_8O_2$ | Isobutyric | 517.4 | 2462.8 |
| $C_4H_8O_2$ | Hydroxyisobutyric *(solid)* | 471.8 | 2245.8 |
| $C_4H_8O_3$ | *dl*-$\beta$-Hydroxybutyric *(solid)* | 487.9 | 2322.4 |
| $C_5H_4O_3$ | Pyromucic *(solid)* | 489.7 | 2331.0 |
| $C_5H_6O_4$ | *cis*-Citraconic (methylmaleic) *(solid)* | 479.1 | 2280.5 |
| $C_5H_6O_4$ | Itaconic (methylenesuccinic) *(solid)* | 475.3 | 2262.4 |
| $C_5H_6O_4$ | *trans*-Mesaconic (methylfumaric) *(solid)* | 476.4 | 2267.7 |
| $C_5H_6O_4$ | $\alpha$, $\alpha$-Trimethylenedicarboxylic *(solid)* | 483 | 2299.1 |
| $C_5H_6O_4$ | $\alpha$, $\beta$-Trimethylenedicarboxylic *(solid)* | 484 | 2303.8 |
| $C_5H_8O_2$ | Allylacetic | 641 | 3051.2 |
| $C_5H_8O_2$ | Angelic *(solid)* | 634 | 3017.8 |
| $C_5H_8O_2$ | $\alpha$, $\beta$-Pentenic *(solid)* | 623 | 2965.5 |
| $C_5H_8O_2$ | $\beta$, $\gamma$-Pentenic *(solid)* | 632 | 3008.3 |
| $C_5H_8O_2$ | Tetramethylenecarboxylic | 640 | 3046.4 |
| $C_5H_8O_2$ | Tiglic *(solid)* | 626 | 2979.8 |
| $C_5H_8O_3$ | Levulinic ($\beta$-acetylpropionic) *(solid)* | 576.8 | 2745.6 |
| $C_5H_8O_4$ | Dimethylmalonic *(solid)* | 515.1 | 2451.9 |
| $C_5H_8O_4$ | Ethylmalonic *(solid)* | 517.7 | 2464.3 |
| $C_5H_8O_4$ | Glutaric *(solid)* | 514.8 | 2450.4 |
| $C_5H_8O_4$ | Methylsuccinic *(solid)* | 515.3 | 2452.8 |
| $C_5H_8O_7$ | Trihydroxyglutaric *(solid)* | 388.3 | 1848.3 |
| $C_5H_{10}O_2$ | *n*-Valeric | 679 | 3232.0 |
| $C_6H_6C_6$ | Aconitic *(solid)* | 475.1 | 2261.5 |
| $C_6H_8O_2$ | Sorbic *(solid)* | 743 | 3536.7 |
| $C_6H_8O_4$ | Allylmalonic *(solid)* | 638 | 3036.9 |

**Appendix C2b. Heat of combustion of organic compounds** *(continued)*

| Formula | Name | kg cal/mol | Btu/scf |
|---|---|---|---|
| | **3. C, H, O, Acids** *(continued)* | | |
| $C_6H_8O_4$ | α, β-Hydromuconic *(solid)* | 628.8 | 2993.1 |
| $C_6H_8O_4$ | β, γ-Hydromuconic *(solid)* | 629.1 | 2994.5 |
| $C_6H_8O_4$ | α, α-Tetramethylenedicarboxylic *(solid)* | 642 | 3055.9 |
| $C_6H_8O_4$ | α, β-Tetramethylenedicarboxylic *(solid)* | 642.1 | 3056.4 |
| $C_6H_8O_4$ | α, γ-Tetramethylenedicarboxylic *(solid)* | 639.4 | 3043.5 |
| $C_6H_8O_6$ | Tricarballylic *(solid)* | 516 | 2456.2 |
| $C_6H_8O_7$ | Citric (anhydr.) *(solid)* | 474.5 | 2258.6 |
| $C_6H_8O_7.H_2O$ | Citric (cryst.) *(solid)* | 471.4 | 2243.9 |
| $C_6H_{10}O_2$ | Hydrosorbic | 795 | 3784.2 |
| $C_6H_{10}O_4$ | Adipic *(solid)* | 668.8 | 3183.5 |
| $C_6H_{10}O_4$ | *sym.*-Dimethylsuccinic *(anti) (solid)* | 674.2 | 3209.2 |
| $C_6H_{10}O_4$ | *sym.*-Dimethylsuccinic *(anti) (solid)* | 673 | 3203.5 |
| $C_6H_{10}O_4$ | *dl-sym.*-Dimethylsuccinic *(solid)* | 671.5 | 3196.3 |
| $C_6H_{10}O_4$ | *sym.*-Dimethylsuccinic *(para) (solid)* | 670.6 | 3192.1 |
| $C_6H_{10}O_4$ | *unsym.*-Dimethylsuccinic *(solid)* | 671 | 3194.0 |
| $C_6H_{10}O_4$ | Ethylsuccinic *(solid)* | 671.2 | 3194.9 |
| $C_6H_{10}O_4$ | Methylethylmalonic *(solid)* | 672 | 3198.7 |
| $C_6H_{10}O_4$ | α-Methylglutaric *(solid)* | 670.6 | 3192.1 |
| $C_6H_{10}O_4$ | Propylmalonic *(solid)* | 674 | 3208.2 |
| $C_6H_{10}O_4$ | Isopropylmalonic *(solid)* | 674 | 3208.2 |
| $C_6H_{10}O_8$ | Allomucic *(solid)* | 494.2 | 2352.4 |
| $C_6H_{10}O_8$ | Mucic *(solid)* | 483.6 | 2301.9 |
| $C_6H_{12}O_2$ | Caproic | 831 | 3955.6 |
| $C_6H_{12}O_2$ | Diethylacetic | 831 | 3955.6 |
| $C_6H_{12}O_2$ | Isobutylacetic | 833 | 3965.1 |
| $C_7H_4O_7$ | Meconic *(solid)* | 490.8 | 2336.2 |
| $C_7H_6O_2$ | Benzoic *(solid)* | 711.2 | 3385.3 |
| $C_7H_6O_3$ | Salicylic *(solid)* | 723.1 | 3442.0 |
| $C_7H_6O_3$ | *m*-Hydrozybenzoic *(solid)* | 725 | 3451.0 |
| $C_7H_6O_3$ | *p*-Hydrozybenzoic *(solid)* | 724 | 3446.2 |
| $C_7H_6O_4$ | 2, 4-Dihydroxybenzoic *(solid)* | 676.5 | 3220.1 |
| $C_7H_6O_5$ | Gallic *(solid)* | 633.7 | 3016.4 |
| $C_7H_6O_5$ | Pyrogallolcarboxylic *(solid)* | 633.4 | 3015.0 |
| $C_7H_6O_3$ | α, α, β, β-Trimethylenetetracarboxylic *(solid)* | 482.7 | 2297.7 |
| $C_7H_{10}O_2$ | $\Delta_1$-Tetrahydrobenzoic *(solid)* | 857 | 4079.3 |
| $C_7H_{10}O_2$ | $\Delta_2$-Tetrahydrobenzoic *(solid)* | 886 | 4217.4 |
| $C_7H_{10}O_4$ | α, β-Pentamethylenedicarboxylic *(solid)* | 781 | 3717.6 |

**Appendix C2b. Heat of combustion of organic compounds** *(continued)*

| Formula | Name | kg cal/mol | Btu/scf |
|---|---|---|---|
| | **3. C, H, O, Acids** *(continued)* | | |
| $C_7H_{10}O_4$ | Teraconic acid (γ-dimethylitaconic ( *(solid)* | 796 | 3789.0 |
| $C_7H_{12}O_2$ | Hexahydrobenzoic | 934 | 4445.8 |
| $C_7H_{12}O_4$ | Diethylmalonic *(solid)* | 829 | 3946.0 |
| $C_7H_{12}O_4$ | Pimelic (isopropylsuccinic) | 827 | 3936.5 |
| $C_7H_{12}O_4$ | Trimethylsuccinic *(solid)* | 829.9 | 3950.3 |
| $C_7H_{14}O_2$ | Ethylpropylacetic | 990 | 4712.4 |
| $C_7H_{14}O_2$ | Heptylic | 986.1 | 4693.8 |
| $C_8H_6O_4$ | *o*-Phthalic *(solid)* | 771 | 3670.0 |
| $C_8H_6O_4$ | Isophthalic *(solid)* | 768 | 3655.7 |
| $C_8H_6O_4$ | Terephthalic *(solid)* | 770 | 3665.2 |
| $C_8H_6O_4$ | Piperonylic *(solid)* | 803 | 3822.3 |
| $C_8H_8O_2$ | Phenylacetic *(solid)* | 930 | 4426.8 |
| $C_8H_8O_2$ | *o*-Toluic *(solid)* | 928.9 | 4421.6 |
| | | 921 | 4384.0 |
| $C_8H_8O_2$ | *m*-Toluic *(solid)* | 928.6 | 4420.1 |
| | | 922.2 | 4389.7 |
| $C_8H_8O_2$ | *p*-Toluic *(solid)* | 926.9 | 4412.0 |
| $C_8H_8O_2$ | *o*-Hydroxymethylbenzoic *(solid)* | 887.3 | 4223.5 |
| $C_8H_8O_3$ | 1, 2, 3-Hydroxytoluic *(solid)*[a] | 878 | 4179.3 |
| $C_8H_8O_3$ | 1, 2, 4-Hydroxytoluic *(solid)*[a] | 877 | 4174.5 |
| $C_8H_8O_3$ | 1, 2, 5-Hydroxytoluic *(solid)*[a] | 879 | 4184.0 |
| $C_8H_8O_3$ | 1, 2, 6-Hydroxytoluic *(solid)*[a] | 882 | 4198.3 |
| $C_8H_8O_3$ | Mandelic *(solid)* | 890 | 4236.4 |
| $C_8H_8O_3$ | *p*-Methoxybenzoic *(solid)* | 894 | 4255.4 |
| $C_8H_8O_3$ | Phenoxyacetic *(solid)* | 903 | 4298.3 |
| $C_8H_8O_4$ | Dihydroterephthalic *(solid)* | 842 | 4007.9 |
| $C_8H_8O_4$ | $\Delta^{1,4}$-Dihydroterephthalic *(solid)* | 835 | 3974.6 |
| $C_8H_8O_4$ | $\Delta^{1,5}$-Dihydroterephthalic *(solid)* | 842 | 4007.9 |
| $C_8H_8O_4$ | $\Delta^{2,5}$-Dihydroterephthalic (fum.) *(solid)* | 844 | 4017.4 |
| $C_8H_{10}O_4$ | $\Delta^{1}$-Tetrahydroterephthalic *(solid)* | 882 | 4198.3 |
| $C_8H_{10}O_4$ | $\Delta^{2}$-Tetrahydroterephthalic *(solid)* | 881 | 4193.6 |
| $C_8H_{12}O_2$ | Amylpropiolic | 1083 | 5155.1 |
| $C_8H_{12}O_2$ | Cyclohexene-1-acetic *(solid)* | 1045 | 4974.2 |
| $C_8H_{12}O_2$ | Cyclohexylideneacetic *(solid)* | 1042 | 4959.9 |
| $C_8H_{12}O_4$ | *trans*-1, 2-Cyclohexanedicarboxylic *(solid)* | 930 | 4426.8 |

[a] The numbers denote the positions of carboxyl, hydroxyl and methyl groups repsectively.

**Appendix C2b. Heat of combustion of organic compounds** *(continued)*

| Formula | Name | kg cal/mol | Btu/scf |
|---|---|---|---|
| | **3. C, H, O, Acids** *(continued)* | | |
| $C_8H_{12}O_4$ | *cis*-Hexahydroterephthalic *(solid)* | 928 | 4417.3 |
| $C_8H_{12}O_4$ | *trans*-Hexahydroterephthalic *(solid)* | 929 | 4422.0 |
| $C_8H_{14}O_2$ | Cycloheptanecarboxylic *(solid)* | 1088 | 5178.9 |
| $C_8H_{14}O_2$ | Cyclooctane (act.) | 1089 | 5183.6 |
| $C_8H_{14}O_2$ | Hexahydro-m-toluic | 1086 | 5169.4 |
| $C_8H_{14}O_4$ | *sym.*-Diethylsuccinic *(solid)* | 987 | 4698.1 |
| $C_8H_{14}O_4$ | *unsym.*-Diethylsuccinic *(solid)* | 985 | 4688.6 |
| $C_8H_{14}O_4$ | *sym.*-Dimethyladipic *(solid)* | 986 | 4693.4 |
| $C_8H_{14}O_4$ | Ethylpropylmalonic *(solid)* | 983 | 4679.1 |
| $C_8H_{14}O_4$ | Suberic *(solid)* | 983.4 | 4681.0 |
| $C_8H_{14}O_4$ | Tetramethylsuccinic *(solid)* | 989 | 4707.6 |
| $C_8H_{14}O_6$ | Dimethyldihydroxyadipic *(solid)* | 889 | 4231.6 |
| $C_8H_{16}O_2$ | Dipropylacetic | 1146 | 5455.0 |
| $C_9H_6O_2$ | Phenylpropiolic *(solid)* | 1021 | 4860.0 |
| $C_9H_6O_6$ | Trimesic *(solid)* | 767 | 3650.9 |
| $C_9H_8O_2$ | *cis*-Allocinnamic (M. P. 58°) *(solid)* | 1047 | 4983.7 |
| $C_9H_8O_2$ | *trans*-Cinnamic *(solid)* | 1040 | 4950.4 |
| $C_9H_8O_2$ | Atropic *(solid)* | 1044 | 4969.4 |
| $C_9H_8O_3$ | *cis-(allo)-p*-Hydroxycinnamic *(solid)* | 996 | 4741.0 |
| $C_9H_8O_3$ | *trans-p*-Hydroxycinnamic *(solid)* | 991 | 4717.2 |
| $C_9H_8O_4$ | Uvitic *(solid)* | 928 | 4417.3 |
| $C_9H_{10}O_2$ | Hydrocinnamic (β-phenylpropionic) *(solid)* | 1085 | 5164.6 |
| $C_9H_{10}O_2$ | Mesitylenic *(solid)* | 1085 | 5164.6 |
| $C_9H_{14}O_2$ | α-Cyclohexene-1-propionic | 1200 | 5712.0 |
| $C_9H_{14}O_2$ | Hexylpropiolic | 1232 | 5864.3 |
| $C_9H_{14}O_2$ | Campholytic *(solid)* | 1366 | 6502.2 |
| $C_9H_{14}O_2$ | Isocampholytic *(solid)* | 1363 | 6487.9 |
| $C_9H_{16}O_4$ | Axelaic *(solid)* | 1141.4 | 5433.1 |
| $C_9H_{16}O_4$ | Dipropylmalonic *(solid)* | 1140 | 5426.4 |
| $C_9H_{18}O_2$ | Heptylacetic | 1287 | 6126.1 |
| $C_{10}H_6O_3$ | Pyromellitic *(solid)* | 776 | 3693.8 |
| $C_{10}H_8O_4$ | *cis*-Allopiperonylacrylic *(solid)* | 1076 | 5121.8 |
| $C_{10}H_8O_4$ | *trans*-Piperonylacrylic *(solid)* | 1067 | 5078.9 |
| $C_{10}H_8O_4$ | Benzalmalonic *(solid)* | 1056 | 5026.6 |
| $C_{10}H_{10}O_2$ | α-Methylcinnamic *(solid)* | 1198 | 5702.5 |
| $C_{10}H_{10}O_2$ | β-Methylcinnamic *(solid)* | 1197 | 5697.7 |
| $C_{10}H_{10}O_2$ | *cis*-Phenylisocrotonic ($C_6H_5$.CH:CH.$CH_2$COOH.) *(solid)* | 1195 | 5688.2 |

**Appendix C2b. Heat of combustion of organic compounds** *(continued)*

| Formula | Name | kg cal/mol | Btu/scf |
|---|---|---|---|
| | **3. C, H, O, Acids** *(continued)* | | |
| $C_{10}H_{10}O_3$ | *cis-(allo)-p*-Methoxycinnamic *(solid)* | 1172 | 5578.7 |
| $C_{10}H_{10}O_3$ | *trans-p*-Methoxycinnamic *(solid)*[b] | 1163 | 5535.9 |
| $C_{10}H_{10}O_3$ | *cis*-Methylcoumarinic *(solid)* | 1167 | 5554.9 |
| $C_{10}H_{10}O_3$ | *trans*-Methylcoumarinic *(solid)* | 1161 | 5526.4 |
| $C_{10}H_{10}O_4$ | Benzylmalonic *(solid)* | 1082 | 5150.3 |
| $C_{10}H_{10}O_4$ | Phenylsuccinic *(solid)* | 1082 | 5150.3 |
| $C_{10}H_{10}O_6$ | Opianic *(solid)* | 1090 | 5188.4 |
| $C_{10}H_{10}O_6$ | Hemipinic *(solid)* | 1025 | 4879.0 |
| $C_{10}H_{12}O_2$ | Cumic (*p*-isopropylbenzoic) *(solid)* | 1238 | 5892.9 |
| $C_{10}H_{16}O_2$ | Geranic | 1379 | 6564.0 |
| $C_{10}H_{16}O_3$ | *α*-Tanacetoneketocarboxylic *(solid)* | 1327 | 6316.5 |
| $C_{10}H_{16}O_4$ | *d*-Camphoric *(solid)* | 1244 | 5921.4 |
| $C_{10}H_{18}O_2$ | Campholic *(solid)* | 1412 | 6721.1 |
| $C_{10}H_{18}O_2$ | Hexahydrocumic *(solid)* | 1396 | 6645.0 |
| $C_{10}H_{18}O_4$ | Heptylmalonic *(solid)* | 1297 | 6173.7 |
| $C_{10}H_{18}O_4$ | Sebacic *(solid)* | 1297 | 6173.7 |
| $C_{10}H_{18}O_4$ | Triethylsuccinic *(solid)* | 1301 | 6192.8 |
| $C_{10}H_{20}O_2$ | Capric *(solid)* | 1453 | 6916.3 |
| $C_{11}H_8O_2$ | *a*-Naphthoic *(solid)* | 1232 | 5864.3 |
| $C_{11}H_8O_2$ | *b*-Naphthoic *(solid)* | 1227 | 5840.5 |
| $C_{11}H_{10}O_2$ | Allocinnamylideneacetic *(solid)* | 1319 | 6278.4 |
| $C_{11}H_{10}O_3$ | Cinnamylideneacetic *(solid)* | 1311 | 6240.4 |
| $C_{11}H_{10}O_4$ | *cis*-Acetylcoumarinic *(solid)* | 1212 | 5769.1 |
| $C_{11}H_{10}O_4$ | *trans*-Acetylcoumarinic *(solid)* | 1208 | 5750.1 |
| $C_{11}H_{10}O_4$ | Phenylparaconic *(solid)* | 1195 | 5688.2 |
| $C_{11}H_{12}O_3$ | *cis*-Ethylcoumarinic *(solid)* | 1323 | 6297.5 |
| $C_{11}H_{12}O_3$ | *trans*-Ethylcoumaric (M.P. 133-134°) *(solid)* | 1316 | 6264.2 |
| $C_{11}H_{18}O_2$ | Undecolic *(solid)* | 1538 | 7320.9 |
| $C_{11}H_{20}O_2$ | Undecylenic *(solid)* | 1580 | 7520.8 |
| $C_{11}H_{20}O_4$ | Nonanedicarboxylic *(solid)* | 1456 | 6930.6 |
| $C_{11}H_{20}O_4$ | n-Octylmalonic *(solid)* | 1453 | 6916.3 |
| $C_{11}H_{22}O_2$ | Undecylic *(solid)* | 1610 | 7663.6 |
| $C_{12}H_6O_{12}$ | Mellitic *(solid)* | 788 | 3750.9 |
| $C_{12}H_8O_4$ | Naphthalic (1, 8-naphthalenedicarboxylic) *(solid)* | 1244 | 5921.4 |
| $C_{12}H_{10}O_4$ | Cinnamylidenemalonic (yellow) *(solid)* | 1319 | 6278.4 |

[b] M. P. 170° (liq. cryst.); clears at 185°.

**Appendix C2b. Heat of combustion of organic compounds** *(continued)*

| Formula | Name | kg cal/mol | Btu/scf |
|---|---|---|---|
| | **3. C, H, O, Acids** *(continued)* | | |
| $C_{12}H_{12}O_3$ | β-Bensallevulinic *(solid)* | 1413 | 6725.9 |
| $C_{12}H_{12}O_3$ | δ-Bensallevulinic *(solid)* | 1410 | 6711.6 |
| $C_{12}H_{12}O_{12}$ | Hexahydromellitic (fum.) *(solid)* | 923 | 4393.5 |
| $C_{12}H_{14}O_3$ | *cis*-Propylcoumarinic *(solid)* | 1476 | 7025.8 |
| $C_{12}H_{14}O_3$ | *trans*-Propylcoumaric *(solid)* | 1470 | 6997.2 |
| $C_{12}H_{22}O_4$ | Decanedicarboxylic *(solid)* | 1611 | 7668.4 |
| $C_{12}H_{22}O_4$ | Tetraethylsuccinic *(solid)* | 1619 | 7706.4 |
| $C_{12}H_{24}O_2$ | Lauric *(solid)* | 1772 | 8434.7 |
| $C_{13}H_{16}O_2$ | *trans-n*-Butylcoumaric *(solid)* | 1631 | 7763.6 |
| $C_{13}H_{16}O_2$ | *cis-n*-Butylcoumarinic *(solid)* | 1637 | 7792.1 |
| $C_{13}H_{24}O_4$ | Brassylic (undecanedicarboxylic) *(solid)* | 1769 | 8420.4 |
| $C_{14}H_{12}O_2$ | Diphenylacetic *(solid)* | 1651 | 7858.8 |
| $C_{14}H_{12}O_3$ | Benzilic *(solid)* | 1618 | 7701.7 |
| $C_{14}H_{18}O_3$ | Isoamylcoumaric *(solid)* | 1790 | 8520.4 |
| $C_{14}H_{18}O_3$ | *cis*-Isoamylcoumarinic *(solid)* | 1791 | 8525.2 |
| $C_{14}H_{28}O_2$ | Myristic *(solid)* | 2086 | 9929.4 |
| $C_{16}H_{14}O_4$ | α-Diphenylsuccinic (anhydr.) (easily sol. form) *(solid)*[c] | 1810 | 8615.6 |
| $C_{16}H_{14}O_4$ | βDiphenylsuccinic (difficulty sol.) *(solid)* | 1087 | 5174.1 |
| $C_{16}H_{16}O_2$ | Dibenzylacetic *(solid)* | 1954 | 9301.0 |
| $C_{16}H_{32}O_3$ | Palmitic *(solid)* | 2380 | 11328.8 |
| $C_{17}H_{16}O_3$ | β-Tolylmethoxycinnamic (stable) *(solid)* | 2035 | 9686.6 |
| $C_{18}H_{14}O_4$ | α-Truxillic *(solid)* | 2083 | 9915.1 |
| $C_{18}H_{32}O_3$ | Stearolic *(solid)* | 2629 | 12514.0 |
| $C_{18}H_{34}O_2$ | Elaidic *(solid)* | 2664 | 12680.6 |
| $C_{18}H_{34}O_2$ | Oleic | 2657 | 12647.3 |
| $C_{18}H_{36}O_2$ | Stearic *(solid)* | 2698 | 12842.5 |
| $C_{19}H_{36}O_4$ | Cetylmalonic *(solid)* | 2893 | 13770.7 |
| $C_{20}H_{40}O_2$ | Arachidic *(solid)* | 3010 | 14327.6 |
| $C_{22}H_{40}O_2$ | Behenolic | 3255 | 15493.8 |
| $C_{22}H_{42}O_3$ | Brassidic *(solid)* | 3290 | 15660.4 |
| $C_{22}H_{42}O_2$ | Erucic *(solid)* | 3297 | 15693.7 |
| $C_{22}H_{44}O_2$ | Behenic *(solid)* | 3338 | 15888.9 |
| $C_{22}H_{44}O_4$ | Dihydroxybehenic *(solid)* | 3236 | 15403.4 |
| $C_{24}H_{30}O_8$ | Cinnamylidenemalonic[d] | 2639 | 12561.6 |

[c] The heat of combustion of the acetone addition product of α-diphenyl-succinic acid (easily sol. form) is given as 2237.9 kg-cal.

[d] Exposed to the action of light.

**Appendix C2b. Heat of combustion of organic compounds** *(continued)*

| Formula | Name | kg cal/mol | Btu/scf |
|---|---|---|---|
| | **4. C, H, O, Adic Anhydrides and Lacyones** | | |
| $C_4H_2O_3$ | Maleic anhydride *(solid)* | 333.9 | 1589.4 |
| $C_4H_4O_3$ | Succinic anhydride *(solid)* | 369.5 | 1758.8 |
| $C_4H_6O_3$ | Acetic anhydride *(gas)* | 458.3 | 2181.5 |
| | *(liquid)* | 431.9 | 2055.8 |
| $C_5H_4O_2$ | Itaconic anhydride *(solid)* | 481.5 | 2291.9 |
| $C_5H_6O_3$ | Glutaric anhydride *(solid)* | 528 | 2513.3 |
| $C_5H_6O_3$ | Methylsuccinic anhydride *(solid)* | 527.7 | 2511.9 |
| $C_6H_8O_3$ | Dimethylsuccinic anhydride *(sym.)* *(solid)* | 680 | 3236.8 |
| $C_6H_8O_3$ | Dimethylsuccinic anhydride *(unsym.)* *(solid)* | 682.6 | 3249.2 |
| $C_6H_8O_3$ | Ethylsuccinic anhydride | 684 | 3255.8 |
| $C_6H_{10}O_3$ | Propionic anhydride | 746 | 3551.0 |
| $C_6H_{10}O_5$ | Saccharic acid lactone *(solid)* | 656 | 3122.6 |
| $C_6H_{10}O_6$ | *l*-Gulonolactone *(solid)* | 614 | 2922.6 |
| $C_6H_{10}O_6$ | *d*-Mannolactone *(solid)* | 618 | 2941.7 |
| $C_6H_{10}O_6$ | *l*-Mannolactone *(solid)* | 616 | 2932.2 |
| $C_7H_{10}O_3$ | Trimethylsuccinic anhydride *(solid)* | 836 | 3979.4 |
| $C_7H_{10}O_4$ | Terebic acid (γ, γ-dimethylparaconic acid) *(solid)* | 778 | 3703.3 |
| $C_7H_{12}O_7$ | Glucoheptonic acid lactone *(solid)* | 726 | 3455.8 |
| $C_8H_4O_3$ | Phthalic anhydride *(solid)* | 781 | 3717.6 |
| $C_8H_6O_2$ | Phthalide *(solid)* | 884 | 4207.8 |
| $C_8H_{10}O_3$ | *cis*-Hexahydrophthalic anhydride (sol.) *(solid)* | 932 | 4436.3 |
| | *trans*-Hexahydrophthalic anhydride (sol.) *(solid)* | 937 | 4460.1 |
| $C_8H_{12}O_3$ | *sym.-dl*-Diethylsuccinic anhydride *(solid)* | 997 | 4745.7 |
| | *anti.*-Diethylsuccinic anhydride *(solid)* | 997 | 4745.7 |
| $C_8H_{12}O_3$ | Diethylsuccinic anhydride *(unsym.)* *(solid)* | 998 | 4750.5 |
| $C_8H_{12}O_3$ | Tetramethylsuccinic anhydride *(solid)* | 993 | 4726.7 |
| $C_8H_{14}O_3$ | Glucooctonic lactone *(solid)* | 836 | 3979.4 |
| $C_{10}H_8O_3$ | Phenylsuccinic anhydride *(solid)* | 1094 | 5207.4 |
| $C_{10}H_{10}O_4$ | Meconine (dimethoxyphthalide) | 1137 | 5412.1 |
| $C_{10}H_{14}O_3$ | Camphoric anhydride *(solid)* | 1252 | 5959.5 |
| $C_{10}H_{16}O_8$ | Triethylsuccinic anhydride *(solid)* | 1310 | 6235.6 |
| $C_{11}H_{12}O_5$ | 4-Methyl opianate *(solid)* | 1263 | 6011.9 |
| $C_{12}H_6O_3$ | Naphthalic anhydride *(solid)* | 1258 | 5988.1 |
| $C_{12}H_{20}O_3$ | Tetraethylsuccinic anhydride *(solid)* | 1621 | 7716.0 |
| $C_{12}H_{22}O_3$ | Diethylacetic anhydride *(solid)* | 1669 | 7944.4 |
| $C_{14}H_{10}O_3$ | Benzoic anhydride *(solid)* | 1555 | 7401.8 |
| $C_{14}H_{26}O_3$ | Heptylic anhydride *(solid)* | 1985 | 9448.6 |
| $C_{16}H_{10}O_3$ | Diphenylmaleic anhydride | 1769 | 8420.4 |

**Appendix C2b. Heat of combustion of organic compounds** *(continued)*

| Formula | Name | kg cal/mol | Btu/scf |
|---|---|---|---|
| | **4. C, H, O, Acid Anhydrides and Lactones** *(continued)* | | |
| $C_{16}H_{12}O_3$ | *dl*-Diphenylsuccinic anhydride *(solid)* | 1816 | 8644.2 |
| $C_{18}H_{14}O_3$ | Cinnamic anhydride *(solid)* | 2091 | 9953.2 |
| $C_{28}H_{22}O_3$ | Diphenylacetic anhydride *(solid)* | 3308 | 15746.1 |
| $C_{22}H_{30}O_3$ | Dibenzylacetic anhydride (glassy) *(solid)* | 3931 | 18711.6 |

| Formula | Name | kg cal/mol | Btu/scf |
|---|---|---|---|
| | **5. C, H, O, Carbohydrates: Cellulose, Starch, Etc.** | | |
| $C_5H_{10}O_3$ | Arabinose *(solid)* | 559 | 2660.8 |
| $C_5H_{10}O_3$ | Xylose *(solid)* | 561 | 2670.4 |
| $C_6H_{10}O_3$ | *l*-Glucosan *(solid)* | 678 | 3227.3 |
| $C_6H_{12}O_5$ | Fucose *(solid)* | 712 | 3389.1 |
| $C_6H_{12}O_5$ | Rhamnose *(solid)* | 718 | 3417.7 |
| $C_6H_{12}O_5$ | Rhamnose *(solid)* | 711 | 3384.4 |
| $C_6H_{12}O_6$ | Galactose *(solid)* | 670 | 3189.2 |
| $C_6H_{12}O_6$ | *d*-Glucose (dextrose) | 673 | 3203.5 |
| $C_6H_{12}O_6$ | *l*-Fructose *(solid)* | 675 | 3213.0 |
| $C_6H_{12}O_6$ | Sorbinose *(d-sorbose) (solid)* | 668 | 3179.7 |
| $C_7H_{14}O_7$ | Glucoheptose *(solid)* | 783 | 3727.1 |
| $C_{12}H_{18}O_8$ | Rhamnose triacetate *(solid)* | 1351 | 6430.8 |
| $C_{12}H_{22}O_{11}$ | Cellobiose (anhydr.) *(solid)* | 1350 | 6426.0 |
| $C_{12}H_{22}O_{11}$ | Lactose (anhydr.) *(solid)* | 1350 | 6426.0 |
| $C_{12}H_{22}O_{11}.H_2O$ | Lactose *(solid)* | 1344 | 6397.4 |
| $C_{12}H_{22}O_{11}$ | Maltose *(solid)* | 1350 | 6426.0 |
| $C_{12}H_{22}O_{22}.H_2O$ | Maltose (cryst.) | 1339 | 6373.6 |
| $C_{12}H_{22}O_{11}$ | Sucrose *(solid)* | 1349 | 6421.2 |
| $C_{12}H_{22}O_{11}$ | Trehalose (cryst.) | 1341 | 6383.2 |
| $C_{12}H_{22}O_{11}$ | Trehalose (mycose) *(solid)* | 1349 | 6421.2 |
| $C_{16}H_{22}O_{11}$ | Galactose pentaacetate *(solid)* | 1726 | 8215.8 |
| $C_{16}H_{22}O_{11}$ | Glucose pentaacetate *(solid)* | 1726 | 8215.8 |
| $C_{18}H_{32}O_{16}.H_2O$ | Melezitose *(solid)* | 2042 | 9719.9 |
| $C_{18}H_{32}O_{16}$ | Raffinose (melitose) *(solid)* | 2025 | 9639.0 |
| $C_{18}H_{32}O_{16}.5H_2O$ | Raffinose (cryst.) | 2018 | 9605.7 |
| $C_{18}H_{32}O_{16}$ | Stachyose (anhyd.) (calc.) *(solid)* | 2709 | 12894.8 |
| $C_{28}H_{38}O_{19}$ | Cellobiose octoacetate *(solid)* | 3032 | 14432.3 |
| $C_{28}H_{38}O_{19}$ | Lactose octoacetate *(solid)* | 3029 | 14418.0 |
| $C_{28}H_{38}O_{19}$ | Maltose octoacetate *(solid)* | 3030 | 14422.8 |
| $C_{28}H_{38}O_{19}$ | Sucrose octoacetate *(solid)* | 3033 | 14437.1 |

**Appendix C2b. Heat of combustion of organic compounds** *(continued)*

| Name | g cal/gram | Btu/lb |
|---|---|---|
| **5. C, H, O, Carbohydrates: Cellulose, Starch, Etc.** *(continued)* | | |
| Starch | 4179 | 7539 |
| Starch acetate | 4499 | 8116 |
| Inulin | 4130 | 7451 |
| | 4190 | 7558 |
| Inulin acetate | 4522 | 8158 |
| Dextrin | 4110 | 7414 |
| Glycogen | 4186 | 7553 |
| Cellulose | 4181 | 7543 |
| Cellulose acetate | 4496 | 8111 |
| Xylan | 4260 | 7685 |
| Xylan acetate | 4548 | 8205 |

| Formula | Name | kg cal/mol | Btu/scf |
|---|---|---|---|
| | **6. Other C, H, O-Compounds (partial list only)** | | |
| $CH_2O$ | Formaldehyde *(gas)* | 134 | 637.8 |
| $1/n\ (CH_2O)_n$ | Paraformaldehyde *(solid)* | 122 | 580.7 |
| $C_2H_4O$ | Acetaldehyde *(gas)* | 280.5 | 1335.2 |
| | | 279.0 | 1328.0 |
| $C_2H_4O_2$ | Methyl formate *(gas)* | 240.2 | 1143.4 |
| | *(liquid)* | 233.1 | 1109.6 |
| $C_2H_6O$ | Dimethyl ether *(gas)* | 347.6 | 1654.6 |
| $C_3H_6O$ | Acetone | 427 | 2032.5 |
| $C_3H_6O_2$ | Ethyl formate *(gas)* | 398.4 | 1896.4 |
| | *(liquid)* | 391.7 | 1864.5 |
| $C_3H_6O_2$ | Methyl acetate *(gas)* | 397.7 | 1893.1 |
| | *(liquid)* | 381.2 | 1814.5 |
| $C_3H_8O$ | Methyl ethyl ether *(gas)* | 503.4 | 2396.2 |
| $C_3H_8O_2$ | Methylal | 462.8 | 2202.9 |
| $C_4H_8O$ | Methyl ethyl ketone | 582 | 2770.3 |
| $C_4H_8O_2$ | Ethyl acetate *(gas)* | 544.4 | 2591.3 |
| | *(liquid)* | 538 | 2560.9 |
| $C_4H_8O_2$ | Methyl propionate *(gas)* | 552.3 | 2628.9 |
| $C_4H_{10}O$ | Diethyl ether *(gas)* | 660.3 | 3143.0 |
| | *(liquid)* | 651.7 | 3102.1 |
| $C_5H_{10}O$ | Diethyl ketone | 736 | 3503.4 |
| $C_5H_{10}O$ | Methyl propyl ketone | 736 | 3503.4 |
| $C_5H_{10}O$ | Methyl isopropyl ketone | 734 | 3493.8 |

**Appendix C2b. Heat of combustion of organic compounds** *(continued)*

| Formula | Name | kg cal/mol | Btu/scf |
|---|---|---|---|
| | **6. Other C, H, O-Compounds** *(continued)* | | |
| $C_5H_{10}O_2$ | Ethyl propionate | 690 | 3284.4 |
| $C_6H_4O_2$ | Quinone *(solid)* | 656 | 3122.6 |
| $C_6H_6O$ | Phenol *(solid)* | 732 | 3484.3 |
| $C_6H_6O_2$ | Pyrocatechol *(solid)* | 685 | 3260.6 |
| $C_6H_6O_2$ | Resorcinol *(solid)* | 683 | 3251.1 |
| $C_6H_6O_2$ | Hydroquinol *(solid)* | 683 | 3251.1 |
| $C_6H_6O_3$ | Pyrogallol *(solid)* | 639 | 3041.6 |
| $C_6H_{10}O_3$ | Ethyl acetoacetate | 753 | 3584.3 |
| $C_6H_{12}O_2$ | Ethyl *n*-butyrate | 851 | 4050.8 |
| $C_6H_{12}O_2$ | Ethyl isobutyrate | 845 | 4022.2 |
| $C_7H_{14}O$ | Dipropyl ketone | 1051 | 5002.8 |
| $C_7H_{14}O_2$ | Amyl acetate | 104 | 495.0 |
| $C_7H_{14}O_2$ | Ethyl valerate | 1017 | 4840.9 |
| $C_8H_8O$ | Acetophenone *(solid)* | 989 | 4707.6 |
| $C_8H_8O_2$ | Methyl benzoate | 943 | 4488.7 |
| $C_8H_8O_3$ | Methyl salicylate | 898 | 4274.5 |
| $C_8H_{16}O$ | Methyl hexyl ketone | 1205 | 5735.8 |
| $C_9H_{10}O_2$ | Ethyl benzoate | 1099 | 5231.2 |
| $C_9H_{10}O_3$ | Ethyl *p*-hydroxybenzoate *(solid)* | 1043 | 4964.7 |
| $C_9H_{10}O_3$ | Ethyl salicylate | 1051 | 5002.8 |
| $C_{10}H_8O$ | $\alpha$-Naphthol *(solid)* | 1185 | 5640.6 |
| $C_{10}H_8O$ | $\beta$-Naphthol *(solid)* | 1187 | 5650.1 |
| $C_{10}H_{16}O$ | Camphor *(solid)* | 1411 | 6716.4 |
| $C_{13}H_{10}O$ | Benzophenone *(solid)* | 1557 | 7411.3 |
| $C_{14}H_8O_2$ | Anthraquinone *(solid)* | 1545 | 7354.2 |
| | **7. Nitrogen Compounds (partial list only)** | | |
| $CH_3NO$ | Formamide | 135 | 642.6 |
| $CH_3NO_2$ | Nitromethane | 169.4 | 806.3 |
| $CH_4N_2O$ | Urea *(solid)* | 152 | 723.5 |
| $CH_5N$ | Methylamine *(gas)* | 259 | 1232.8 |
| | *(liquid)* | 256 | 1218.6 |
| $C_2N_2$ | Cyanogen *(gas)* | 260 | 1237.6 |
| $C_2H_3N$ | Acetonitrile *(gas)* | 310.4 | 1477.5 |
| | *(liquid)* | 302.4 | 1439.4 |
| $C_2H_3NO$ | Glycolic nitrile | 256.7 | 1221.9 |
| $C_2H_3NO$ | Methyl isocyanate | 269.4 | 1282.3 |

**Appendix C2b. Heat of combustion of organic compounds** *(continued)*

| Formula | Name | kg cal/mol | Btu/scf |
|---|---|---|---|
| | **7. Nitrogen Compounds** *(continued)* | | |
| $C_2H_3NO_3$ | Oxamic acid *(solid)* | 130 | 618.8 |
| $C_2H_4N_2O_2$ | Oxamide *(solid)* | 203 | 966.3 |
| $C_2H_5NO$ | Acetamide *(solid)* | 283 | 1347.1 |
| $C_2H_5NO_2$ | Nitroethane | 322.2 | 1533.7 |
| $C_2H_5NO_2$ | Ethyl nitrite *(gas)* | 333 | 1585.1 |
| $C_2H_5NO_2$ | Glycine (aminoacetic acid) *(solid)* | 234 | 1113.8 |
| $C_2H_5NO_3$ | Ethyl nitrate *(gas)* | 322 | 1532.7 |
| $C_2H_7N$ | Dimethylamine *(gas)* | 422 | 2008.7 |
| | *(liquid)* | 417 | 1984.9 |
| $C_2H_7N$ | Ethylamine *(gas)* | 413 | 1965.9 |
| | *(liquid)* | 408.5 | 1944.5 |
| $C_2H_8N_2.H_2O$ | Ethylenediamine | 452.6 | 2154.4 |
| $C_3H_2N_2$ | Malononitrile *(solid)* | 394.8 | 1879.2 |
| $C_3H_2N_2O_3$ | Parabanic acid *(solid)* | 212.4 | 1011.0 |
| $C_3H_4N_2O_2$ | Hydantoin *(solid)* | 311.7 | 1483.7 |
| $C_3H_5N$ | Propionitrile | 456.4 | 2172.5 |
| $C_3H_5NO$ | Ethyl isocyanate | 424.5 | 2020.6 |
| $C_3H_5N_3O_9$ | Trinitroglycerol | 432.4 | 2058.2 |
| $C_3H_6N_2O_2$ | Malonamide *(solid)* | 359 | 1708.8 |
| $C_3H_6N_2O_3$ | Hydantoic acid *(solid)* | 308.6 | 1468.9 |
| $C_3H_7N$ | Allylamine *(gas)* | 528.1 | 2513.8 |
| | *(liquid)* | 524.8 | 2498.0 |
| $C_3H_7NO$ | Propionamide *(solid)* | 440 | 2094.4 |
| $C_3H_7NO_2$ | Alanine *(solid)* | 388 | 1846.9 |
| $C_3H_7NO_2$ | *d*-Alanine *(solid)* | 387 | 1842.1 |
| $C_3H_7NO_2$ | Nitropropane | 477.9 | 2274.8 |
| $C_3H_7NO_2$ | Sarcosine *(solid)* | 401 | 1908.8 |
| $C_3H_7NO_2$ | Urethane *(solid)* | 397 | 1889.7 |
| $C_3H_7NO_3$ | Isoserine *(solid)* | 343 | 1632.7 |
| $C_3H_8N_2O$ | Ethylurea *(solid)* | 472 | 2246.7 |
| $C_3H_9N$ | Propylamine *(gas)* | 572.3 | 2724.1 |
| | *(liquid)* | 558.3 | 2657.5 |
| $C_3H_9N$ | Trimethylamine | 578.6 | 2754.1 |
| $C_4N_2$ | Carbon subnitride (acetylenedicarboxylic acid nitrile) *(solid)* | 515 | 2451.4 |
| $C_4H_2N_2O_4.H_2O$ | Alloxan *(solid)* | 276.3 | 1315.2 |
| $C_4H_5N$ | Allyl cyanide | 575 | 2737.0 |
| $C_4H_5N$ | Trimethylene nitrile | 581 | 2765.6 |

**Appendix C2b. Heat of combustion of organic compounds** *(continued)*

| Formula | Name | kg cal/mol | Btu/scf |
|---|---|---|---|
| | **7. Nitrogen Compounds** *(continued)* | | |
| $C_4H_5N$ | Pyrrole | 567.7 | 2702.3 |
| $C_4H_5NO_2$ | Succinimide *(solid)* | 438 | 2084.9 |
| $C_4H_6N_2O_2$ | Diketopiperazine *(solid)* | 474.6 | 2259.1 |
| $C_4H_7N$ | *n*-Butyronitrile | 613.3 | 2919.3 |
| $C_4H_7NO_4$ | Aspartic acid *(solid)* | 384.9 | 1832.1 |
| $C_4H_7N_3O$ | Creatinine *(solid)* | 563.4 | 2681.8 |
| $C_4H_8N_2O_2$ | Succinamide *(solid)* | 509.2 | 2423.8 |
| $C_4H_8N_2O_3$ | Asparagine *(solid)* | 463.3 | 2205.3 |
| $C_4H_8N_2O_3.H_2O$ | Asparagine (cryst.) | 459.7 | 2188.2 |
| $C_4H_9NO$ | *n*-Butyramide *(solid)* | 596 | 2837.0 |
| $C_4H_9NO$ | Isobutyramide *(solid)* | 596 | 2837.0 |
| $C_4H_9NO_2$ | Isobutyl nitrite *(gas)* | 644.6 | 3068.3 |
| $C_4H_9N_3O_2$ | Creatine (anhydr.) *(solid)* | 558 | 2656.1 |
| $C_4H_9N_3O_2.H_2O$ | Creatine (cryst.) | 553.1 | 2632.8 |
| $C_4H_{11}N$ | *n*-Butylamine | 710 | 3379.6 |
| $C_4H_{11}N$ | Isobutylamine | 713 | 3393.9 |
| $C_4H_{11}N$ | *sec.*-Butylamine | 713 | 3393.9 |
| $C_4H_{11}N$ | *tert.*-Butylamine | 716 | 3408.2 |
| $C_4H_{11}N$ | Diethylamine *(gas)* | 731 | 3479.6 |
| | *(liquid)* | 722.8 | 3440.5 |
| $C_5H_4N_4O_3$ | Uric acid *(solid)* | 460.2 | 2190.6 |
| $C_5H_5N$ | Pyridine | 660 | 3141.6 |
| $C_5H_5N_5O$ | Guanine *(solid)* | 586 | 2789.4 |
| $C_5H_6N_2O_2$ | 4-Methyluracil *(solid)* | 566 | 2694.2 |
| $C_5H_6N_2O_2$ | 5-Methyluracil *(solid)* | 565 | 2689.4 |
| $C_5H_6N_4O_4$ | Pseudouric acid *(solid)* | 454.2 | 2162.0 |
| $C_5H_8N_2O_2$ | 4-Methylhydrouracil *(solid)* | 618 | 2941.7 |
| $C_5H_9NO_4$ | Glutamic acid (act.) *(solid)* | 542.4 | 2581.8 |
| $C_5H_{11}N$ | Piperidine | 826.6 | 3934.6 |
| $C_5H_{11}NO_2$ | *d-l-α*-Aminoisovaleric acid *(solid)* | 701 | 3336.8 |
| $C_5H_{13}N$ | Isoamylamine | 867 | 4126.9 |
| $C_6H_3N_3O_6$ | 1, 2, 4-Trinitrobenzene *(solid)* | 674 | 3208.2 |
| $C_6H_3N_3O_6$ | 1, 3, 5-Trinitrobenzene *(solid)* | 564 | 2684.6 |
| $C_6H_3N_3O_7$ | 2, 4, 6-Trinitrophenol *(solid)* | 620 | 2951.2 |
| $C_5H_4N_2O_4$ | *o*-Dinitrobenzene *(solid)* | 703.2 | 3347.2 |
| $C_6H_4N_2O_4$ | *m*-Dinitrobenzene *(solid)* | 697 | 3317.7 |
| $C_6H_4N_2O_4$ | *p*-Dinitrobenzene *(solid)* | 695 | 3308.2 |
| $C_6H_4N_2O_5$ | 2, 4-Dinitrophenol *(solid)* | 648 | 3084.5 |
| $C_6H_5NO_2$ | Nitrobenzene | 739 | 3517.6 |

**Appendix C2b. Heat of combustion of organic compounds** *(continued)*

| Formula | Name | kg cal/mol | Btu/scf |
|---|---|---|---|
| | **7. Nitrogen Compounds** *(continued)* | | |
| $C_6H_5NO_3$ | *o*-Nitrophenol *(solid)* | 689 | 3279.6 |
| $C_6H_5NO_3$ | *m*-Nitrophenol *(solid)* | 684 | 3255.8 |
| $C_6H_5NO_3$ | *p*-Nitrophenol *(solid)* | 686 | 3265.4 |
| $C_6H_6N_2O_2$ | *o*-Nitroaniline *(solid)* | 766 | 3646.2 |
| $C_6H_6N_2O_2$ | *m*-Nitroaniline *(solid)* | 765 | 3641.4 |
| $C_6H_6N_2O_2$ | *p*-Nitroaniline *(solid)* | 761 | 3622.4 |
| $C_6H_6N_2O_3$ | *m*-Nitrophenylhydroxylamine | 766 | 3646.2 |
| $C_6H_7N$ | Aniline | 812 | 3865.1 |
| $C_6H_7N$ | α-Picoline | 815 | 3879.4 |
| $C_6H_7N$ | β-Picoline | 812 | 3865.1 |
| $C_6H_7N$ | γ-Picoline | 816 | 3884.2 |
| $C_6H_7NO$ | Phenylhydroxylamine | 804 | 3827.0 |
| $C_6H_8N_2$ | Phenylhydrazine *(solid)* | 875 | 4165.0 |
| | | 805 | 3831.8 |
| $C_6H_8N_2$ | *p*-Phenylenediamine *(solid)* | 843 | 4012.7 |
| $C_6H_{12}N_2O_3$ | d-Alanine anhydride *(solid)* | 786 | 3741.4 |
| $C_6H_{12}N_4$ | Hexamethylenetetramine *(solid)* | 1006.7 | 4791.9 |
| $C_6H_{13}NO_2$ | Leucine *(solid)* | 856 | 4074.6 |
| $C_6H_{15}N$ | Hexylamine | 1022 | 4864.7 |
| $C_6H_{15}N$ | Triethylamine | 1037 | 4936.1 |
| $C_7H_5N$ | Benzonitrile | 866 | 4122.2 |
| $C_7H_5NO_3$ | *m*-Nitrobenzaldehyd *(solid)* | 800 | 3808.0 |
| $C_7H_5NO_4$ | *o*-Nitrobenzoic acid *(solid)* | 735 | 3498.6 |
| $C_7H_5NO_4$ | *m*-Nitrobenzoic acid *(solid)* | 729 | 3470.0 |
| $C_7H_5NO_4$ | *p*-Nitrobenzoic acid *(solid)* | 728 | 3465.3 |
| $C_7H_5N_3O_6$ | 2, 3, 4-Trinitrotoluene | 833 | 3965.1 |
| $C_7H_5N_3O_6$ | 2, 3, 5-Trinitrotoluene *(solid)* | 824 | 3922.2 |
| $C_7H_5N_3O_6$ | 2, 3, 5-Trinitrotoluene | 825 | 3927.0 |
| $C_7H_5N_3O_6$ | 2, 4, 6-Trinitrotoluene *(solid)* | 821 | 3908.0 |
| $C_7H_5N_3O_6$ | 3, 4, 5-Trinitrotoluene *(solid)* | 828 | 3941.3 |
| $C_7H_5N_3O_6$ | 3, 4, 6-Trinitrotoluene *(solid)* | 826 | 3931.8 |
| $C_7H_5N_5O_3$ | Tetryl *(solid)* | 842 | 4007.9 |
| $C_7H_6N_2O_4$ | 2, 3-Dinitrotoluene *(solid)* | 859 | 4088.8 |
| $C_7H_6N_2O_4$ | 2, 4-Dinitrotoluene *(solid)* | 853 | 4060.3 |
| $C_7H_6N_2O_4$ | 2, 5-Dinitrotoluene *(solid)* | 855 | 4069.8 |
| $C_7H_6N_2O_4$ | 2, 6-Dinitrotoluene *(solid)* | 854 | 4065.0 |
| $C_7H_6N_2O_4$ | 3, 4-Dinitrotoluene *(solid)* | 860 | 4093.6 |
| $C_7H_6N_2O_4$ | 3, 5-Dinitrotoluene *(solid)* | 853 | 4060.3 |

**Appendix C2b. Heat of combustion of organic compounds** *(continued)*

| Formula | Name | kg cal/mol | Btu/scf |
|---|---|---|---|
| | **7. Nitrogen Components** *(continued)* | | |
| $C_7H_7NO$ | Benzamide *(solid)* | 848 | 4036.5 |
| $C_7H_7NO$ | Formanilide *(solid)* | 861 | 4098.4 |
| $C_7H_7NO_2$ | *o*-Nitrotoluene | 897 | 4269.7 |
| $C_7H_7NO_2$ | *m*-Nitrotoluene | 893 | 4250.7 |
| $C_7H_7NO_2$ | *p*-Nitrotoluene *(solid)* | 889 | 4231.6 |
| $C_7H_7N_3O_4$ | 2, 4-Dinitromethylaniline *(solid)* | 884.5 | 4210.2 |
| $C_7H_8N_2O_2$ | *p*-Nitromethylaniline *(solid)* | 924 | 4398.2 |
| $C_7H_8N_4O_2$ | Theobromine *(solid)* | 845 | 4022.2 |
| $C_7H_9N$ | Benzylamine | 967 | 4602.9 |
| $C_7H_9N$ | Methylaniline | 973 | 4631.5 |
| $C_7H_9N$ | *o*-Toluidine | 964 | 4588.6 |
| $C_7H_9N$ | *m*-Toluidine | 965 | 4593.4 |
| $C_7H_9N$ | *p*-Toluidine | 958 | 4560.1 |
| $C_7H_{17}N$ | Heptylamine | 1179 | 5612.0 |
| $C_8H_5NO$ | Benzoyl cyanide *(solid)* | 940 | 4474.4 |
| $C_8H_5NO_2$ | Isatin *(solid)* | 868 | 4131.7 |
| $C_8H_5NO_2$ | Phthalimide *(solid)* | 850 | 4046.0 |
| $C_8H_7N$ | Benzyl cyanide | 1024 | 4874.2 |
| $C_8H_7N$ | Indole *(solid)* | 1022 | 4864.7 |
| $C_8H_7NO_2$ | Dioxindol *(solid)* | 916 | 4360.2 |
| $C_8H_7N_5O_8$ | Methyltetryl *(solid)* | 1009 | 4802.8 |
| $C_8H_8N_2O_3$ | *o*-Nitroacetanilide *(solid)* | 974 | 4636.2 |
| $C_8H_8N_2O_3$ | *m*-Nitroacetanilide *(solid)* | 970 | 4617.2 |
| $C_8H_8N_2O_3$ | *p*-Nitroacetanilide *(solid)* | 968 | 4607.7 |
| $C_8H_9NO$ | Acetanilide *(solid)* | 1010 | 4807.6 |
| $C_8H_9NO_2$ | Phenylglycine (phenylaminoacetic acid) *(solid)* | 955 | 4545.8 |
| $C_8H_9NO_3$ | *o*-Nitrophenetole | 1021 | 4860.0 |
| $C_8H_9NO_3$ | *m*-Nitrophenetole | 1009 | 4802.8 |
| $C_8H_9NO_3$ | *p*-Nitrophenetole | 1006 | 4788.6 |
| $C_8H_{10}N_4O_2$ | Caffeine *(solid)* | 1014 | 4826.6 |
| $C_8H_{11}N$ | Dimethylaniline | 114 | 542.6 |
| $C_8H_{11}N$ | Ethylaniline | 1122 | 5340.7 |
| $C_8H_{12}N_2O$ | Veronal (5,5-diethylbarbituric acid) *(solid)* | 983 | 4679.1 |
| $C_8H_{17}N$ | Coniine | 1275 | 6069.0 |
| $C_8H_{19}N$ | Diisobutylamine | 1348 | 6416.5 |
| $C_9H_7N$ | Quinoline | 1123 | 5345.5 |
| $C_9H_9N$ | 3-Methylindole | 117 | 556.9 |

**Appendix C2b. Heat of combustion of organic compounds** *(continued)*

| Formula | Name | kg cal/mol | Btu/scf |
|---|---|---|---|
| | **7. Nitrogen Components** *(continued)* | | |
| $C_9H_9N$ | α-Methylindole | 1169 | 5564.4 |
| $C_9H_9NO_3$ | Hippuric acid *(solid)* | 1012 | 4817.1 |
| $C_9H_{11}NO$ | Propionanilide *(solid)* | 1168 | 5559.7 |
| $C_9H_{11}NO_2$ | Phenylalanine *(solid)* | 1111 | 5288.4 |
| $C_9H_{11}NO_3$ | Tyrosine *(solid)* | 1070 | 5093.2 |
| $C_9H_{13}N$ | Benzylethylamine | 1290 | 6140.4 |
| $C_{10}H_9N$ | α-Naphthylamine *(solid)* | 1264 | 6016.6 |
| $C_{10}H_9N$ | β-Naphthylamine *(solid)* | 1261 | 6002.4 |
| $C_{10}H_9NO_4$ | Hemipinimide *(solid)* | 1099 | 5231.2 |
| $C_{10}H_{13}N$ | Tetrahydroquinaldine | 1382 | 6578.3 |
| $C_{10}H_{13}NO_2$ | Phenacetin *(solid)* | 1285 | 6116.6 |
| $C_{10}H_{14}N_2$ | Nicotine | 1428 | 6797.3 |
| $C_{10}H_{15}N$ | Diethylaniline | 145 | 690.2 |
| $C_{10}H_{23}N$ | Diisoamylamine | 1660 | 7901.6 |
| $C_{11}H_7N$ | α-Naphthonitrile *(solid)* | 1326 | 6311.8 |
| $C_{11}H_7N$ | β-Naphthonitrile *(solid)* | 1321 | 6288.0 |
| $C_{12}H_9N$ | Carbazole *(solid)* | 1475 | 7021.0 |
| $C_{12}H_{10}N_2$ | Azobenzene *(solid)* | 1546 | 7359.0 |
| $C_{12}H_{11}N$ | Diphenylamine *(solid)* | 153 | 728.3 |
| $C_{12}H_{12}N_2$ | Hydrazobenzene *(solid)* | 1597 | 7601.7 |
| | | 1599 | 7611.2 |
| $C_{12}H_{12}N_2$ | Benzidine *(solid)* | 1561 | 7430.4 |
| $C_{12}H_{27}N$ | Triisobutylamine | 1974 | 9396.2 |
| $C_{13}H_{11}NO$ | Benzanilide *(solid)* | 1576 | 7501.8 |
| $C_{13}H_{12}N_2O$ | *sym.*-Diphenylurea *(solid)* | 1612 | 7673.1 |
| $C_{13}H_{12}N_2O$ | *unsym.*-Diphenylurea *(solid)* | 1614 | 7682.6 |
| $C_{14}H_{15}N$ | Dibenzylamine *(solid)* | 1853 | 8820.3 |
| $C_{15}H_{33}N$ | Triisoamylamine | 2459 | 11704.8 |
| $C_{16}H_{10}N_2O_2$ | Indigo *(solid)* | 1815 | 8639.4 |
| $C_{16}H_{13}N$ | Phenyl-α-naphthylamine *(solid)* | 2004 | 9539.0 |
| $C_{16}H_{13}N$ | Phenyl-β-naphthylamine *(solid)* | 1998 | 9510.5 |
| $C_{17}H_{19}NO_3.H_2O$ | Morphine *(solid)* | 2146 | 10215.0 |
| $C_{18}H_{15}N$ | Triphenylamine *(solid)* | 226 | 1075.8 |
| $C_{18}H_{21}NO_3.H_2O$ | Codeine *(solid)* | 2328 | 11081.3 |
| $C_{19}H_{21}NO_3$ | Thebaine *(solid)* | 2441 | 11619.2 |
| $C_{20}H_{21}NO_4$ | Papaverine *(solid)* | 2478 | 11795.3 |
| $C_{20}H_{27}NO_{11}$ | Amygdalin *(solid)* | 2348 | 11176.5 |
| $C_{21}H_{21}N$ | Tribenzylamine *(solid)* | 2762 | 13147.1 |

**Appendix C2b. Heat of combustion of organic compounds** *(continued)*

| Formula | Name | kg cal/mol | Btu/scf |
|---|---|---|---|
| | **7. Nitrogen Compounds** *(continued)* | | |
| $C_{21}H_{22}N_2O_2$ | Strychnine *(solid)* | 2686 | 12785.4 |
| $C_{22}H_{23}NO_7$ | Narcotine *(solid)* | 1644 | 7825.4 |
| $C_{23}H_{26}N_2O_4$ | Brucine *(solid)* | 2933 | 13961.1 |
| $C_{23}H_{27}NO_8.2H_2O$ | Narceine *(solid)* | 2803 | 13342.3 |
| | **8. Halogen Components (partial list only)** | | |
| $CCl_4$ | Carbon tetrachloride *(gas)* | 44.5 | 211.8 |
| | *(liquid)* | 37.3 | 177.5 |
| $CHCl_3$ | Chloroform *(gas)* | 70.3 | 334.6 |
| | *(liquid)* | 89.2 | 424.6 |
| $CHI_3$ | Iodoform *(solid)* | 162 | 771.1 |
| $CH_2Cl_2$ | Methylene chloride *(gas)* (HCI gas) | 106.8 | 508.4 |
| $CH_2I_2$ | Methylene iodide | 178 | 847.3 |
| $CH_3Br$ | Methyl bromide *(gas)* | 184 | 875.8 |
| | | 180 | 856.8 |
| $CH_3Cl$ | Methyl chloride | 164 | 780.6 |
| | | 173 | 823.5 |
| $CH_3I$ | Methyl iodide *(gas)* | 200.5 | 954.4 |
| | *(liquid)* | 194.7 | 926.8 |
| $C_2HCl_3O_2$ | Trichloroacetic acid *(solid)* | 92.8 | 441.7 |
| $C_2H_3ClO$ | Chloroacetaldehyde | 234.4 | 1115.7 |
| $C_2H_3ClO_2$ | Chloroacetic acid *(solid)* | 171 | 814.0 |
| $C_2H_4Cl_2$ | Ethylene chloride *(gas)* | 271 | 1290.0 |
| $C_2H_4Cl_2$ | Ethylidene chloride *(gas)* | 271.1 | 1290.4 |
| | *(liquid)* (HCI gas) | 267.1 | 1271.4 |
| $C_2H_5Br$ | Ethyl bromide *(gas)* | 340 | 1618.4 |
| | | 329 | 1566.0 |
| $C_2H_5Cl$ | Ethyl chloride *(gas)* | *316.7* | *1507.5* |
| | | *326.9* | *1556.0* |
| *C{H¶I* | *Ethyl iodide (gas)* | 358 | 1704.1 |
| | *(liquid)* | 356 | 1694.6 |
| $C_3H_5Cl$ | Allyl chloride *(gas)* | 440.8 | 2098.2 |
| $C_3H_7Br$ | Propyl bromide *(gas)* | 497 | 2365.7 |
| $C_3H_7Cl$ | Propyl chloride *(gas)* | 478.3 | 2276.7 |
| $C_3H_7I$ | *n*-Propyl iodide | 514.3 | 2448.1 |
| $C_3H_7I$ | Isopropyl iodide | 509.1 | 2423.3 |

**Appendix C2b. Heat of combustion of organic compounds** *(continued)*

| Formula | Name | kg cal/mol | Btu/scf |
|---|---|---|---|
| | **8. Halogen Components** *(continued)* | | |
| $C_4H_6Cl_2O_2$ | Ethyl dichloroacetate | 463.4 | 2205.8 |
| $C_4H_7ClO_2$ | Ethyl chloroacetate | 493.9 | 2351.0 |
| $C_4H_9Cl$ | Isobutyl chloride | 635.5 | 3025.0 |
| $C_6Cl_4O_2$ | Tetrachloroquinone *(solid)* | 517.7 | 2464.3 |
| $C_6HCl_3O_2$ | Trichloroquinone *(solid)* | 546.4 | 2600.9 |
| $C_6H_2Cl_2O_2$ | 2, 6-Dichloroquinone *(solid)* | 578.9 | 2755.6 |
| $C_6H_2Cl_2O_4$ | Chloroanilic acid *(solid)* | 484.9 | 2308.1 |
| $C_6H_2Cl_4O_2$ | Tetrachlorohydroquinol *(solid)* | 563 | 2679.9 |
| $C_6H_3ClO_2$ | Chloroquinone *(solid)* | 616.6 | 2935.0 |
| $C_6H_3Cl_3O_2$ | Trichlorohydroquinol *(solid)* | 593 | 2822.7 |
| $C_6H_4Cl_2$ | *o*-Dichlorobenzene *(solid)* | 671.8 | 3197.8 |
| $C_6H_4Cl_2O_2$ | 2, 6-Dichlorohydroquinol | 614 | 2922.6 |
| $C_6H_5ClO_2$ | Chlorohydroquinol *(solid)* | 646 | 3075.0 |
| $C_6H_5I$ | Phenyl iodide | 771 | 3670.0 |
| $C_7H_5ClO$ | Benzoyl chloride | 783 | 3727.1 |
| $C_7H_7Cl$ | Benzyl chloride | 886 | 4217.4 |
| $C_8H_4Cl_2O_2$ | Phthalyl chloride *(solid)* | 802 | 3817.5 |
| **9. Sulfur Compounds, final state: dilute $H_2SO_4$ aq., except as otherwise noted.** | | | |
| COS | Carbonyl sulfide *(gas)* (Gaseous $SO_2$) | 130.5 | 621.2 |
| $CS_2$ | Carbon disulfide (Gaseous $SO_2$) | 246.6 | 1173.8 |
| $CH_4N_2S$ | Thiourea *(solid)* ($H_2SO_4.200H_2O$.) | 342.8 | 1631.7 |
| $CH_4S$ | Methylmercaptan *(gas)* (Gaseous $SO_2$) | 297.6 | 1416.6 |
| $C_2H_3NS$ | Methyl thiocyanate *(gas)* (Gaseous $SO_2$) | 397.4 | 1891.6 |
| | *(liquid)* | 453.1 | 2156.8 |
| $C_2H_3NS$ | Methyl isothiocyanate *(gas)* (Gaseous $SO_2$) | 390.5 | 1858.8 |
| | *(solid)* | 442.9 | 2108.2 |
| $C_2H_6S$ | Dimethyl sulfide *(gas)* (Gaseous $SO_2$) | 455.6 | 2168.7 |
| $C_2H_6S$ | Ethylmercaptan *(gas)* (Gaseous $SO_2$) | 452 | 2151.5 |
| | *(liquid)* | 517.2 | 2461.9 |
| $C_2H_7NO_3S$ | Taurine *(solid)* | 382.9 | 1822.6 |
| $C_3H_4N_2OS$ | Thiohydantoin *(solid)* | 503 | 2394.3 |
| $C_3H_5NS$ | Ethyl thiocyanate | 613.8 | 2921.7 |
| $C_3H_5NS$ | Ethyl isothiocyanate | 604.1 | 2875.5 |
| $C_3H_6N_2O_2S$ | Thiohydantoic acid *(solid)* | 498.5 | 2372.9 |
| $C_4H_4S$ | Thiophene *(liquid)* | 670.5 | 3191.6 |
| | *(gas)* | 608.2 | 2895.0 |

**Appendix C2b. Heat of combustion of organic compounds** *(concluded)*

| Formula | Name | kg cal/mol | Btu/scf |
|---|---|---|---|
| | **9. Sulfur Compounds** *(continued)* | | |
| $C_4H_5NS$ | Allyl mustart oil *(liquid)* | 733 | 3489.1 |
| | *(gas)* (Gaseous $SO_2$) | 673 | 3203.5 |
| $C_4H_8N_2S$ | Allythiourea *(solid)* | 792 | 3769.9 |
| $C_4H_9NS_2$ | Dimethyl *N*-methylcarbimidedithiolate | 969 | 4612.4 |
| $C_4H_9NS_2$ | *S*-Methyl *N*-dimethyldithiocarbamate (sol.) | 954 | 4541.0 |
| $C_4H_9NS_2$ | Methyl formothialdine | 964 | 4588.6 |
| $C_4H_{10}S$ | Diethyl sulfide *(gas)* (Gaseous $SO_2$) | 769 | 3660.4 |
| | *(liquid)* | 829 | 3946.0 |
| $C_5H_4O_2S$ | α-Thiophenecarboxylic acid *(solid)* | 646 | 3075.0 |
| $C_5H_8O_2S$ | Tetrahydro-α-thiophenecarboxylic acid *(solid)* | 755 | 3593.8 |
| $C_5H_{10}N_2S_2$ | Dimethyl formocarbothialdine (sol.) | 1098 | 5226.5 |
| $C_5H_{10}N_2S_2$ | Carbothialdine (sol.) | 1086 | 5169.4 |
| $C_5H_{10}N_2S_2$ | Pentamethylenediamine disulfine (sol.) | 1113 | 5297.9 |
| $C_5H_{11}NS_2$ | *S*-Ethyl *N*-dimethyldithiocarbamate | 1122 | 5340.7 |
| $C_5H_{11}NS_2$ | Dimthyl *N*-ethylcarbimidedithiolate | 1130 | 5378.8 |
| $C_6H_{12}N_2O_4S_2$ | Cystine *(solid)* (Gaseous $SO_2$) | 994 | 4731.4 |
| $C_6H_{13}NS_2$ | *S*-Methyl *N*-diethyldithiocarbamate | 1272 | 6054.7 |
| $C_6H_{13}NS_2$ | Diethyl *N*-methylcarbimidedithiolate | 1289 | 6135.6 |
| $C_6H_{13}NS_2$ | Thialdine (sol.) | 1264 | 6016.6 |
| $C_7H_5NS$ | Phenyl isothiocyanate | 1024 | 4874.2 |
| $C_9H_{11}NS_2$ | Dimethyl *N*-phenylcarbimidedithiolate | 1545 | 7354.2 |
| $C_{18}H_{14}N_2O_2S$ | Benzonaphthoquinonethiazine *(solid)* | 2278 | 10843.3 |

**Appendix C2c. Calorific (heating) values.** (Adapted from Reference 14)

| **Substance**[(a)] | **Formula** | (b) | (c) | **kcal/kg**[(c)] | **Btu/lb**[(c)] |
|---|---|---|---|---|---|
| Acetaldehyde | $CH_3CHO$ | daf | n | 6340 | 11410 |
| Acetic Acid | $CH_3CO_2H$ | daf | n | 3490 | 6280 |
| Acetone | $(CH_3)_2CO$ | daf | n | 7360 | 13250 |
| Acetylene | CHCH | daf | n | 12000 | 21600 |
| Aluminum | (to $Al_2O_3$) | daf | g | 7420 | 13350 |
| Aniline | $C_6H_5NH_2$ | daf | n | 8730 | 15710 |
| Asphalt | | ar | n | 9530 | 17190 |
| Bagasse, 12m | | ar | n | 4050 | 7300 |
| Bamboo, 10m | | ar | n | 4110 | 7410 |
| Benzaldehyde | $C_6H_5CHO$ | daf | n | 7940 | 14290 |
| Benzene | $C_6H_6$ | daf | n | 10030 | 18050 |
| Books, 24a | | daf | g | 4200 | 7560 |
| Brown Paper, 6m, 1a | | ar | g | 4030 | 7260 |
| Brown Paper, 6m, 1a | | d | g | 4280 | 7710 |
| Brown Paper, 6m, 1a | | daf | g | 4330 | 7800 |
| Brown Peanut Skins | | d | g | 5800 | 10430 |
| Brush, 40m, 5a | | ar | g | 2640 | 4740 |
| Brush, 40m, 5a | | d | g | 4390 | 7900 |
| Brush, 40m, 5a | | daf | g | 4780 | 8600 |
| iso-Butyl Alcohol | $(CH_3)_2CH_2CH_2OH$ | daf | n | 8510 | 15320 |
| n-Butyl Alcohol | $C_4H_9OH$ | daf | n | 8630 | 15530 |
| Butyl sole composition, 1m, 30a | | ar | g | 6060 | 10900 |
| Carbon | | ar | g | 7830 | 14090 |
| Carbon Disulfide | $CS_2$ | daf | n | 3240 | 5840 |
| Carbon Tetrachloride | $CCl_4$ | daf | n | 240 | 440 |
| Cardboard | | daf | g | 4650 | 8370 |
| Castor Oil | | d | g | 8860 | 15950 |
| Castor Oil | | daf | g | 8860 | 15950 |
| Charcoal, 4m | | ar | n | 7260 | 13090 |
| Chloroform | $CHCl_3$ | daf | n | 750 | 1340 |
| Citrus Rinds & Seeds, 79m, 1a | | ar | g | 950 | 1710 |
| Citrus Rinds & Seeds, 79m, 1a | | d | g | 4450 | 8020 |
| Citrus Rinds & Seeds, 79m, 1a | | daf | g | 4610 | 8300 |
| Coal, Sub-bituminous B, 15m, 7a | | ar | g | 5690 | 10240 |
| Coal, Sub-bituminous B, 15m, 7a | | d | g | 6720 | 12100 |
| Coal, Sub-bituminous B, 15m, 7a | | daf | g | 7200 | 12960 |
| Coal, Bituminous-high volatile B, 9m, 8a | | ar | g | 6800 | 12240 |
| Coal, Bituminous-high volatile B, 9m, 8a | | d | g | 7440 | 13390 |
| Coal, Bituminous-high volatile B, 9m, 8a | | daf | g | 8190 | 14740 |
| Coal, Bituminous-volatile, 4m, 5a | | ar | g | 8030 | 14450 |
| Coal, Bituminous-volatile, 4m, 5a | | d | g | 8330 | 14990 |
| Coal, Bituminous-volatile, 4m, 5a | | daf | g | 8780 | 15800 |

*(continued)*

(a) a = % ash by weight
m = % moisture by weight

(b) ar = as received
d = dry
daf = dry & ash-free

(c) g = gross or higher HV
n = net or lower HV

**Appendix C2c.** *(continued)*

| Substance[a] | Formula | (b) | (c) | kcal/kg[c] | Btu/lb[c] |
|---|---|---|---|---|---|
| Coal, Anthracite, 5m, 14a | | ar | g | 6170 | 11110 |
| Coal, Anthracite, 5m, 14a | | d | g | 6460 | 11630 |
| Coal, Anthracite, 5m, 14a | | daf | g | 7600 | 13680 |
| Coffee Grounds, 20m, 2a | | d | g | 5590 | 10060 |
| Copper | (to CuO) | daf | g | 600 | 1090 |
| Corn Cobs, 5m, 3a | | ar | g | 4440 | 8000 |
| Corrugated Boxes, 5.2m | | ar | g | 3910 | 7040 |
| Corrugated Boxes, 5.2m | | d | g | 4130 | 7430 |
| Corrugated Boxes, 5.2m | | daf | g | 4360 | 7850 |
| Cotton Seed Hulls, 10m, 2a | | ar | g | 4440 | 8000 |
| Cresol (av. o, m, p) | $CH_3C_6H_4OH$ | daf | n | 8170 | 14700 |
| Ethane | $C_2H_6$ | daf | n | 12280 | 22100 |
| Ethyl Acetate | $CH_3CO_2C_2H_5$ | daf | n | 6100 | 10980 |
| Ethyl Alcohol | $C_2H_5OH$ | daf | n | 7120 | 12820 |
| Ethylene (ethene) | $CH_2CH_2$ | daf | n | 11840 | 21320 |
| Evergreen Shrubs, 69m, 1a | | ar | g | 1500 | 2710 |
| Evergreen Shrubs, 69m, 1a | | d | g | 4850 | 8740 |
| Evergreen Shrubs, 69m, 1a | | daf | g | 4980 | 8960 |
| Excelsior, 0.77a | | daf | g | 4790 | 8620 |
| Fats (Animal) | | ar | n | 9500 | 17130 |
| Fats, Cooking | | daf | g | 9000 | 16200 |
| Fats, Fried | | ar | g | 9150 | 16470 |
| Fats, Fried | | d | g | 9150 | 16470 |
| Fats, Fried | | daf | g | 9150 | 16470 |
| Flowering Plants, 54m, 2a | | ar | g | 2050 | 3700 |
| Flowering Plants, 54m, 2a | | d | g | 4460 | 8030 |
| Flowering Plants, 54m, 2a | | daf | g | 4700 | 8460 |
| Fuel Oil, -2 | | ar | g | 10870 | 19570 |
| Fuel Oil, -2 | | d | g | 10870 | 19570 |
| Fuel Oil, -2 | | daf | g | 10870 | 19570 |
| Fuel Oil, -6, 2m, 0.1a | | ar | g | 10150 | 18260 |
| Fuel Oil, -6, 2m, 0.1a | | d | g | 10300 | 18540 |
| Fuel Oil, -6, 2m, 0.1a | | daf | g | 10380 | 18680 |
| Grass, Dirt, Leaves, 21-62m | | d | g | 3490 | 6280 |
| Grass, Dirt, Leaves, 21-62m | | daf | g | 4990 | 8990 |
| Glycerol (glycerin) | $(CH_2OH)_2$ | daf | n | 6400 | 11530 |
| n-Heptane | $C_7H_{16}$ | daf | n | 11500 | 20700 |
| n-Hexane | $C_6H_{14}$ | daf | n | 11510 | 20720 |
| Household Dirt, 3m, 70a | | ar | g | 2040 | 3670 |
| Household Dirt, 3m, 70a | | d | g | 2110 | 3790 |
| Household Dirt, 3m, 70a | | daf | g | 7580 | 13650 |
| Hydrogen | | ar | g,n | 33890 | 61000 |

*(continued)*

(a) a = % ash by weight
m = % moisture by weight

(b) ar = as received
d = dry
daf = dry & ash-free

(c) g = gross or higher HV
n = net or lower HV

**Appendix C2c.** *(continued)*

| Substance[a] | Formula | (b) | (c) | kcal/kg[c] | Btu/lb[c] |
|---|---|---|---|---|---|
| Iron | (to $Fe_2O_3$) | daf | g | 1760 | 3160 |
| Iron | (to $FeO_{.947}$) | daf | g | 1200 | 2160 |
| Iron | (to $Fe_3O_4$) | daf | g | 1590 | 2870 |
| Kerosene, 0.5a | | ar | g | 10500 | 18900 |
| Latex | | ar | g | 5560 | 10000 |
| Lead | (to PbO) | daf | g | 250 | 450 |
| Leather, 10m, 9a | | ar | g | 4420 | 7960 |
| Leather, 10m, 9a | | d | g | 4920 | 8850 |
| Leather, 10m, 9a | | daf | g | 5470 | 9850 |
| Lignin | | daf | g | 5850 | 10530 |
| Lignite | | ar | g | 3930 | 7070 |
| Lignite | | d | g | 6130 | 11030 |
| Lignite | | daf | g | 7560 | 13610 |
| Linoleum, 2m, 27a | | ar | g | 4530 | 8150 |
| Linoleum, 2m, 27a | | d | g | 4620 | 8310 |
| Linoleum, 2m, 27a | | daf | g | 6360 | 11450 |
| Logs, Green, 50m, 0.5a | | ar | g | 1170 | 2100 |
| Logs, Green, 50m, 0.5a | | d | g | 2340 | 4200 |
| Logs, Green, 50m, 0.5a | | daf | g | 2360 | 4250 |
| Magazines, 4m, 22a | | ar | g | 2920 | 5250 |
| Magazines, 4m, 22a | | d | g | 3040 | 5480 |
| Magazines, 4m, 22a | | daf | g | 3970 | 7150 |
| Magnesium | (to MgO) | ar | n | 4730 | 8530 |
| Magnesium | (to MgO) | daf | g | 5910 | 10640 |
| Meat Scraps, 39m, 3a | | ar | g | 4240 | 7620 |
| Meat Scraps, 39m, 3a | | d | g | 6910 | 12440 |
| Meat Scraps, 39m, 3a | | daf | g | 7280 | 13110 |
| Methane | $CH_4$ | daf | n | 13180 | 23720 |
| Methyl Alcohol | $CH_3OH$ | daf | n | 5340 | 9610 |
| Methyl Ethyl Ketone | $CH_3COC_2H_5$ | daf | n | 8090 | 14560 |
| Methylene Chloride | $CH_2Cl_2$ | daf | n | 1260 | 2260 |
| Naphtha | | ar | g | 8330 | 15000 |
| Newsprint, 6m, 1.4a | | ar | g | 4430 | 7970 |
| Newsprint, 6m, 1.4a | | d | g | 4710 | 8480 |
| Newsprint, 6m, 1.4a | | daf | g | 4780 | 8600 |
| Oats | | d | | 4440 | 8000 |
| n-Octane | $C_8H_{18}$ | daf | n | 11430 | 20570 |
| Oil, cotton seed | | ar | n | 9500 | 17130 |
| Oils, Paints, 16a | | ar | g | 7440 | 13400 |
| Oils, Paints, 16a | | d | g | 7440 | 13400 |
| Oils, Paints, 16a | | daf | g | 8890 | 16000 |
| Paper Food Cartons | | ar | g | 4030 | 7260 |

*(continued)*

(a) a = % ash by weight
m = % moisture by weight

(b) ar = as received
d = dry
daf = dry & ash-free

(c) g = gross or higher HV
n = net or lower HV

**Appendix C2c.** *(continued)*

| Substance[a] | Formula | (b) | (c) | kcal/kg[c] | Btu/lb[c] |
|---|---|---|---|---|---|
| Paper Food Cartons | | d | g | 4290 | 7730 |
| Paper Food Cartons | | daf | g | 4580 | 8250 |
| Paraffin | | ar | n | 10340 | 18650 |
| Peat | | ar | g | 1800 | 3240 |
| Peat | | d | g | 5030 | 9050 |
| Peat | | daf | g | 7000 | 12590 |
| Phenol | $C_6H_5OH$ | daf | n | 7790 | 14020 |
| Phenol formaldehyde | | daf | g | 6220 | 11190 |
| Phthalic Acid | $C_6H_4(CO_2H)_2$ | daf | n | 4640 | 8360 |
| Pitch | | ar | n | 8400 | 15150 |
| Plastic Coated Paper, 4.71m | | ar | g | 4080 | 7340 |
| Plastic Coated Paper, 4.71m | | d | g | 4280 | 7700 |
| Plastic Coated Paper, 4.71m | | daf | g | 4410 | 7940 |
| Plastic Film, 3-20m | | d | g | 7690 | 13850 |
| Plastic Film, 3-20m | | daf | g | 8260 | 14870 |
| Plastics, Mixed, 2m, 10a | | ar | g | 7830 | 14100 |
| Plastics, Mixed, 2m, 10a | | d | g | 7980 | 14370 |
| Plastics, Mixed, 2m, 10a | | daf | g | 8890 | 16000 |
| Polyethylene, 0.2m, 1a | | ar | g | 10930 | 19680 |
| Polyethylene, 0.2m, 1a | | d | g | 10960 | 19730 |
| Polyethylene, 0.2m, 1a | | daf | g | 11110 | 20000 |
| Polypropylene | | daf | g | 11080 | 19950 |
| Polystyrene, 0.2m, 0.5a | | ar | g | 9120 | 16420 |
| Polystyrene, 0.2m, 0.5a | | d | g | 9140 | 16450 |
| Polystyrene, 0.2m, 0.5a | | daf | g | 9170 | 16510 |
| Polyurethane, 0.2m, 4a | | ar | g | 6220 | 11200 |
| Polyurethane, 0.2m, 4a | | d | g | 6240 | 11220 |
| Polyurethane, 0.2m, 4a | | daf | g | 6520 | 11730 |
| Polyurethane (foamed) | | ar | g | 7220 | 13000 |
| Polyurethane (foam) | | daf | g | 5700 | 10260 |
| Polyvinyl Chloride, 0.2m, 2a | | ar | g | 5420 | 9750 |
| Polyvinyl Chloride, 0.2m, 2a | | d | g | 5430 | 9780 |
| Polyvinyl Chloride, 0.2m, 2a | | daf | g | 5560 | 10000 |
| Propane | $C_3H_8$ | daf | n | 11960 | 21530 |
| iso-Propyl Alcohol | $(CH_3)_2CHOH$ | daf | n | 7910 | 14240 |
| n-Propyl Alcohol | $C_3H_7OH$ | daf | n | 8010 | 14420 |
| Propylene | $CH_3CHCH_2$ | daf | n | 11670 | 21010 |
| Rags, 10m, 2a | | ar | g | 3830 | 6900 |
| Rags, 10m, 2a | | d | g | 4250 | 7650 |
| Rags, 10m, 2a | | daf | g | 4360 | 7840 |
| Rags, cellulose acetate, 2.19a | | daf | g | 4440 | 8000 |
| Rags, cotton | | daf | g | 4000 | 7200 |

*(continued)*

(a) a = % ash by weight
m = % moisture by weight

(b) ar = as received
d = dry
daf = dry & ash-free

(c) g = gross or higher HV
n = net or lower HV

**Appendix C2c.** *(continued)*

| Substance[a] | Formula | (b) | (c) | kcal/kg[c] | Btu/lb[c] |
|---|---|---|---|---|---|
| Rags, linen | | d | g | 3960 | 7130 |
| Rags, mixed, 2.19a | | daf | g | 4191 | 7540 |
| Rags, nylon | | daf | g | 7330 | 13190 |
| Rags, rayon | | daf | g | 420 | 750 |
| Rags, silk | | d | g | 4660 | 8390 |
| Rags, wool | | daf | g | 5440 | 9800 |
| Rubber, 1m, 10a | | ar | g | 6220 | 11200 |
| Rubber, 1m, 10a | | d | g | 6290 | 11330 |
| Rubber, 1m, 10a | | daf | g | 7000 | 12600 |
| Rubber Waste | | ar | g | 5560 | 10000 |
| Sawdust (pine) | | d | g | 5380 | 9680 |
| Sawdust (fir) | | d | g | 4580 | 8250 |
| Shellac | | daf | g | 7540 | 13580 |
| Shoe, Heel & Sole, 1m, 30a | | ar | g | 6060 | 10900 |
| Shoe, Heel & Sole, 1m, 30a | | d | g | 6130 | 11030 |
| Shoe, Heel & Sole, 1m, 30a | | daf | g | 8770 | 15790 |
| Starch | | d | g | 4180 | 7520 |
| Starch | | daf | g | 4180 | 7520 |
| Street Sweepings, 20m, 20a | | ar | g | 2670 | 4800 |
| Street Sweepings, 20m, 20a | | d | g | 3330 | 6000 |
| Street Sweepings, 20m, 20a | | daf | g | 4440 | 8000 |
| Styrene-butadiene copolymer | | daf | g | 9830 | 17700 |
| Sugar (sucrose) | | d | g | 3940 | 7100 |
| Sugar (sucrose) | | daf | g | 3940 | 7100 |
| Sulfur (rhombic) | | ar | n | 2200 | 3970 |
| Tar or asphalt, 1m | | ar | g | 9440 | 17000 |
| Tar 1/3, Paper 2/3, 1m, 2a | | ar | g | 6110 | 11000 |
| Textiles, 15-31m | | d | g | 4460 | 8040 |
| Textiles, 15-31m | | daf | g | 4610 | 8300 |
| Tin | (to $SnO_2$) | daf | g | 1170 | 2100 |
| Tires, 1m, 7a | | ar | g | 7670 | 13800 |
| Tires, 1m, 7a | | d | g | 7730 | 13910 |
| Tires, 1m, 7a | | daf | g | 8280 | 14900 |
| Toluene | $CH_3C_6H_5$ | daf | n | 10150 | 18280 |
| Turpentine | | ar | g | 9440 | 17000 |
| Upholstery, 7m, 3a | | ar | g | 3870 | 6960 |
| Upholstery, 7m, 3a | | d | g | 4160 | 7480 |
| Upholstery, 7m, 3a | | daf | g | 4270 | 7690 |
| Urea | $(NH_2)_2CO$ | daf | n | 2530 | 4550 |
| Vegetable Food Waste, 78m | | ar | g | 1000 | 1790 |
| Vegetable Food Waste, 78m | | d | g | 4590 | 8270 |
| Vegetable Food Waste, 78m | | daf | g | 4830 | 8700 |

*(continued)*

[a] a = % ash by weight
m = % moisture by weight

[b] ar = as received
d = dry
daf = dry & ash-free

[c] g = gross or higher HV
n = net or lower HV

**Appendix C2c.** *(concluded)*

| Substance[a] | Formula | (b) | (c) | kcal/kg[c] | Btu/lb[c] |
|---|---|---|---|---|---|
| Vinyl chloride/acetate copolymer | | daf | g | 4910 | 8830 |
| Waste, Type 0, 5a, 10m | | ar | g | 4720 | 8500 |
| Waste, Type 1, 10a, 25m | | ar | g | 3610 | 6500 |
| Waste, Type 2, 7a, 50m | | ar | g | 2390 | 4300 |
| Waste, Type 3, 5a, 70m | | ar | g | 1390 | 2500 |
| Waste, Type 4, 5a, 85m | | ar | g | 560 | 1000 |
| Waxed Cartons | | daf | g | 6670 | 12000 |
| Wax paraffin | | ar | g | 10350 | 18620 |
| Wheat | | d | g | 4180 | 7530 |
| Wood and Bark, 20m, 0.8a | | ar | g | 3830 | 6900 |
| Wood and Bark, 20m, 0.8a | | d | g | 4790 | 8610 |
| Wood and Bark, 20m, 0.8a | | daf | g | 4830 | 8700 |
| Wood, Balsam, Spruce 74m, 1a | | ar | g | 1360 | 2450 |
| Wood, Balsam, Spruce 74m, 1a | | d | g | 5300 | 9540 |
| Wood, Balsam, Spruce 74m, 1a | | daf | g | 5470 | 9850 |
| Wood, beech, 13m | | d | g | 3640 | 6540 |
| Wood, birch, 11.8m | | d | g | 3720 | 6690 |
| Wood, Demolition Softwood, 7.7m, .8a | | ar | g | 4060 | 7300 |
| Wood, Demolition Softwood, 7.7m, .8a | | d | g | 4400 | 7920 |
| Wood, Demolition Softwood, 7.7m, .8a | | daf | g | 4440 | 8000 |
| Wood, Furniture, 6m, 1a | | ar | g | 4080 | 7350 |
| Wood, Furniture, 6m, 1a | | d | g | 4340 | 7810 |
| Wood, Furniture, 6m, 1a | | daf | g | 4410 | 7940 |
| Wood, oak, 13m | | ar | n | 3990 | 7200 |
| Wood, pine, 12m | | ar | n | 4420 | 7970 |
| Wood, Rotten Timbers, 27m, 2a | | ar | g | 2620 | 4710 |
| Wood, Rotten Timbers, 27m, 2a | | d | g | 3540 | 6370 |
| Wood, Rotten Timbers, 27m, 2a | | daf | g | 3640 | 6560 |
| Wood, Waste Hardwood, 12m, 0.5a | | ar | g | 3570 | 6430 |
| Wood, Waste Hardwood, 12m, 0.5a | | d | g | 4060 | 7300 |
| Wood, Waste Hardwood, 12m, 0.5a | | daf | g | 4080 | 7340 |
| Xylene (av. o, m, p) | $(CH_3)_2C_6H_4$ | daf | n | 10280 | 18500 |
| Zinc | (to NzO) | daf | g | 1280 | 2300 |

(a) a = % ash by weight
m = % moisture by weight

(b) ar = as received
d = dry
daf = dry & ash-free

(c) g = gross or higher HV
n = net or lower HV

**Appendix C3. Flammability Limits and Ignition Temperatures.** (Courtesy of Reference 14)

| Substance | Flammability Limit % volume (at 20°C, 1 atm) in air | | | Ignition Temperature F/C | |
|---|---|---|---|---|---|
| | State | Lean | Rich | In Air | In $O_2$ |
| Acetic Acid | gas | 5.4 | | | |
| Acetone | gas | 2.55 | 12.80 | 1042/561 | |
| Acetylene | gas | 2.50 | 8.00 | 581/305 | |
| Ammonia | gas | 15.50 | 27.00 | | |
| Benzene | gas | 1.40 | 7.10 | 1364/740 | 1224/662 |
| Butane | gas | 1.86 | 8.41 | 806/430 | |
| Butylene | gas | 1.65 | 9.95 | 829/443 | |
| Carbon, fixed (anthracite coal) | solid | | | 838-1114/448-601 | |
| Carbon fixed (bituminous coal) | solid | | | 765/407 | |
| Carbon, fixed (semibituminous coal) | solid | | | 869/465 | |
| Carbon disulfide | gas | 1.25 | 50.00 | 248/120 | |
| Carbon monoxide | gas | 12.50 | 71.20 | 1191-1216/644-658 | 1179-1216/637-658 |
| Charcoal | solid | | | 1202/650 | |
| Ethane | gas | 3.00 | 12.50 | 968-1166/520-630 | 968-1166/520-630 |
| Ethyl alcohol | gas | 3.28 | 18.95 | 738/392 | |
| Ethylene | gas | 2.75 | 28.60 | 1008-1017/542-547 | 932-966/500-519 |
| Ethylene oxide | gas | 3.00 | 80.00 | 804/429 | |

*(continued)*

**Appendix C3.** *(concluded)*

| Substance | Flammability Limit % volume (at 20°C, 1 atm) in air | | | Ignition Temperature F/C | |
|---|---|---|---|---|---|
| | State | Lean | Rich | In Air | In $O_2$ |
| Gasoline | liquid | | | 500-799/260-426 | |
| Heptane (n-) | gas | 1.10 | 6.70 | 477/247 | |
| Hexane (n-) | gas | 1.18 | 7.40 | 500/260 | |
| Hydrogen | gas | 4.00 | 74.20 | 1076-1094/580-590 | 1076-1094/580-590 |
| Hydrogen sulfide | gas | 4.30 | 45.50 | 554/290 | |
| Kerosene | liquid | | | 489-559/254-293 | |
| Methane | gas | 5.00 | 15.00 | 1202-1382/650-750 | 1033-1292/556-700 |
| Methyl alcohol | gas | 6.72 | 36.50 | 878/470 | |
| Propane | gas | 12.12 | 9.35 | 939/504 | 914-1058/490-570 |
| Propylene | gas | 2.00 | 11.10 | 1036/558 | |
| Sulfur | solid | | | 469/243 | |
| Toluene | gas | 1.27 | 6.75 | 1490/810 | 1026/552 |
| Xylene | gas | 1.00 | 6.00 | | |

**Appendix C4. Comparison of various fuel oils.** Local legal limits may supersede these figures.

| ASTM[a] grades | Flash point, F (C), min | Pour point, F (C), max | Water and sediment, % by volume, max | Carbon residue on 10% bottoms, %, max | Ash % by weight, max | Distillation temperatures, F (C): 10% point, max | Distillation temperatures, F (C): 90% point, min | Distillation temperatures, F (C): 90% point, max | Kinematic viscosity: Saybolt seconds Universal at 100 F (38 C), min | Saybolt seconds Universal at 100 F (38 C), max | Saybolt seconds Furol at 122 F (100 C), min | Saybolt seconds Furol at 122 F (100 C), max | mm²/s, Centistokes at 100 F (38 C), min | mm²/s, Centistokes at 100 F (38 C), max | mm²/s Centistokes at 212 F (100 C), min | mm²/s Centistokes at 212 F (100 C), max | Gravity, deg API | Sulfur, %, max |
|---|---|---|---|---|---|---|---|---|---|---|---|---|---|---|---|---|---|---|
| No. 1 | 100 (38) | 0 (−18) | trace | 0.15 | — | 420 (216) | — | 550 (288) | — | — | — | — | 1.3 | 2.1 | — | — | 35 min. | 0.5 |
| No. 2 | 100 (38) | 20 (−7) | 0.05 | 0.35 | — | — | 540 (282) | 640 (338) | 32.6 | 37.93 | — | — | 1.9 | 3.4 | — | — | 30 min. | 0.5 |
| No. 4 (light) | 100 (38) | 28 (−6) | 0.50 | — | 0.05 | — | — | — | 32.6[b] | 45.0[b] | — | — | 1.9 | 5.5 | — | — | 30 max. | — |
| No. 4 | 130 (55) | 20 (−7) | 0.50 | — | 0.10 | — | — | — | 45 | 125 | — | — | >5.0 | 24.0 | — | — | — | — |
| No. 5 (light) | 130 (55) | — | 1.00 | — | 0.10 | — | — | — | >125[b] | 300[b] | — | — | >24.0[b,c] | 58[b] | 5 | 8.9 | — | — |
| No. 5 (heavy) | 130 (55) | — | 1.00 | — | 0.10 | — | — | — | >300[b] | 900[b] | 23[b] | 40[b] | >58[b] | 168[b] | 9.0 | 14.9 | — | — |
| No. 6 | 150 (65) | — | 2.00 | — | — | — | — | — | >900[b] | 9000[b] | >45[b] | 300[b] | 92[b,c] | 638[b,c] | 15.0 | 50.0 | — | — |

[a] This is not a reproduction of ASTM specifications--merely a comparison of various fuels. For complete ASTM specifications, purchase a copy of standard D396 from: American Society for testing and Materials, 1916 Race Street, Philadelphia, PA 19103.

[b] Approximate, not limiting.

[c] At 122 F (50 C).

**Appendix C5. Bulk Densities of Waste Substances** (Abstracted from References 14, 15. See also Appendix C7.)

| Substance | kg/m³ | lb/ft³ |
|---|---|---|
| Agricultural pen sweepings | 650 | 41 |
| Agricultural waste, mixed | 565 | 35 |
| Alley cleanings | 150 | 9 |
| Appliances, major | 180 | 11 |
| Auto bodies | 130 | 8 |
| Average bulky waste | 190 | 12 |
| Bark | 64 | 4 |
| Bark slabs | 400 | 25 |
| Battery cases and miscellaneous auto | 715 | 45 |
| Beans or grain waste | 775 | 48 |
| Brush | 30 | 2 |
| Bunkered refuse, 10% moisture (8-11 m high) | 155 | 10 |
| Bunkered refuse, 20% moisture (8-11 m high) | 180 | 11 |
| Bunkered refuse, 30% moisture (8-11 m high) | 225 | 14 |
| Bunkered refuse, 40% moisture (8-11 m high) | 275 | 17 |
| Bunkered refuse, 50% moisture (8-11 m high) | 345 | 22 |
| Catch-basin cleaning | 1445 | 90 |
| Cement industry waste | 1425 | 89 |
| Chaff | 60 | 4 |
| Chemical waste, dry | 640 | 40 |
| Chemical waste, wet | 965 | 60 |
| Cinders | 900 | 56 |
| Construction waste | 160-965 | 10-60 |
| Construction waste, burnable | 255 | 16 |
| Dead animals | 355 | 22 |
| Demolition waste, nonburnable | 1430 | 89 |
| Demolition waste, burnable | 360 | 22 |
| Earth, sand, gravel | 1445 | 90 |
| Fiberboard | 231 | 14 |
| Fly ash | 1285 | 80 |
| Fruit waste, agricultural | 355 | 22 |
| Furniture | 50 | 3 |

(continued)

**Appendix C5.** *(continued)*

| Substance | $kg/m^3$ | $lb/ft^3$ |
|---|---|---|
| Garbage, kitchen waste | 300-450 | 19-28 |
| Glass | 2307-2996 | 144-187 |
| Glass and bottles, unbroken | 415 | 26 |
| Glass, broken | 1190 | 74 |
| Grass and trimmings | 130 | 8 |
| Household bulky waste | 100 | 6 |
| Leather scraps | 180 | 11 |
| Limbs and leaves | 160-190 | 10-12 |
| Logs and stumps | 400 | 25 |
| Logs, green | 320-800 | 20-50 |
| Meat-packing waste | 1030 | 64 |
| Metal cans | 95 | 6 |
| Metals, mixed | 120 | 7 |
| Metal, scrap, heavy | 2410 | 150 |
| Metal, scrap, light | 800 | 50 |
| Mixed sludges (wet) | 1190 | 74 |
| Oils, tars, asphalts | 965 | 60 |
| Paper | 140 | 9 |
| Paper and cardboard | 110 | 7 |
| Paper and cardboard, compacted | 215 | 13 |
| Paper, wet | 165 | 10 |
| Paunch manure | 1030 | 64 |
| Pavement, broken, sidewalk | 1520 | 95 |
| Plastics | 30 | 2 |
| Potato-processing waste | 670 | 42 |
| Rags | 115 | 7 |
| Refractories, firebrick | 615-2403 | 38-150 |
| Refuse with garbage | 195 | 12 |
| Refuse without garbage (72% void) | 273 | 17 |
| Residential | 60-255 | 4-16 |
| Residential in paper sacks | 80-160 | 5-10 |
| Residential, damp | 90-215 | 6-13 |
| Residential, dry | 50-145 | 3-9 |
| Residential, wet | 115-255 | 7-16 |

*(continued)*

**Appendix C5.** *(concluded)*

| Substance | kg/m³ | lb/ft³ |
|---|---|---|
| Sawdust | 290 | 18 |
| Sewage screenings | 950 | 59 |
| Sewage skim (grease) | 950 | 59 |
| Sewage-sludge solids | 1040 | 65 |
| Shells, offal, paunch, fleshings | 300 | 19 |
| Street dirt | 1370 | 86 |
| Street sweepings, litter | 250 | 16 |
| Textile wastes | 180 | 11 |
| Tires and rubber products | 240 | 15 |
| Tree cuttings | 135 | 8 |
| Turpentine | 859 | 54 |
| Vegetable waste, agricultural | 355 | 22 |
| Wire | 320 | 20 |
| Wood chips (50% void) | 280 | 17 |
| Wood crates | 110 | 7 |
| Wood pallets, driftwood | 210 | 13 |
| Wood shavings | 240 | 15 |
| Wood trimmings | 580 | 36 |

**Appendix C6. Air Required for Fuels and Refuse.** (Derived from data in References 14 and 15)

| Fuel | Air/fuel weight ratio | kg air/ kcal[a] | $sm^3$ air/ kcal[a] | $ft^3$ air/ Btu[a] |
|---|---|---|---|---|
| Coal, anthracite | 9.92 | 1.40 E-3 | 1.15 E-3 | 10.2 E-3 |
| Coal, bit, hi-volatile | 9.08 | 1.34 E-3 | 1.13 E-3 | 10.2 E-3 |
| Coal, sub-bitum B | 7.58 | 1.22 E-3 | 1.12 E-3 | 9.97 E-3 |
| Fuel oil, #2 | 14.35 | 1.35 E-3 | 1.11 E-3 | 9.88 E-3 |
| Fuel oil, #6 | 13.44 | 1.33 E-3 | 1.09 E-3 | 9.69 E-3 |
| Lignite | 5.27 | 1.34 E-3 | 1.13 E-3 | 10.1 E-3 |
| Natural gas | 15.73 | 1.29 E-3 | 1.06 E-3 | 9.43 E-3 |
| Peat | 2.33 | 1.30 E-3 | 1.10 E-3 | 9.79 E-3 |
| Refuse | 3.22 | 1.30 E-3 | 1.10 E-3 | 9.79 E-3 |
| Wood, non-resinous | 4.90 | 1.40 E-3 | 1.14 E-3 | 10.2 E-3 |

[a] Note that the ratios of air requirements to gross heat releases are within a fairly narrow range for all fossil fuels, including refuse. In the absence of measured data, designers may be forced to use this to make approximations.

**Appendix C7. Properties of Some Common Solvents** (Courtesy of Reference 17b)

| | (a) | (b) | (c) | (d) | (e) |
|---|---|---|---|---|---|
| FC-113 | 188 | 87 | 63 | 35 | 0.21 |
| Methylene chloride | 104 | 40 | 142 | 79 | 0.28 |
| Methyl chloroform (k) | 165 | 74 | 102 | 57 | 0.25 |
| Perchloro-ethylene | 250 | 121 | 90 | 50 | 0.21 |
| Trichloro-ethylene (k) | 188 | 87 | 103 | 57 | 0.23 |

| | (f) | (g) | (h) | (i) |
|---|---|---|---|---|
| FC-113 | 13.2 | 1590 | 0.459 | 7.35 |
| Methylene chloride | 11 | 1320 | 0.206 | 3.30 |
| Methyl chloroform (k) | 11 | 1320 | 0.279 | 4.47 |
| Perchloro-ethylene | 12.5 | 1500 | 0.326 | 5.22 |
| Trichloro-ethylene (k) | 12.1 | 1450 | 0.278 | 4.45 |

(a) Boiling point, F
(b) Boiling point, C
(c) Latent heat of vaporization, Btu/lb
(d) Latent heat of vaporization, kcal/kg
(e) Specific heat of liquid, Btu/lb°F or kcal/kg°C
(f) Density, liquid, lb/USgallon
(g) Density, liquid, kg/m$^3$
(h) Density, vapor at boiling point, lb/ft$^3$
(i) Density, vapor at boiling point, kg/m$^3$
(k) 1,1,1-Trichloroethane

## Appendix D. Typical analyses and combustion data for some waste materials.[g]

| Analysis, % wt | | Sulfite paper[a] | | Average wood[b] | | Douglas fir[c] | | Garbage[d] | |
|---|---|---|---|---|---|---|---|---|---|
| Carbon (C) | | 44.34 | | 49.56 | | 52.30 | | 52.78 | |
| Hydrogen (H) | | 6.27 | | 6.11 | | 6.30 | | 6.27 | |
| Nitrogen (N) | | | | 0.07 | | 0.10 | | | |
| Oxygen (O) | | 48.39 | | 43.83 | | 40.50 | | 39.95 | |
| Ash | | 1.00 | | 0.42 | | 0.80 | | 1.00 | |
| Gross heating value (dry basis), Btu/lb | | 7590 | | 8517 | | 9050 | | 8820 | |
| **Air and flue gas per pound of waste** | | **scf[e]** | **lb** | **scf** | **lb** | **scf** | **lb** | **scf** | **lb** |
| Theoretical stp[e] dry air/lb waste | | 67.58 | 5.16 | 77.30 | 5.90 | 84.16 | 6.43 | 85.12 | 6.50 |
| ditto with 40% relative humidity | | 68.05 | 5.18 | 77.84 | 5.93 | 84.75 | 6.46 | 85.72 | 6.53 |
| Flue gas with theoretical air | $CO_2$ | 13.99 | 1.62 | 15.64 | 1.81 | 16.51 | 1.91 | 16.66 | 1.93 |
| | $N_2$ | 53.40 | 3.94 | 61.10 | 4.51 | 66.53 | 4.91 | 67.23 | 4.97 |
| | $H_2O$ formed | 11.78 | 0.56 | 11.48 | 0.54 | 11.84 | 0.56 | 11.88 | 0.56 |
| | $H_2O$ | 0.47 | 0.02 | 0.53 | 0.02 | 0.58 | 0.02 | 0.59 | 0.02 |
| Total flue gas (w/0% xsa)[f] | | 79.65 | 6.15 | 88.77 | 6.90 | 95.46 | 7.42 | 96.37 | 7.49 |
| Flue gas with excess air (xsa) | 50% | 113.44 | 8.74 | 127.42 | 9.86 | 137.55 | 10.64 | 139.24 | 10.77 |
| | 100% | 147.23 | 11.32 | 166.07 | 12.81 | 179.63 | 13.86 | 182.00 | 14.04 |
| | 150% | 181.26 | 13.91 | 204.99 | 15.78 | 222.01 | 17.09 | 224.86 | 17.21 |
| | 200% | 215.28 | 16.51 | 243.91 | 18.75 | 264.38 | 20.12 | 267.72 | 20.58 |
| | 300% | 283.33 | 21.70 | 321.75 | 24.68 | 349.13 | 26.58 | 353.44 | 27.12 |

[a] Constituents of sulfite paper: 84% Cellulose, $C_6H_{10}O_5$
8% Hemicellulose, $C_5H_{10}O_5$
6% Lignin, $C_6H_{10}O_5$
2% Resin, $C_6H_{10}O_5$
1% Ash, $C_{20}H_{30}O_2$

[b] Kent, 1936.

[c] Kent, 1961.

[d] Estimated probably higher paper content now.

[e] At 60 F and 14.7 psia.

[f] xsa = excess air

[g] Adapted from USEPA: "Air Pollution Engineering Manual" (formerly AP40), 2nd Edition, 1973, Edited by J. A. Danielson, Los Angeles County Air Pollution Control District.

## Appendix E1. Formulas for conversion of viscosity units *(from Reference 15).*

**$\mu$ and $\nu$ formulas**

abs visc in cp = kin visc in cs × specific gravity
abs visc in #m/sec ft = kin visc in ft$^2$/sec × mass density in #m/ft$^3$
abs visc in #f sec/ft$^2$ = kin visc in ft$^2$/sec × (mass density in #m/ft$^3$) ÷ 32.17

**$\mu$, absolute viscosity units**

1 #f sec/ft$^2$ = 32.17 #m/sec ft
1 poise = 100 cp = 1 dyne·s/cm$^2$† = 1 g/s·cm
1 #m/sec ft = 1488 cp = 1.488 Pa·s
1 #m/hr ft = 0.413 cp = 1.488 kgm/hr·m
1 #m/sec in. = 17 850 cp = 178.5 poise
1 #f sec/ft$^2$ = 47 850 cp = 0.4883 kgf·s/m$^2$
1 #f hr/ft$^2$ = 172 260 000 cp = 0.4883 kgf·hr/m$^2$
1 #f sec/in.$^2$ = 6 890 000 cp = 1 Reyn

**$\nu$, kinematic viscosity units**

1 ft$^2$/sec = 92 900 cs = 0.0929 m$^2$/s
1 ft$^2$/hr = 25.81 cs = 0.0929 m$^2$/hr
1 in.$^2$/sec = 645.2 cs = 0.000 645 m$^2$/s
1 stoke = 100 cs = 0.0001 m$^2$/s
kin visc in centistokes = (0.266 × SSU) − (195/SSU), for SSU 32 to 100
kin visc in centistokes = (0.220 × SSU) − (135/SSU), for SSU >100
kin visc in centistokes = (2.24 × SSF) − (184/SSF), for SSF 25 to 40
kin visc in centistokes = (2.16 × SSF) − (60/SSF), for >40
kin visc in centistokes = (0.26 × SR1) − (179/SR1), for SR1 34 to 100
kin visc in centistokes = (0.247 × SR1) − (50/SR1), for SR1 >100
kin visc in centistokes = (8.0 × °E) − (8.64/°E), for °E 1.35 to 3.2
kin visc in centistokes = (7.6 × °E) − (4.0/°E), for °E >3.2

**definitions**

abs visc is absolute viscosity, $\mu$; kin visc is kinematic viscosity, $\nu$
cp is centipoise (abs visc); cs is centistokes (kin visc)
specific gravity is relative to water, = (mass density in #m/ft$^3$) ÷ 62.35
#m is pounds mass; #f is pounds force
kgm = kilograms mass; kgf = kilograms force
SSU is kinematic viscosity in seconds, Saybolt Universal
SSF is kinematic viscosity in seconds, Saybolt Furol
SR1 is kinematic viscosity in seconds, Redwood No. 1 (or Redwood Standard)
°E is kinematic viscosity in degrees Engler. 1°E = 51.28 seconds Engler

**At 60 F & 14.7 psia, 15.6 C & 760 mm Hg**

| Water* | Air | Nat'l Gas |
|---|---|---|
| $\mu$ = 1.124 cp | $\mu$ = 0.0180 cp | $\mu$ = 0.011 cp |
| = 2.72 #m/hr·ft | = 0.0436 #m/hr·ft | = 0.0266 #m/hr·ft |
| = 2.349 × 10$^{-5}$ #f·s/ft$^2$ | = 3.763 × 10$^{-7}$ #f·s/ft$^2$ | = 2.30 × 10$^{-7}$ #f·s/ft$^2$ |
| $\nu$ = 1.130 cs | $\nu$ = 14.69 cs | $\nu$ = 14.92 cs |
| = 32 SSU | = 1.469 × 10$^{-5}$ m$^2$/s | = 1.49 × 10$^{-5}$ m$^2$/s |
| = 0.043 78 ft$^2$/hr | = 0.5691 ft$^2$/hr | = 0.578 ft$^2$/hr |

* Approximate viscosity of water at 70 F (21 C) is 1 cp and 1 cs.
† A dot between symbols means "times" or "multiply by".

**Appendix E2. Alignment chart for comparing viscosity scales at the same temperature.** Use a straight-edge to read horizontally across all scales -- far left and far right scales are identical. See Appendix E1. *(From Reference 15.)*

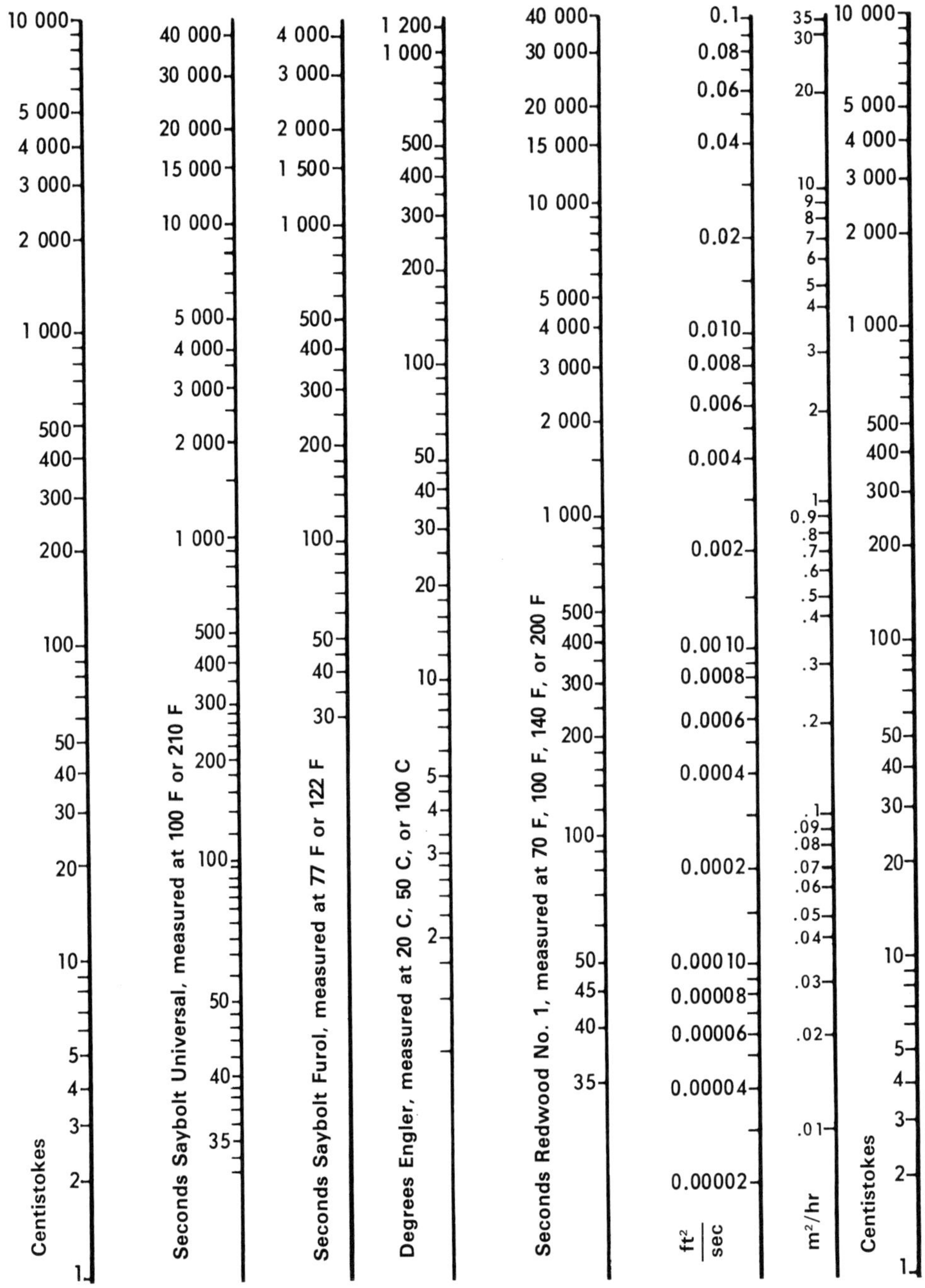

**Appendix E3. Viscosity-temperature relations for typical fuel oils.** The diagonal lines are samples, not standards, for the designated grades of fuel oils. Source of the crude determines the slope of the line. This ASTM chart may be used for plotting properties of other fuel oils.

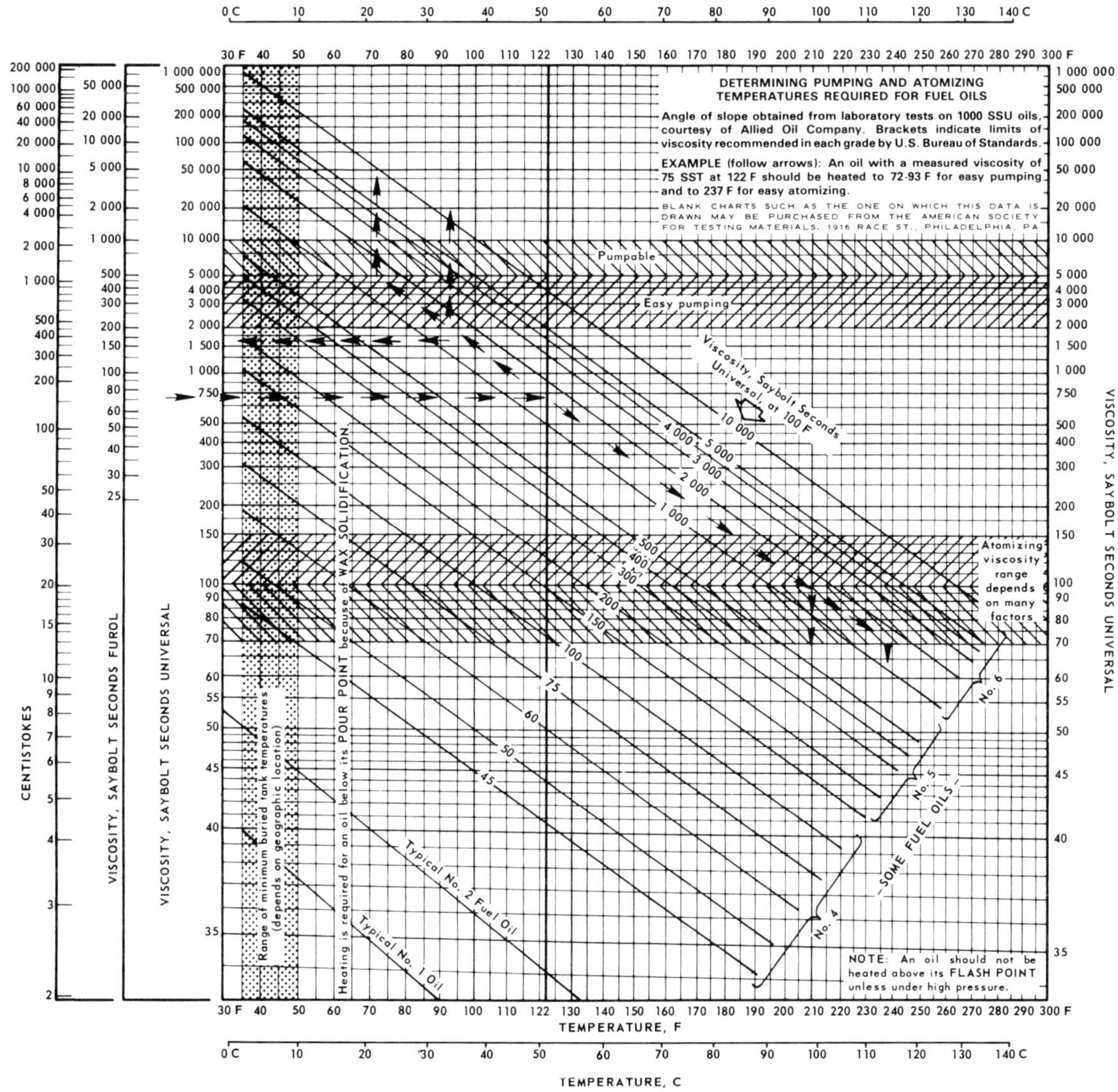

**Appendix F. Thermal stability-based incinerability hierarchy rankings of POHCs.** *(From Appendix H [University of Dayton Research Institute] of Reference 21.)*

| **Class 1. Principal organic hazardous constituents** | **T99** | **Rank** |
|---|---|---|
| **Acetonitrile (Ethanenitrile)** | **1000 C** | **18-19** |
| **Acrylonitrile (2-Propenenitrile)** | 985 | 21 |
| Benzanthracene (1,2) (Benz[a]anthracene) | 1060 | 7-11 |
| **Benzene** | 1150 | 3 |
| **Benzo(a)pyrene (1,2-Benzopyrene)** | 1025 | 12 |
| **Benzo(b)fluoranthene (2,3-Benzofluoranthene)** | 1060 | 7-11 |
| Benzo(j)fluoranthene (7,8-Benzofluoranthene) | 1060 | 7-11 |
| **Bromomethane (Methylbromide)** | 935 | 33-35 |
| **Chlorobenzene** | 990 | 20 |
| **Chlorodibenzo-P-Dioxin (2-)** | 925 | 36 |
| **Chlorodibenzo-P-Dioxin (1-)** | 960 | 26-27 |
| **Chloromethane (Methyl Chloride)** | 950 | 30-31 |
| **Chloronaphthalene (1-)** | 975 | 22-23 |
| *Chlorotrifluoromethane*[c,d] | 1010 | 17 |
| Chrysene (1,2-Benzphenanthrene) | 1060 | 7-11 |
| Cyanogen Bromide (Bromine Cyanide) | 970 | 24-25 |
| Cyanogen Chloride (Chlorine Cyanide) | 1000 | 18-19 |
| Cyanogen (Ethanedinitrile) | 1500 | 2 |
| Dibenz(a,h)anthracene (1,2,5,6-Dibenzanthracene) | 1020 | 13-16 |
| Dibenzo(a,e)pyrene (1,2,4,5-Dibenzopyrene) | 1020 | 13-16 |
| Dibenzo(a,h)pyrene (1,2,5,6-Dibenzopyrene) | 1020 | 13-16 |
| Dibenzo(a,i)pyrene (1,2,7,8-Dibenzopyrene) | 1020 | 13-16 |
| **Dichlorobenzene (1,2-Dichlorobenzene)** | 970 | 24-25 |
| **Dichlorobenzene (1,3-Dichlorobenzene)** | 960 | 26-27 |
| Dichlorobenzene (1,4-Dichlorobenzene) | 975 | 22-23 |
| **Dichlorodibenzo-P-Dioxin (2,7-)** | 920 | 37 |
| Fluoranthene (Benzo[j,k]fluorene) | 1062 | 6 |
| **Hexachlorobenzene** | 935 | 33-35 |
| **Hydrogen Cyanide (Hydrocyanic Acid)**[a] | >1500 | 1 |
| Indeno(1,2,3-cd)pyrene (1,10-[1,2-Phenylene]pyrene) | 1060 | 7-11 |
| **Naphthalene** | 1070 | 5 |
| **Pentachlorobenzene** | 935 | 33-35 |
| Sulfur Hexafluoride[b] | 1090 | 4 |
| **Tetrachlorobenzene (1,2,3,5-Tetrachlorobenzene)**[c] | 955 | 28-29 |
| **Tetrachlorobenzene (1,2,4,5-Tetrachlorobenzene)** | 940 | 32 |
| **Trichlorobenzene (1,2,4-Trichlorobenzene)** | 955 | 28-29 |
| **Trichlorobenzene (1,3,5-Trichlorobenzene)**[c] | 950 | 30-31 |

*(continued)*

See footnotes on page 260.

**Appendix F.** *(continued)*

| **Class 2. Principal organic waste constituents** | **T 99** | **Rank** |
|---|---|---|
| Acetylaminofluorene (2-) (Acetamide,n-[9H-Fluoren-2-yl]-) | 800 C | 67-70 |
| Acrylamide (2-Propenamide) | 820 | 61-62 |
| Aniline | 850 | 50-53 |
| **Chloroaniline (Chlorobenzenamine)** | 810 | 65 |
| Chloro-1,3-Butadiene (2-) (Chloroprene) | 800 | 67-70 |
| DDE (1,1-Dichloro-2,2-Bis (4-Chlorophenylethylene) | 875 | 40-41 |
| **Dichlorodibenzo-P-Dioxin (2,3-)** | 875 | 40-41 |
| **Dichloroethene (1,1-)** | 860 | 46-48 |
| **Dichloroethene (trans-1,2-)** | 835 | 59 |
| **Dichloromethane (Methylene Chloride)** | 815 | 63-64 |
| Dimethylbenz)a)anthracene (7,12-) | 855 | 49 |
| Dimethylphenethylamine (alpha,alpha-) | 840 | 55-58 |
| Diphenylamine (N-Phenylbenzenamine) | 800 | 67-70 |
| Fluoroacetamide (2-) | 830 | 60 |
| Fluoroacetic Acid | 860 | 46-48 |
| Formaldehyde (Methylene Oxide) | 850 | 50-53 |
| Formic Acid (Methanoic Acid) | 870 | 42-43 |
| **Malononitrile (Propanedinitrile)** | 840 | 55-58 |
| **Methacrylonitrile (2-Methyl-2-Propenenitrile)** | 815 | 63-64 |
| Methyl Chlorocarbonate (Carbonochloridic Acid, Methyl Ester) | 850 | 50-53 |
| Methylcholanthrene (3-) | 805 | 66 |
| Methyl Isocyanate (Methylcarbylamine) | 850 | 50-53 |
| Methyl Methacrylate (2-Propenoic Acid, 2-Methyl-,Methyl Ester) | 820 | 61-62 |
| Naphthylamine (1-) | 840 | 55-58 |
| Naphthylamine (2-) | 840 | 55-58 |
| Phosgene (Carbonyl Chloride) | 870 | 42-43 |
| Pronamide (3,5-Dichloro-N-[1,1-Dimethyl-2-Propynyl]Benzamide) | 800 | 67-70 |
| **Tetrachlorodibenzo-P-Dioxin (1,2,3,4-)** | 845 | 54 |
| Tetrachlorodibenzo-P-Dioxin (2,3,7,8-) | 865 | 44-45 |
| **Tetrachloroethene** | 890 | 39 |
| **Toluene (Methylbenzene)** | 895 | 38 |
| Trichlorodibenzo-P-Dioxin (1,2,4-) | 860 | 46-48 |
| **Trichloroethene** | 865 | 44-45 |

*(continued)*

See footnotes on page 260.

**Appendix F.** *(continued)*

| **Class 3. POHCs** | **T 99** | **Rank** |
|---|---|---|
| Acetonylbenzyl-4-Hydroxycoumarin (3-alpha-) (Warfarin) | 760 C | 100-101 |
| Acetophenone (Ethanone, 1-Phenyl-) | 775 | 86-90 |
| Acetyl Chloride (Ethanoyl Chloride) | 765 | 94-99 |
| Aminobiphenyl (4-) ([1,1′-Biphenyl]-4-Amine) | 796 | 71 |
| Benz(c)acridine (3,4-Benzacridine) | 775 | 86-90 |
| Benzenethiol (Thiophenol) | 725 | 120 |
| Benzidine ([1,1′-Biphenyl]-4,4′Diamine) | 794 | 74 |
| Benzoquinone (1,4-Cyclohexadienedione) | 770 | 91-93 |
| **Chlorophenol (2-)** | 795 | 72-73 |
| **Chlorophenol (3-)** | 785 | 80-81 |
| **Chlorophenol (4-)** | 755 | 102-103 |
| Chlorocresol (4-Chloro-3-Methylphenol) | 709 | 124 |
| Cresol (1,2-) (Methylphenol) | 746 | 115 |
| Cresol (1,3-) (Methylphenol) | 749 | 114 |
| **Cresol (1,4-) (Methylphenol)** | 745 | 116 |
| **Crotonaldehyde (2-Butenal)** | 710 | 122-123 |
| Dibenz(a,h)acridine (1,2,5,6-Dibenzacridine) | 765 | 94-99 |
| Dibenz(a,j)acridine (1,2,7,8-Dibenzacridine) | 765 | 94-99 |
| Dibenzo(c,g)carbazole (7H-) (3,4,5,6-Dibenzcarbazole) | 750 | 105-113 |
| Dichlorobenzidine (3,3′-) | 795 | 72-73 |
| **Dichlorodifluoromethane (Freon 12)** | 775 | 86-90 |
| **Dichloropropene (1,1-)** | 780 | 82-85 |
| Dichloropropene (2,3-) | 754 | 104 |
| Dichloropropene (trans-1,2-) | 750 | 105-113 |
| Dichlorophenol (2,4-) | 710 | 122-123 |
| Dichlorophenol (2,6-) | 740 | 117 |
| **Dichloro-2-Butene (1,2-)** | 755 | 102-103 |
| Diethylstilbesterol | 730 | 118-119 |
| Dihydroxy-Alpha-(Methylamino) Methyl Benzyl Alcohol (3,4-) (Adrenaline) | 705 | 125-126 |
| Dimethylbenzidine (3,3′-) | 793 | 75 |
| Dimethylphenol (2,4-) | 700 | 127 |
| **Dimethyl Phthalate** | 765 | 94-99 |
| **Ethyl Cyanide (Propionitrile)** | 770 | 91-93 |
| **Hexachlorobutadiene (trans-1,3)** | 765 | 94-99 |
| **Isobutyl Alcohol (2-Methyl-1-Propanol)** | 715 | 121 |

*(continued)*

See footnotes on page 260.

**Appendix F.** *(continued)*

| **Class 3. POCHs** *(continued)* | **T 99** | **Rank** |
|---|---|---|
| Maleic Anhydride (2,5-Furandione) | 760 C | 100-101 |
| Methylactonitrile (2-) (Propanenitrile, 2-Hydroxy-2-Methyl) | 705 | 125-126 |
| **Methyl Ethyl Ketone (2-Butanone)** | 730 | 118-119 |
| Naphthoquinone (1,4-) (1,4-Naphthalenedione) | 765 | 94-99 |
| n-Propylamine (1-Propanamine) | 789 | 79 |
| **Phenol (Hydroxybenzene)** | 775 | 86-90 |
| Phenylenediamine (1,2-) (Benzenediamine) | 791 | 77 |
| Phenylenediamine (1,3-) (Benzenediamine) | 792 | 76 |
| Phenylenediamine (1,4) (Benzenediamine) | 790 | 78 |
| Picoline (2-) (Pyridine, -2-Methyl-) | 780 | 82-85 |
| **Pyridine** | 785 | 80-81 |
| Resorcinol (1,3-Benzenediol) | 750 | 105-113 |
| Thioacetamide (Ethanethioamide) | 780 | 82-85 |
| Toluenediamine (2,4-) (Diaminotoluene) | 750 | 105-113 |
| Toluenediamine (2,4-) (Diaminotoluene) | 750 | 105-113 |
| Toluenediamine (2,5-) (Diaminotoluene) | 750 | 105-113 |
| Toluenediamine (2,6-) (Diaminotoluene) | 750 | 105-113 |
| Toluenediamine (3,4-) (Diaminotoluene) | 750 | 105-113 |
| Toluenediamine (3,5-) (Diaminotoluene) | 750 | 105-113 |
| **Trichlorofluoromethane (Freon 11)** | 775 | 86-90 |
| **Trichloro(1,2,2-)Trifluoroethane(1,1,2-) (Freon 113)**[c] | 780 | 82-85 |
| **Vinyl Chloride (Chloroethene)** | 770 | 91-93 |

| **Class 4. POCHs** | | |
|---|---|---|
| **Acrolein (2-Propenal)** | 670 | 138-141 |
| Aldrin | 632 | 167-169 |
| **Allyl Alcohol (2-Propen-1-Ol)** | 665 | 142 |
| Auramine | 618 | 188-189 |
| **Benzal Chloride (Alpha, Alpha-Dichlorotoluene)** | 630 | 170-174 |
| Benzotrichloride (Trichloromethylbenzene) | 600 | 204-205 |
| **Benzyl Chloride (Chloromethylbenzene)** | 685 | 131-133 |
| bis(2-Chloroethoxy) Methane | 610 | 197-201 |
| **bis(2-Chloroethyl) Ether** | 615 | 190-194 |
| Bromoacetone (1-Bromo-2-Propanone) | 670 | 138-141 |
| Butyl-4,6-Dinitrophenol (2-sec-) (DNBP) | 613 | 195-196 |

*(continued)*

See footnotes on page 260.

**Appendix F.** *(continued)*

| **Class 4. POCHs** *(continued)* | **T 99** | **Rank** |
|---|---|---|
| Chloral (Trichloroacetaldehyde) | 610 C | 197-201 |
| Chlorambucil | 659 | 144 |
| Chloroacetaldehyde | 630 | 170-174 |
| **Chlorodifluoromethane (Freon 22)**[c] | 645 | 155-157 |
| *Chloroethane (Ethyl Chloride)*[d] | 670 | 138-141 |
| **Chloropropene (3-) (Allyl Chloride)** | 695 | 128 |
| **Chloropropionitrile (3-) (3-Chloropropanenitrile)** | 655 | 145-146 |
| Chloro-2,3-Epoxypropane (1-) (Oxirane, 2-Chloromethyl-) | 615 | 190-194 |
| Cyclohexyl-4,6-Dinitrophenol (2-) | 613 | 195-196 |
| DDD (Dichlorodiphenyldichloroethane) | 654 | 147-148 |
| DDT (Dichlorodiphenyltrichloroethane) | 620 | 184-187 |
| **Dibromomethane (Methylene Bromide)** | 685 | 131-133 |
| Dibromoethane (1,2-) (Ethylene Dibromide) | 630 | 170-174 |
| **Dichloroethane (1,1-) (Ethylidene Dichloride)** | 610 | 197-201 |
| **Dichloroethane (1,2-)** | 680 | 134 |
| **Dichlorofluoromethane (Freon 21)**[c] | 640 | 158-163 |
| *Dichloropropane (1,1-)* | 615 | 190-194 |
| *Dichloropropane (1,2-) (Propylene Dichloride)* | 620 | 184-187 |
| *Dichloropropane (1,3-)* | 630 | 170-174 |
| Dichloropropene (3,3-) | 690 | 129-130 |
| Dichloropropene (cis-1,3-) | 625 | 175-183 |
| **Dichloropropene (trans-1,3-)** | 625 | 175-183 |
| Dichloro-1-Propanol (2,3-) | 625 | 175-183 |
| Dichloro-2-Propanol (1,1-) | 654 | 146-148 |
| Dichloro-2-Propanol (1,3-) | 652 | 149 |
| Dieldrin | 632 | 167-169 |
| Diepoxybutane (1,2,3,4-) (2,2′-Bioxirane) | 604 | 203 |
| Dimethylcarbamoylchloride | 620 | 184-187 |
| Dinitrobenzene (1,2-) | 635 | 164-166 |
| Dinitrobenzene (1,3-) | 640 | 158-163 |
| Dinitrobenzene (1,4-) | 635 | 164-166 |
| Dinitrocresol (4,6-) (Phenol,2,4-Dinitro-6-Methyl-) | 610 | 197-201 |
| Dinitrophenol (2,4-) | 615 | 190-194 |
| Dinitrotoluene (2,4-) | 625 | 175-183 |
| Dinitrotoluene (2,6-) | 625 | 175-183 |
| **Dioxane (1,4-) (1,4-Diethylene Oxide)** | 660 | 143 |

*(continued)*

See footnotes on page 260.

**Appendix F.** *(continued)*

| **Class 4. POCHs** *(continued)* | **T 99** | **Rank** |
|---|---|---|
| *Ethylene Oxide (Oxirane)* | 625 C | 175-183 |
| Glycidyaldehyde (1-Propanol-2,3-Epoxy) | 620 | 184-187 |
| Heptachlor | 618 | 188-189 |
| Heptachlor Epoxide | 609 | 202 |
| **Hexachlorocyclohexane (Lindane)** | 640 | 158-163 |
| Hexachlorocyclopentadiene | 625 | 175-183 |
| Hexachlorophene (2,2′-Methylenebis[3,4,6-Trichlorophenol]) | 670 | 138-141 |
| *Hydrazine (Diamine)* | 685 | 131-133 |
| Isodrin | 632 | 167-169 |
| Methapyrilene | 600 | 204-205 |
| Methyl Parathion | 650 | 150-154 |
| Mustard Gas (bis[2-Chloroethyl]-Sulfide) | 679 | 135-137 |
| Nitroaniline (4-Nitrobenzenamine) | 640 | 158-163 |
| **Nitrobenzene** | 655 | 145-146 |
| Nitrogen Mustard | 679 | 135-137 |
| Nitrophenol (4-) | 650 | 150-154 |
| Nitrotoluidine (5-) (Benzenamine, 2-Methyl-5-Nitro-) | 630 | 170-174 |
| N,N-bis (2-Chloroethyl) 2-Naphthylamine (Chlornaphazine) | 679 | 135-137 |
| **Pentachloroethane** | 640 | 158-163 |
| *Pentachlorophenol* | 640 | 158-163 |
| Phthalic Anhydride (1,2-Benzenedicarboxylic Acid Anhydride) | 650 | 150-154 |
| **Tetrachloroethane (1,1,2,2-)** | 690 | 129-130 |
| **Tetrachloromethane (Carbon Tetrachloride)** | 645 | 155-157 |
| Tetrachlorophenol (2,3,4,6-) | 645 | 155-157 |
| **Trichloroethane (1,1,2-)** | 635 | 164-166 |
| **Trichloromethane (Chloroform)** | 625 | 175-183 |
| Trichloromethanethiol | 610 | 197-201 |
| Trichlorophenol (2,4,5-) | 650 | 150-154 |
| **Trichlorophenol (2,4,6-)** | 650 | 150-154 |
| **Trichloropropane (1,2,3-)** | 625 | 175-183 |
| Trinitrobenzene (1,3,5-Trinitrobenzene) | 615 | 190-194 |

*(continued)*

See footnotes on page 260.

**Appendix F.** *(continued)*

| **Class 5. POCHs** | **T 99** | **Rank** |
|---|---|---|
| Aflatoxins | 590 C | 208 |
| Amitrole (1H-1,2,4-Triazol-3-Amine) | 580 | 215-216 |
| Aramite | 500 | 244-248 |
| bis-(Chloromethyl) Ether (Methane-Oxybis[2-Chloro-]) | 530 | 231-233 |
| bis(2-Chloroisopropyl) Ether | 520 | 236-237 |
| **Bromoform (Tribromomethane)** | 585 | 209-210 |
| Bromophenyl Phenyl Ether (4-) (Benzene, 1-Bromo-4-Phenoxy-) | 523 | 235 |
| Brucine (Strychnidin-10-One,2,3-Dimethoxy-) | 470 | 254-255 |
| Chhlordane (Alpha and Gamma Isomers) | 533 | 230 |
| Chlorobenzilate | 584 | 211-214 |
| **Chloroethylvinylether (2-) (Ethane,[2-Chloroethoxy]-)** | 565 | 218-220 |
| Chloromethylmethyl Ether (Chloromethoxymethane) | 550 | 226-228 |
| Diallate (S-[2,3-Dichloroallyl]Diisopropyl Thiocarbamate) | 500 | 244-248 |
| Dibromo-3-Chloropropane (1,2-) | 561 | 221 |
| Dichlorophenoxyacetic Acid (2,4-) (2,4-D) | 565 | 218-220 |
| *Dichloropropane (2,2-)* | 530 | 231-233 |
| Diethylphthalate | 400 | 265-266 |
| Dihydrosafrole (1,2-Methylenedioxy-4-Propylbenzene) | 520 | 236-237 |
| Dimethoate | 500 | 244-248 |
| Dimethylaminoazobenzene | 405 | 264 |
| Dimethoxygenzidine (3,3′-) | 453 | 259 |
| *Dimethylhydrazine (1,1-)* | 555 | 224-225 |
| Dimethylhydrazine (1,2-) | 550 | 226-228 |
| Dimethyl-1-Methylthio-2-Butanone, O-([Methylamino]-Carbonyl) | 550 | 226-228 |
| Diphenylhydrazine (1,2-) | 450 | 260 |
| Ethyl Carbamate (Urethan) (Carbamic Acid, Ethyl Ester) | 584 | 211-214 |
| Ethyleneimine (Aziridine) | 500 | 244-248 |
| Ethyl Methacrylate (2-Propenoic Acid, 2-Methyl-, Ethyl Ester) | 584 | 211-214 |
| **Hexachloroethane** | 585 | 209-210 |
| **Hexachloropropene** | 505 | 243 |
| Iodomethane (Methyl Iodide) | 567 | 217 |

*(continued)*

See footnotes on page 260.

**Appendix F.** *(continued)*

| **Class 5. POCHs** *(continued)* | **T 99** | **Rank** |
|---|---|---|
| Isosafrole (1,2-Methylenedioxy-4-Allylbenzene) | 460 C | 256-258 |
| Kepone | 470 | 254-255 |
| Lasiocarpine | 584 | 211-214 |
| Maleic Hydrazide (1,2-Dihydro-3,6-Pyridazinedione) | 525 | 234 |
| Methoxychlor | 490 | 252-253 |
| Methylaziridine (2-) (1,2-Propylenimine) | 490 | 252-253 |
| Methylene Bis (2-Chloroaniline) (4,4-) | 565 | 218-220 |
| *Methyl Hydrazine* | 593 | 206-207 |
| Methyl Methanesulfonate (Methanesulfonic Acid, Methyl Ester) | 515 | 238 |
| Methyomyl | 510 | 241-242 |
| Methyl-2-Methylthio-Propionaldehyde-O-(Methylcarbonyl) Oxime (2-) | 510 | 241-242 |
| Muscimol (5-Aminomethyl-3-Isoazotol) | 580 | 215-216 |
| **n-Butylbenzyl Phthalate** | 415 | 262 |
| N,N-Diethylhydrazine (1,2-Diethylhydrazine) | 555 | 224-225 |
| O,O-Diethylphosphoric Acid, O-p-Nitrophenyl Ester | 425 | 261 |
| O,O-Diethyl-O-2-Pyrazinyl Phosphorothioate | 410 | 263 |
| Oxime (3,3-) (Thiofanox) | — | — |
| Parathion | 530 | 231-233 |
| Pentachloronitrobenzene (PCNB) | 500 | 244-248 |
| Phenacetin (N-[4-Ethoxyphenyl]Acetamide) | 593 | 206-207 |
| Propane Sulfone (1,3-) (1,2-Oxathiolane, 2,2-Dioxide) | 513 | 239 |
| **Propyn-1-Ol (2-) (Propargyl Alcohol)** | 560 | 222-223 |
| Saccharin (1,2-Benzoisothiazolin-3-One, 1,1-Dioxide) | 511 | 240 |
| Safrole (1,2-Methylene-4-Allylbenzene) | 460 | 256-258 |
| **Tetrachloroethane (1,1,1,2-)** | 560 | 222-223 |
| **Trichloroethane (1,1,1-) (Methyl Chloroform)** | 545 | 229 |
| Trichlorophenoxyacetic Acid (2,4,5-) (2,4,5-T) | 495 | 249-250 |
| Trichlorophenoxypropionic Acid (2,4,5-) (2,4,5-TP) (Silvex) | 495 | 249-250 |
| tris(1-Azridinyl) Phosphine Sulfide | 460 | 256-258 |
| tris(2,3-Dibromopropyl) Phosphate | 492 | 251 |

| **Class 6. POHCs** | | |
|---|---|---|
| bis(2-Ethylhexyl) Pgthalate | 370 | 278-279 |

*(continued)*

See footnotes on page 260.

**Appendix F.** *(continued)*

| **Class 6. POCHs** *(continued)* | **T 99** | **Rank** |
|---|---|---|
| Citrus Red No. 2 (2-Naphthol, 1-[(2,5-Dimethoxyphenyl]Aso)) | 395 C | 267-268 |
| Cyclophosphamide | 300 | 282-285 |
| Diisopropylfluorophosphate (DFP) | 390 | 270-274 |
| Di-n-Butyl Phthalate | 390 | 270-274 |
| **Di-n-Octyl Phthalate** | 380 | 276 |
| Disulfoton | 390 | 270-274 |
| Ethyl Methanesulfonate (Methanesulfonic Acid, Ethyl Ester) | 390 | 270-274 |
| Methylthiouracil | 370 | 278-279 |
| Nicotine ([S]-3-[1-Methyl-2-Pyrrolidinyl]Pyridine) | 300 | 282-285 |
| Octamethylpyrophosphoramide (Octamethyldiphosphoramide) | 374 | 277 |
| O,O-Diethyl S-([Ethylthio]Methyl) Ester of Phosphorodithioic Acid | 395 | 267-268 |
| O,O-Diethyl-S-Methyl Ester of Phosphoric Acid | 395 | 265-266 |
| O,O,O-Triethyl Phosphorothioate | 390 | 270-274 |
| *Paraldehyde (2,4,6-Trimethyl-1,3,5-Trioxane)* | 380 | 275 |
| Propylthiouracil | 360 | 280 |
| Reserpine | 300 | 282-285 |
| Strychnine (Strychnidin-10-One) | 320 | 281 |
| Toluidine Hydrochloride (2-Methyl-Benzenamine Hydrochloride) | 300 | 282-285 |
| Trypan Blue | 392 | 269 |
| **Class 7. POHCs** | | |
| Acetyl-2-Thiourea (1-) (Acetamide,N-[Aminothioxomethyl]-) | 240 | 295-299 |
| Azaserine (L-Serine, Diazoacetate[Ester]) | 200 | 306 |
| Butanone Peroxide (2-) (Methyl Ethyl Ketone, Peroxide) | 260 | 288 |
| Chlorophenyl Thiourea (1-) (Thiourea,[2-Chlorophenyl]-) | 240 | 295-299 |
| Cycasin (beta-D-Glucopyranoside, [Methyl-Onn-Azoxy]Methyl-) | 170 | 308-310 |
| Daunomycin | 230 | 300-301 |

*(continued)*

See footnotes on page 260.

**Appendix F.** *(continued)*

| **Class 7. POCHs** *(continued)* | **T 99** | **Rank** |
|---|---|---|
| Di-n-Propylnitrosamine (N-Nitroso-Di-n-Propylamine) | 130 C | 312-327 |
| Dithiobiuret (2,4-) (Thioimidodicarbonic Diamide) | 220 | 304-305 |
| Endosulfan | 100 | 329 |
| Endrin | 285 | 287 |
| EthylenebisDithiocarbamic Acid | 250 | 291-292 |
| Ethylene Thiourea (2-Imidazolidinethione) | 230 | 300-301 |
| Hexaethyl Tetraphosphate | 190 | 307 |
| Melphalan (Alanine,3-[p-bis(2-Chloroethyl)Amino] Phenyl-,L-) | 225 | 302-303 |
| Naphthyl-2-Thiourea (1-) (Thiourea, 1-Naphthalenyl-) | 240 | 295-299 |
| Nitrogen Mustard N-Oxide | 170 | 308-310 |
| *Nitroglycerine (Trinitrate-1,2,3-Propaneltriol)* | 255 | 290 |
| Nitroquinoline-1-Oxide (4-) | 170 | 308-310 |
| Nitrosopyrrolidine (N-Nitrosotetrahydropyrrole) | 130 | 312-327 |
| N-Methyl-N′-Nitro-N-Nitrosoguanidine | 130 | 312-327 |
| N-Nitroso-Di-Ethanolamine ([2,2′-Nitrosoimino]bisEthanol) | 130 | 312-327 |
| N-Nitrosodiethylamine (N-Ethyl-N-Nitrosoethanamine) | 130 | 312-327 |
| N-Nitrosodimethylamine (Dimethylnitrosamine) | 130 | 312-327 |
| N-Nitroso-Di-N-Butylamine (N-Butyl-N-Nitroso-1-Butanamine) | 130 | 312-327 |
| N-Nitrosomethylethylamine (N-Methyl-N-Nitrosoethanamine) | 130 | 312-327 |
| N-Nitrosomethylvinylamine (N-Methyl-N-Nitrosoethenamine) | 130 | 312-327 |
| N-Nitrosomorpholine | 130 | 312-327 |
| N-Nitroso-N-Ethylurea (N-Ethyl-N-Nitrosocarbamide) | 130 | 312-327 |
| N-Nitroso-N-Methylurea (N-Methyl-N-Nitrosocarbamide) | 130 | 312-327 |
| N-Nitroso-N-Methylurethane | 130 | 312-327 |
| N-Nitrosonornicotine | 130 | 312-327 |
| N-Nitrosopiperidine (Hexahydro-N-Nitrosopyridine) | 130 | 312-327 |
| N-Nitrososarcosine | 130 | 312-327 |
| N-Phenylthiourea | 240 | 295-299 |
| Oxabicyclo(2.2.1)Heptane-2,3-Dicarboxylic Acid (7-) (Endothal) | 120 | 328 |

*(continued)*

See footnotes on page 260.

**Appendix F.** *(concluded)*

| **Class 7. POCHs** *(concluded)* | **T 99** | **Rank** |
|---|---|---|
| Streptozotocin | 155 C | 311 |
| Tetraethyldithiopyrophosphate | 250 | 291-292 |
| Tetraethylpyrophosphate | 255 | 289 |
| *Tetranitromethane* | 245 | 293 |
| Thiosemicarbazide (Hydrazinecarbothioamide) | 225 | 302-303 |
| Thiourea (Thiocarbamide) | 240 | 295-299 |
| Thiuram (bis[Dimethylthiocarbamoyl]Disulfide) | 220 | 304-305 |
| Tolyene Diisocyanate (1,3-Diisocyanatomethylbenzene) | 290 | 286 |
| Uracil Mustard (5-[bis(2-Chloroethyl)Amino]Uracil) | 246 | 294 |

---

[a] Boldface print indicates compound thermal stability is fully evaluated; ranking based on UDRI experimental mixture data coupled with reaction kinetic theory.

[b] Non-appendix VIII compound.

[c] N.O.S. listing; ranking is presented based on either UDRI or literature experimental data coupled with reaction kinetic theory.

[d] Italicized print indicates compound thermal stability is fully evaluated based on UDRI or literature data coupled with reaction kinetic theory.

# INDEX